The Illustrated Encyclopedia of AMMUNITION

The Illustrated Encyclopedia of AMMUNITION

IAN V. HOGG

Quantum Books

A QUANTUM BOOK

This book is produced by
Quantum Publishing Ltd.
6 Blundell Street
London N7 9BH

This edition printed 2003

ISBN 1-86160-728-8

QUMANI

Printed in Singapore by
Star Standard Industries Pte. Ltd.

Page 7

The lead of a Boxer shrapnel shell showing the fuze socket, the wooden head lightly attached to the body of the shell and the central flash tube which is filled with perforated gunpowder pellets to boost the fuze flash down to the expelling charge at the bottom of the shell.

CONTENTS

FOREWORD

THERE IS AN OLD SAYING among artillerymen that their weapon is not the gun; it is the shell, and the gun is only the last stage in transportation from the factory to the target. The same applies to all firearms; they are merely devices for discharging bullets, shells, bombs — projectiles of one sort or another — which are the things destined to have the desired effect on the enemy. Without ammunition the finest firearm is merely an expensive club or, at best, a handle for a bayonet, while a piece of artillery with no ammunition is no more than an ornament.

But for all its importance, ammunition is usually taken for granted; received, loaded, fired, and if for some reason it fails to work the firer gets most aggrieved even if he doesn't know why it failed. Yet it is a fascinating subject in its own right. Many a weapon which appeared to have reached the end of its usefulness has been revitalized and given a new lease of life by nothing more than re-designing the ammunition for it, and many weapons have had their effectiveness enhanced by new and improved ammunition. It is impossible to have a complete understanding of a weapon unless there is complete understanding of its ammunition as well.

Although ammunition has been in existence since the first days of firearms in the 14th century, it remained at a fairly static stage of development for the first 400 years. It was not until the Industrial Revolution got into its stride in the mid-19th century that gun design began to show some advances, and with these came improvements in ammunition. The activity of designers and manufacturers in the 1880s and 1890s showed its effect in the South African and Russo-Japanese wars of the early 20th century. These wars led to improvements and innovations which were seen in use during the 1914-18 war, and it was this war which gave ammunition development its greatest impetus to that time. The First World War produced new tactical situations and new techniques of warfare which demanded new types of ammunition. The arrival of military aircraft, for example, brought the first aerial bombs, but it also led to the development of special ammunition for shooting at the aircraft that were dropping these bombs.

The advances made during the First World War were consolidated and refined during the 1920s and 1930s to become the standard ammunition types with which the Second World War was begun. But, once again, advances in military tactics and technology brought new forms of warfare and new types of ammunition. Improvements in the tank brought about improvements in anti-tank projectiles, higher-flying aircraft demanded new types of shell and fuze, portable infantry anti-tank weapons had to be given special projectiles capable of defeating tanks, and so on.

This book cannot, for reasons of space, cover every facet of ammunition; we have, therefore, concentrated on the principal projectile weapons of war — pistols, rifles, machine-guns, mortars, artillery and grenades. We have had to ignore many other interesting areas — mines, torpedoes, aerial bombs and missiles — but the general principles explained in the body of this book apply to every type of ammunition, and with the knowledge gained from these pages it should not be difficult to understand other types of explosive device.

Finally, and for those of a nervous disposition, may I point out that this is not a terrorist's handbook; there are no plans of how to make bombs or any other type of ammunition, and the illustrations are no more explicit than will be found in any encyclopedia or manufacturer's brochure. Nor should anything said in this book be taken as being sufficiently detailed to warrant anyone attempting to dismantle a souvenir; ammunition is designed to kill, and it is entirely colour-blind, taking no note of what uniform, if any, the tampering hand is wearing. So remember: Never tamper with ammunition — it can kill you.

Ian V. Hogg, 1985

FROM CANNON-BALL TO COPPERHEAD

An engraving showing 'Black Berthold', the Mysterious Monk of Freiburg, discovering the principle of the cannon. Charming though this legend may be, there is no evidence that a monk called Berthold Schwartz ever existed. And there is plenty of evidence to suggest that cannon was in use well before the supposed date of his fabled experiment.

Right: The famous *De Officiis Regnum* written by Walter de Milemete. Although there is no mention of firearms in the text, the illustration at the bottom of the manuscript clearly shows a primitive cannon being fired — the touch hole of a bulbous tube, which has an arrow protruding from its muzzle, is being ignited by a man with a hot iron.

ALTHOUGH THE TERM 'ammunition' can be interpreted broadly to cover the stones, prisoners and dead horses flung by catapults into besieged cities in ancient times, the word is usually taken to mean explosive ammunition of some sort, and therefore its history must begin with the first explosive — gunpowder.

Just who invented gunpowder, and where and when, is forever lost in history. At various times it has been credited to the Chinese, the Arabs, possibly even the Lost Tribes of Atlantis, but there is no firm evidence for any of them. The Chinese certainly had pyrotechnic compositions of sorts — fireworks — long before anything of the kind was known in Europe, but this is no warrant for claiming that they invented guns and gunpowder — although for all we know to the contrary they may very well have done.

The first firm signpost on the gunpowder trail can be said to be the publication entitled *De Mirabili Potestate Artis et Naturae* (On the Marvellous Power of Art and Nature), written by Roger Bacon in AD1242. Bacon was a Franciscan monk who prepared several books and papers on philosophical subjects. In this book, at a point in the text where Bacon appears to be skirting the subject of explosives, he includes a string of apparently meaningless words — an aberration that was ignored for centuries, until in the early years of the 20th century a British artillery officer, Colonel Hime, hit upon the answer. The words were an anagram which, when the letters were re-arranged and punctuated, read 'But of saltpetre take seven parts, five of young hazel twig and five of sulphur, and so thou wilt call up thunder and destruction if thou know the art.' If the 'five of young hazel twig' is read instead as charcoal, this provides a workable formula for gunpowder.

But why the anagram? Why not write it plainly for all the world to see and to credit Bacon with, if not the discovery, at least the first announcement of gunpowder? Because Bacon was a monk, and in 1139 the Second Lateran Council had laid under anathema any person who made 'fiery compositions' for military purposes. Had Bacon made an open claim or announcement, he might well have forfeited his life. As it was, due to his outspokenness on various subjects he was removed from his teaching post at Oxford in 1257 and ordered into cloisters in Paris, where he remained out of sight for the next ten years. Then in 1266 he was requested to write a summary of philosophical knowledge for the Pope, after which he was permitted to return to Oxford and resume his teaching.

Among the collection of papers submitted to the Pope was one entitled *Opus Tertius*, a copy of which was discovered in the Bibliothèque Nationale in Paris in 1909. One passage from this makes interesting reading:

> From the flaming and flashing of certain igneous mixtures and the terror inspired by their noise, wonderful consequences ensue which no one can guard against or endure. As a simple example may be mentioned the noise and flame generated by the powder, known in divers places, composed of saltpetre, charcoal and sulphur. When a quantity of this powder no bigger than a man's finger be wrapped in a piece of parchment and ignited, it explodes with a blinding flash and a stunning noise. If a larger quantity were used, or the case made of some more solid material, the explosion would be much more violent and the noise altogether unbearable ... These compositions can be used at any distance we please, so that the operators escape all hurt from them, while those against whom they are employed are filled with confusion..

Bacon makes no claim to having discovered the composition but refers to its use in 'divers places', suggesting that by 1266 the material was common knowledge and he was merely reporting it as a fact.

If we can assume, then, that gunpowder was known to initiates in 1242 and was general knowledge by 1266, the next question to arise is when and where was it first applied to propel something from a gun? (Obviously it could not be called 'gunpowder' until it was used with a gun; exactly what it *was* called in those far-off days is another unsolved mystery.)

There are more spurious claims and legends which must be rapidly dismissed here. One of the most persistent is that of Black Berthold, the Mysterious Monk of Freiburg. Legend says (and is supported by an engraving dated 1643) that one day, experimenting with some powder in a cast-iron vessel, he ignited a charge and thus blew off the lid, and from this deduced the principle of confining a charge in a tube and propelling a shot. A charming legend, but in the first place there is no evidence that Berthold Schwarz ever existed, and in the second place the engraver of the scene gave it a date of 1380, and there is plenty of evidence to show that cannon were in use well before that.

One difficulty is the use of the word 'artillery' in medieval papers; this is often taken to mean 'guns', but the term was in use for many centuries before the gun appeared, in connection with catapults and other engines of war, and its transference to guns went unremarked in the archives. There are various other misleading reports that would be a waste of time to examine. The first positive record, written in AD 1326, reports on the manufacture of brass cannon and iron balls for 'the defence of the commune, camps and territory' of Florence.

Pre-dating this by one year is the famous document *De Officiis Regnum*, written by Walter de Milemete, in which there is an illustration showing the firing of a cannon. The picture is not explained by any text but shows a bulbous tube, with an arrow protruding from the muzzle, being ignited by a man with a hot iron.

From these early records we can see that the first 'ammunition' consisted of the gunpowder used to propel the missile, and either balls or arrows as the missiles them-

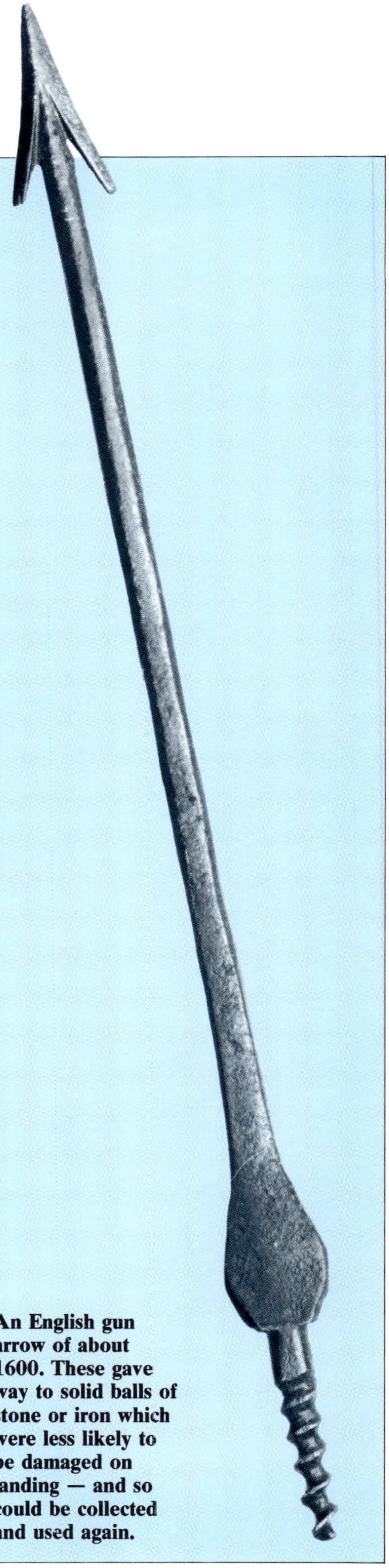

An English gun arrow of about 1600. These gave way to solid balls of stone or iron which were less likely to be damaged on landing — and so could be collected and used again.

Artillery manufacture in Germany in 1500 — Emperor Maximilian and his cannon makers.

H B

The manufacture of gunpowder circa 1450. This engraving shows the material being ground. Note the hourglass. The ingredients had to be ground together for a fixed length of time.

selves. Arrows were well known, and it seems logical that the soldiers of the time would have attempted to adapt the type of missile they knew to the new system of discharge, wrapping a binding of leather around the arrow shaft so as to make it fit tightly into the barrel of the cannon. But making arrows was a skilled business, and using them as cannon projectiles would have been a wasteful affair, so it must have been an obvious step to turn to the use of solid balls, either of stone or cast from iron. These had the advantage that they were unlikely to be damaged on landing and could be salvaged, tidied up, and used again and again (whereas an arrow would probably be damaged beyond repair by a single firing).

Even so, this type of projectile survived for a long time: Sir Francis Drake, in a return to the Government, dated 30 March, 1588, mentions gun-arrows among the stores on his ship.

Stone balls were preferred in the earliest days; indeed, the reference to iron balls in use at Florence is rather remarkable and suggests that the guns were quite small. The argument for this is rather involved, and has to do with the strength and weaknesses of both the guns and the gunpowder. The Milemete drawing shows the cannon in a colour that suggests bronze as the material — a not unreasonable assumption in that the casting of bronze was a well understood art at that time. But it was also expensive, and as the call came for larger cannon, other methods were adopted, notably that of making the barrel from strips of iron bound tightly with straps of metal and rope. Although this was cheaper and easier to make, it was also not as strong as a cast barrel. At the same time, however, the original gunpowder was, by today's standards, weak stuff, so the gun and its powder were more or less well matched. Even so, putting a heavy iron ball into one of these guns and touching off a charge of powder behind it was asking for trouble; the weight of the ball would resist movement so that the explosion of the powder would build up considerable pressure in the gun and might well burst it. A stone ball of the same size would be much lighter — about one-third of the weight of an iron ball — would move more quickly, and would thus not allow such a dangerous build-up of pressure.

Horizontal gun boring. Foreground — a man is removing the mould after the cannon has been cast. Background — the waterwheel is boring the inside of the barrel.

Another point was that the powder of the day, as well as being weak by formula, was weak by manufacture. Bacon's formula was for 41 per cent saltpetre, 29.5 per cent sulphur and 29.5 per cent charcoal, and doubtless the materials of this day were not to any great degree of purity. Modern gunpowder uses percentage proportions of 75:10:15, together with high standards of purity.

The early powder was mixed by grinding the separate ingredients finely and then mixing them by hand. This dusty mixture was loaded simply by scooping it out of a barrel and pouring it into the gun loosely; there it would settle and consolidate in the chamber with the result that the ignition flame might have difficulty in quickly penetrating through the fine mass to ignite it all at the same time. This led to uncertainty of action, variable power from shot to shot, and also meant that some of the charge was blown out of the muzzle either still burning or unburned — a serious defect at a time when gunpowder cost a staggering amount, in comparison to the price of iron and lead.

The main drawback with gunpowder was its susceptibility to damp. Saltpetre attracts moisture, and the records of the 14th and 15th centuries are full of complaints about damp powder. It was a very serious problem in that if the powder was unserviceable, the army's teeth were drawn before it got into battle. In 1372-74 there are reports of the purchase of faggots for drying the English powder. Scottish records in 1459 mention keeping powder in waxed cloth bags. Even as late as in 1759 a Royal Navy report of an action off Grenada complained that the shot would not reach the French ships because of damp powder.

Another drawback was that during prolonged travelling, when barrels were being bounced around in carts, the various components of the powder tended to separate, the heavy sulphur sifting through the lighter charcoal. The solution to this problem was generally to carry the three ingre-

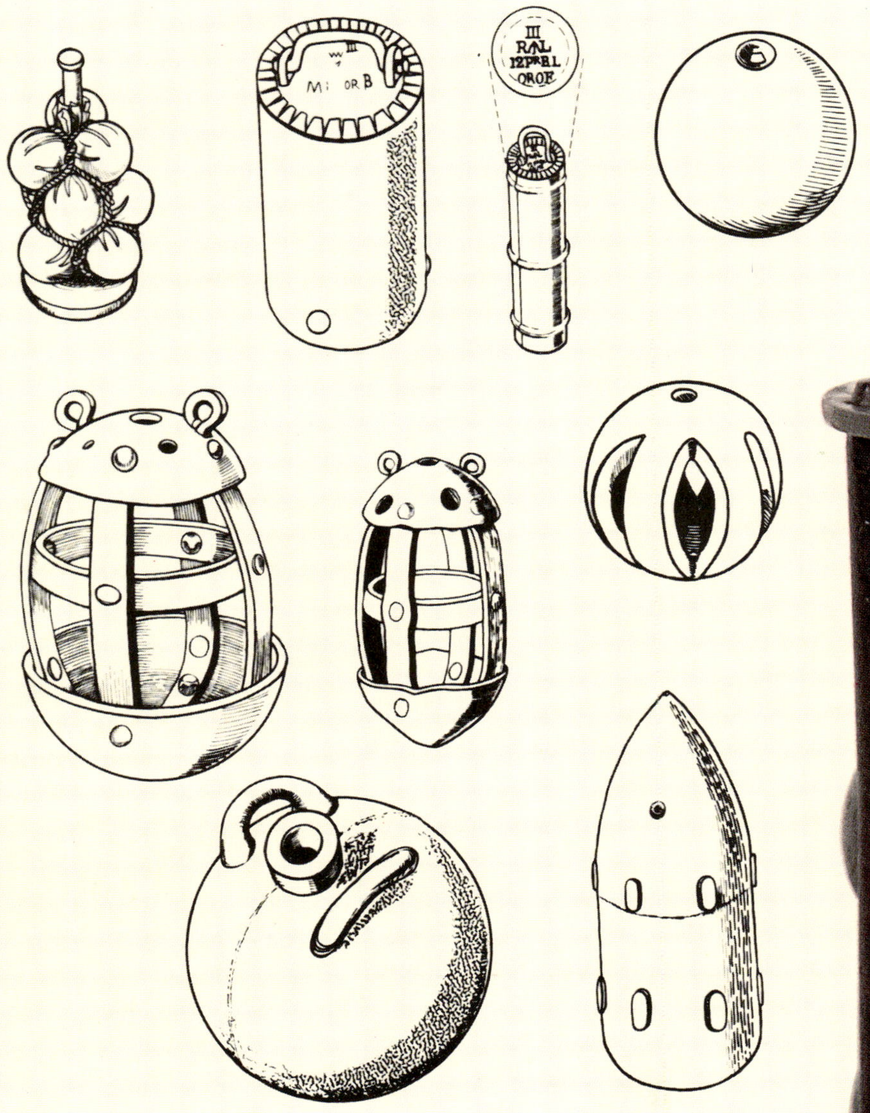

A selection of projectiles from the past.
Left: Top row — grape shot, case shot, a spherical shell; middle row — the framework of two carcases for smoke or light balls, another variety of carcase; bottom row — a mortar bomb and a studded shell.
Below and right: Early grapeshot. Far right: Top row — an iron ball, two projectiles for Manby's life-saving mortar, a bar for cutting rigging; middle row — grape shot, a spherical shell with its bottom attached, grapnel shot; bottom row — a stone ball, a light ball.

dients separately and mix them on the spot, before the battle began (although this was a hazardous enterprise).

The first improvement is as difficult to pin down in location and date as was the origin of gunpowder, but it appears to have developed in France where it was first mentioned in 1429. The improvement was to mix the ingredients in a wet state, which made for better incorporation and less likelihood of accidental explosion, and then to allow it to dry into a cake which was then broken up; the resulting grains were then run through sieves of varying meshes so as to select powder grains for different purposes. The powder was called 'corned powder', and it showed considerable advantages over the original, which then became known as 'serpentine powder'. The granular form allowed flame to penetrate the charge more easily, leading to more certain and regular burning — so much so that corned powder was considered to be about one-third more powerful than the same amount of serpentine. It was also less susceptible to moisture, it no longer separated during transport, and it left a lot less residue and fouling in the bore after firing.

Such is the nature of improvements that these advantages were not attained without some drawbacks. The two principal ones were that corned powder was so powerful it

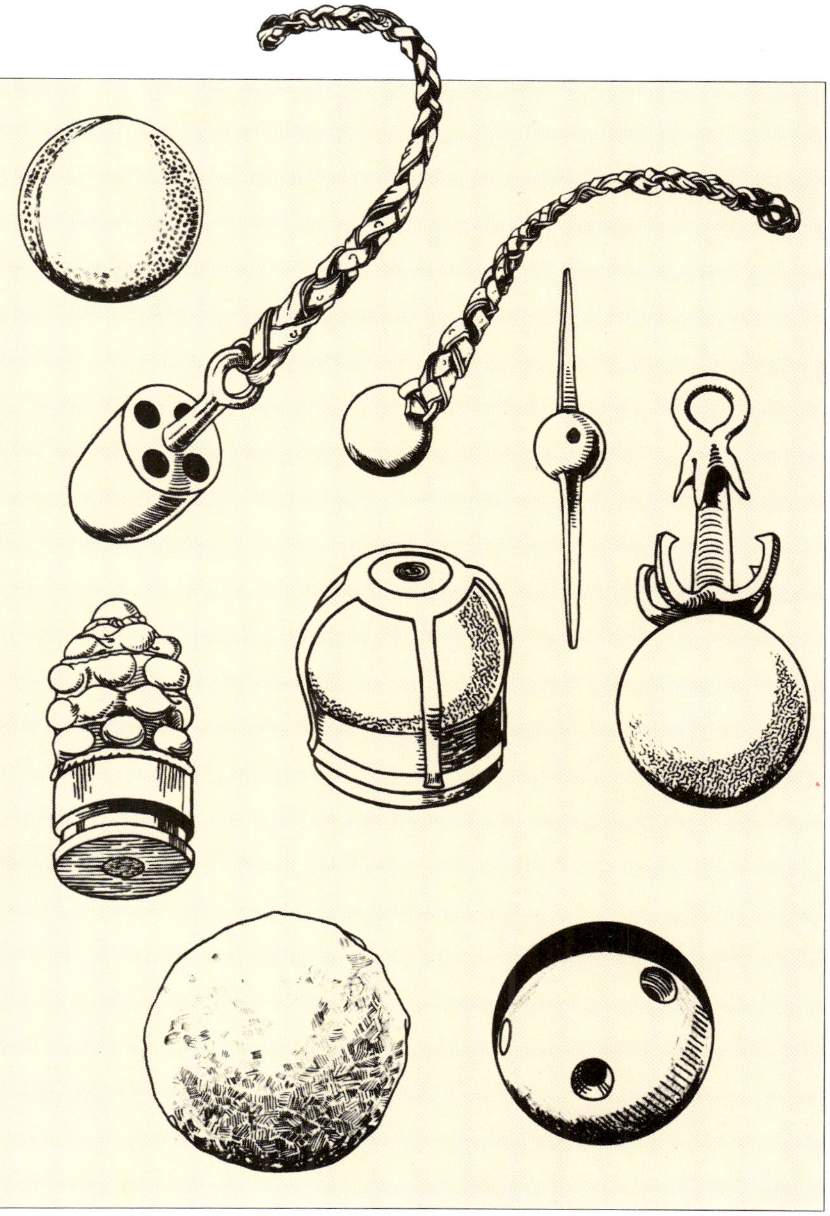

tended to burst the older guns; and that because of the involved method of manufacture it was even more expensive. As a result it was some time before corned powder gained universal acceptance, but it did have the effect of speeding up the development of cannon, in order that the guns might be strong enough to get the full benefit from the new powder. Cast-iron cannon began to appear in the middle of the 16th century, and remained the standard pattern for the next 300 years.

Most of the problems with gunpowder centred on the use of saltpetre (potassium nitrate). Until the middle of the 19th century the only source of supply was from natural deposits, and in Europe these were rare — the material appeared as a surface coating in underground caves, cellars and stables. The major deposits were in the East, and Oriental merchants soon realized the value of their commodity and charged accordingly. In Europe, governments sought to conserve every possible source of supply, so as to reduce their dependence upon the East. In France, 'Saltpetre Commissioners' were appointed in 1540 and had the right of entry into any stable, pigeon-loft, cattle shed or sheep-pen in order to gather saltpetre.

In the early years of the 17th century the East India Company began importing saltpetre into England and built its own

Stefan Batory (1533-88), the Polish king responsible for red-hot shot and forming the Cossacks into a militia.

gunpowder factory in Surrey. Among the terms for the renewal of the Company's Charter in 1693 was a requirement to supply 500 tons of saltpetre to the Board of Ordnance every year. But merchants had been quick to recognize the profitability of gunpowder many years before that; there was a powder mill in Augsburg in 1340, one in Spandau in 1344. The first powder used by an English army was imported from the continent. This importation continued for many years until the threatening attitude of Spain during Elizabeth I's reign prompted the issue of patents for the manufacture of powder. In 1555 a mill appeared near Rotherhithe, followed by others, and in 1561 a factory was built at Waltham Abbey which grew into the Royal Gunpowder Factory and, in the 20th century, a Royal Ordnance Factory.

The projectile remained the stone or cast-iron ball. The propelling charge was a matter of argument between gunners of the early days, but seems to have been more or less agreed as one-ninth of the weight of the stone shot. As guns increased in strength so the charge was increased to one-fifth or even one-quarter, and this rule was carried over to iron shot, since the difference in weight automatically produced a heavier charge for the heavier projectile.

The only alternative to solid shot was 'langridge', used as an anti-personnel short-range measure. To use 'langridge' simply meant loading a charge and a wad, and then shovelling down the barrel anything and everything which could hurt when fired out again — horse-shoe nails, stones, gravel, scrap metal, flints — anything at all: it was a formidable producer of casualties. Later, the idea of placing this collection into a convenient case was devised, and the result was hence known as 'case shot', which first appears on record during the defence of Constantinople in the middle of the 15th century. An alternative method was to fill a cloth bag with langridge and bind it with cord before loading; this binding caused it to bulge and resemble a bunch of grapes, so that shot in a cloth bag became 'grape shot'.

Ignition of the charge was originally carried out using a hot iron, as evidenced by the Milemete manuscript, but hot irons are an inconvenient apparatus to have around the battlefield and a better method soon appeared. The gun was provided with a 'vent' above the chamber into which fine powder was dribbled, to be set off by the hot iron, or by the newly invented alternative, 'quickmatch'. This was simply cord soaked in a solution of saltpetre and then rolled in fine gunpowder and allowed to dry; when ignited it burned very quickly, and a length of this, inserted into the gun vent, proved to be efficient in firing the gun. Unfortunately, it tended to foul the vent, and it became necessary to 'rime' the hole periodically to clean it so that a fresh piece of quickmatch could be used. Some gunners preferred to stick to using fine powder, from their powder horn, because it often found a way past the fouling and reduced the frequency at which riming was needed.

And so, also to replace the hot iron, 'slow match' was devised. This was similar to quickmatch except that it was thicker and dispensed with the gunpowder, so that when lit it merely glowed and burned slowly. It could be revived by blowing on it or whirling it through the air and, attached to a long staff or 'linstock', could be applied to the gun vent.

The first real innovation in projectiles came in 1573 when Master Gunner Zimmermann, a German, invented 'hail shot'. This was a lead cylinder filled half with gunpowder and half with musket balls; a short length of quickmatch passed through a hole in the bottom of the container. As the shot was fired from the gun, so the quickmatch was ignited, burned through and fired the gunpowder inside the casing so that it exploded and blew the musket balls in all directions. This had a range of about a kilometre (600 yards), so that it was a very efficient method of reducing the strength of an opposing army long before it got within hand-to-hand fighting distance.

Very shortly afterwards, in 1579, Stefan Batory, the King of Poland, instituted the use of red-hot shot. Heated balls of clay had often been used with catapults, flung into towns in order to raise fires in the thatched roofs, but the application to guns was slow to come about — which, considering the problems facing the gunners, was not surprising. At first acquaintance the prospect of pushing a red-hot cast-iron ball down on to a charge of gunpowder is not attractive, after all, but Stefan Batory's innovation was firstly to ram a dry wad of cloth down on top of the powder, then a second wad of soaking wet cloth, after which the red-hot ball was tipped into the muzzle and rammed and the gunner touched off the gun as quickly as he could before the hot shot dried out the wet wad.

Solid shot, hot or cold, varied in effect. Against solid targets, such as castles or towns or fortresses, projectiles could do severe damage to the defences. Against individuals on the field of battle they were less reliable — an astute soldier could often see a shot approaching and jump clear, although this was difficult in those days of serried ranks, shoulder-to-shoulder, and with second and third ranks pressing close behind. Moreover, a solid ball did not instantly lose all its energy when it landed; after impact it would often roll and skip for a considerable distance in an erratic manner, retaining sufficient energy to damage anyone it hit. Undoubtedly the gunner's favourite target was a close-packed mass of men, for minor errors in

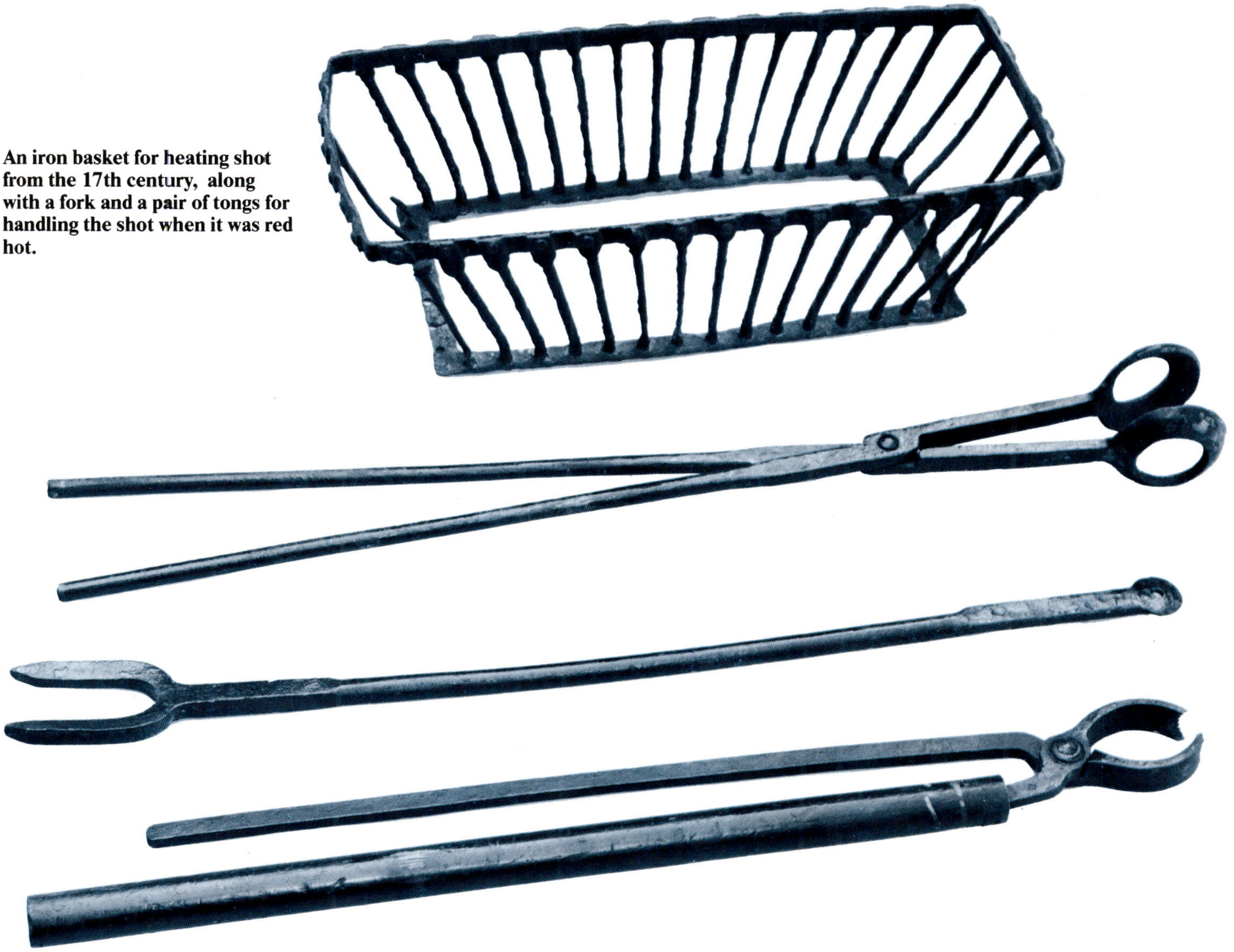

An iron basket for heating shot from the 17th century, along with a fork and a pair of tongs for handling the shot when it was red hot.

aiming, or in the performance of the gun or the powder charge, or in the flight of the ball, would be nullified by the sheer density and vulnerability of the target.

But there were, from the earliest days of artillery, those who sought diligently after a projectile which would contain some explosive effect, and thus have even greater damaging power at the target. Technical problems abounded, however. In the first place the explosive filling had to be set off at the correct place; in the second there was a great danger that the explosive might be jarred into action by the impulse of being shot from the gun, so disposing of the gun and gunners instead of the enemy.

The first moderately successful explosive shells appear to have been fired by the Venetians as early as 1421. The shells were formed of two hollow hemispheres of bronze, joined by a hoop of iron, and with a fuze consisting of a length of quickmatch inside a sheet-iron tube riveted to one of the hemispheres. The interior was filled with a charge of coarse gunpowder. The projectile became known as a *granata* or 'grenade', because the grains of gunpowder were thought to resemble the seeds in a pomegranate. (The word 'shell' appeared later, from the German word *Schall* for 'outside rind or bark'.) This was loaded into a short-barrelled mortar with the fuze against the gunpowder charge so that when the charge was fired to propel the shell, it also lit the fuze. The fuze burned through and, at just about the end of the shell's flight, ignited the gunpowder filling, bursting the shell into fragments.

The business was of course fraught with danger. The construction of the shell was prone to a less than tight joint between the hemispheres, through which the charge's flash could penetrate and burst the shell inside the gun. Or the quickmatch might not thoroughly occupy the iron tube, leaving room for flash to pass down alongside. Or the choice and manufacture of the quickmatch might be faulty so that it flashed down and fired the shell perhaps as it left the muzzle. In any event, the result was unpopular with the gunners and shells never became common as an article of war in the 15th century.

It was not until the 17th century that shells became popular in Europe. The reason for the delay was simply that a shell was not a practical device without an accurate fuze, and there could not be an accurate fuze until there was an accurate and convenient method of measuring short intervals of time. When the first practical watches appeared, in the 1670s, the problem was at last amenable to solution.

The fuzes developed for shells were simple enough devices: tapered cylinders of beech wood (chosen for its hardness and impermeable grain) through which a central hole was bored and filled with a composition of gunpowder and 'spirits of

Muzzle-loading shells with bronze studs angled to the gun's rifling pitch.

wine' — alcohol. This, rammed in to a suitable density, burned relatively slowly, and the wooden body of the fuze was marked off in distances. The fuze was physically cut off at the appropriate distance by means of a knife, and then driven into a hole in the shell; when fired, the flash from the charge lit the exposed outer end of the train of gunpowder, which then burned through until it lit the filling of the shell. Interestingly the expression 'to cut a fuze' is still in use with some artillery forces to this day, even though the setting of the modern fuze is achieved by a totally different method.

The method of igniting the fuze still posed problems. Loading the shell with the fuze towards the charge was one way; another was to use a gun with two vents, one to ignite the fuze and the other to ignite the charge (a rare system, as might be imagined); and a third was to load the shell with the fuze towards the muzzle and then, shortly before firing, reach down through the muzzle with a length of slow match and ignite the fuze. (It can be appreciated that this latter system was far from popular with the unfortunate who was to do the lighting.) But in about 1740 some unknown genius suddenly realized that there was enough air-space around the shell to permit it to be loaded with the fuze towards the muzzle and still be lit efficiently by the charge flame washing around the shell. That there was such space followed from the necessity for the shell to be slightly smaller than the bore of the gun in order that during ramming air would not be trapped underneath it. And where air could escape, flame could pass. And so 'single fire' became the vogue in the mid-18th century.

Another desirable effect which inventors pursued was the use of incendiary materials in shells, both to ignite material and also to provide illumination during night-time warfare. Various expedients were tried with little or no success until the 'carcass' was invented in 1672, reputedly by a Master Gunner in the service of Christopher van Galen, the Prince-Bishop of Münster. The carcass was a spherical iron framework wrapped in cloth and cord and filled with a mixture of saltpetre,

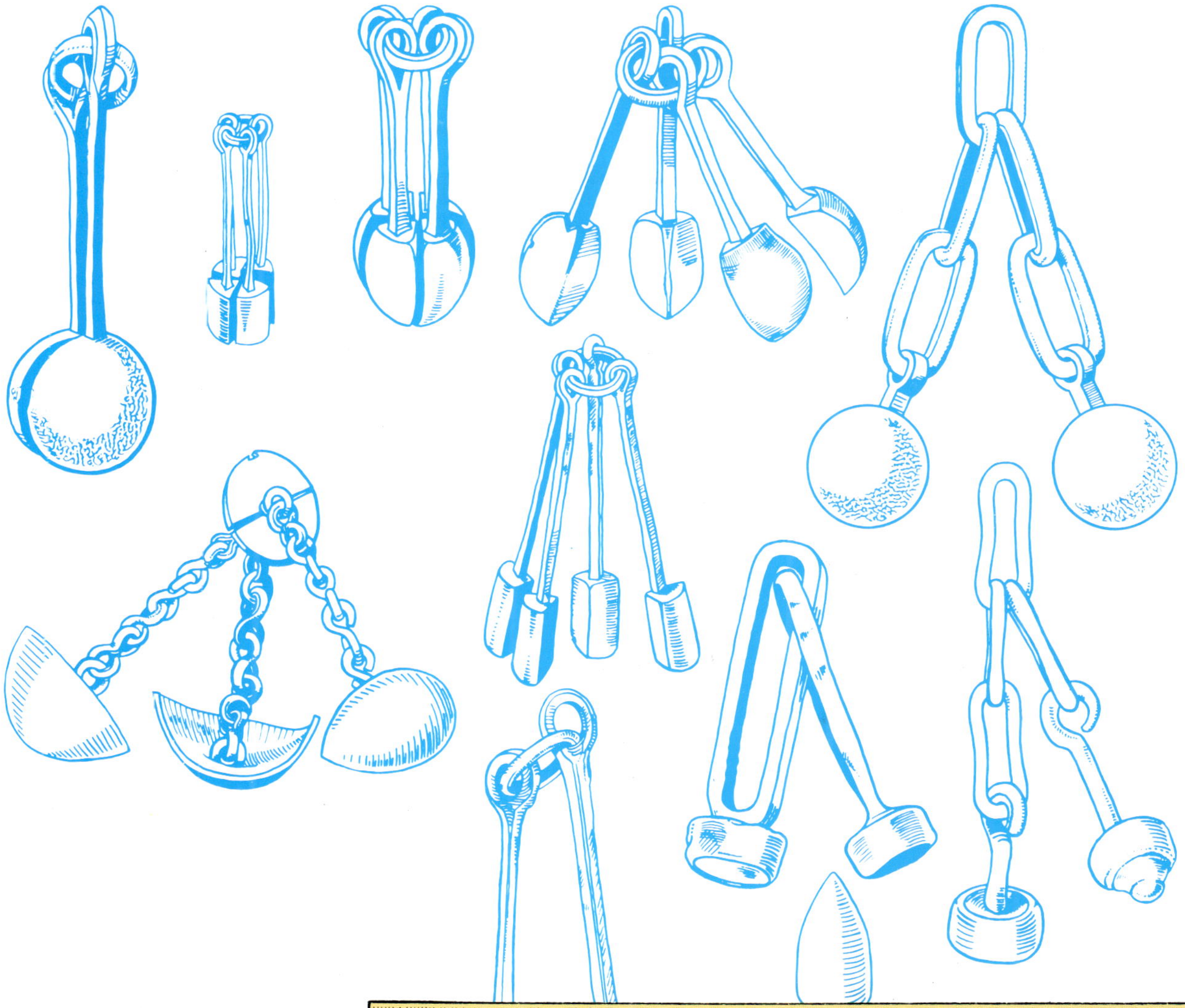

Above: Various types of expanding shot for cannon used particularly by naval guns for cutting rigging.

Above right: An early light ball. The fuze is coiled around the body and burns in flight.

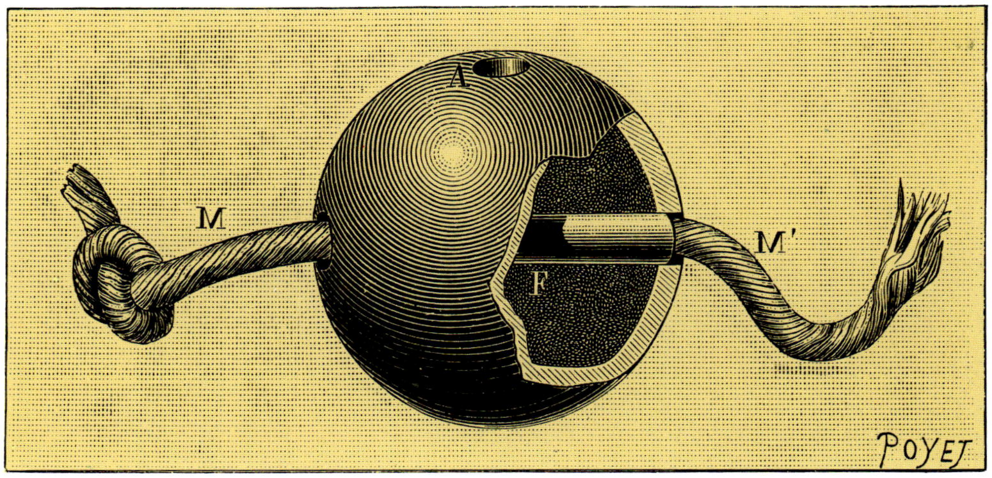

Right: An early type of grenade which was thrown by means of a rope.

sulphur, Venetian turpentine, tallow and resin. This villainous compound, ignited by the normal type of fuze, would then burst asunder, scattering the flaming liquid in all directions; because of its composition it would stick to surfaces and ignite them too, proving very difficult to fight with nothing more sophisticated than water. The only drawback in practice was the fact that in order to get the maximum filling into each carcass, the construction was somewhat flimsy; unless precautions were taken it was all too possible for them to burst in the bore of the gun through the shock of the

Casting cannon balls circa 1800. The man on the right is cutting off the cast marks to make the cannon ball round.

charge being fired behind them. As a result, it was normal to reduce the charge and pack a thick wad of turf or some similarly absorbent substance on top of the charge to act as a damper between the explosion and the carcass.

A variant of the carcass which also appeared in the 17th century was the 'smoke ball' which, according to a contemporary document, 'during their combustion cast forth a noisome smoke and that in such abundance that it is impossible to bear it'. In view of the contents of the smoke ball — saltpetre, coal, pitch, tar, resin, sawdust, crude antimony and sulphur — this seems a valid claim, and probably marks it as the forerunner of chemical warfare.

And so, with the artillerymen amply supplied with gunpowder and their assortment of projectiles, let us leave them for a while and go back again to the 14th century to trace the second line of development: ammunition for hand weapons.

The cannon came first; the 'handgonne' came some time afterwards and was simply a shrunken version of the contemporary cannon that was mounted on a wooden stave which could be tucked under the firer's arm to give support and some direction to the weapon. The design was gradually refined to a more manageable size, and split into two — the shoulder arm and the hand gun — each of which passed through stages of development without moving from the muzzle-loading concept.

The breech of an early breech-loading gun which belonged to Henry VIII.

These three musketeers are carrying arquebuses complete with serpentine locks. This development allowed the gunman to hold the firearm firmly with both hands while firing.

The history of small-arms ammunition during the period is relatively meagre in that it is simply the history of lead balls and gunpowder, neither of which changed very much once the definitive forms had been arrived at in the 16th century. The only important change lay in the method of ignition which, properly, is a matter for the history of the gun and not of the ammunition. Nevertheless that history does eventually lead into the sphere of ammunition, and an understanding of the problem is necessary in order to understand why a solution was so desirable.

The first small arms took the contemporary method of firing a cannon and extended it; the 'hand-gonner' tucked the stave under his arm, aimed — one way or another, since in those days there were no such things as sights — and touched a burning slow match to the vent, which he had primed with gunpowder. Experience soon showed that it was desirable to be gripping the weapon with both hands when it went off, and very quickly the 'serpentine lock' was invented: simply a Σ-shaped piece of metal pivoted on the stave, with the forward end holding the match and the rear end in a convenient place to be pulled by the firer so as to lower the match to the vent while he held the gun securely.

The serpentine lock was refined into the matchlock, in which the apparatus was fixed to a small plate at the side of the gun and controlled by a spring and catch. The arm holding the match was pulled back and retained, against a spring, by a stud which was released by being pushed by the firer, so that the match flew forward on to the vent. Since the lock was at the side, it made sense to alter the vent so that it too came out at the side of the barrel and ended in a pan into which loose powder could be sprinkled. At the same time the form of the weapon changed: the barrel became longer and a shoulder stock appeared, generally taking on the form that is recognized today.

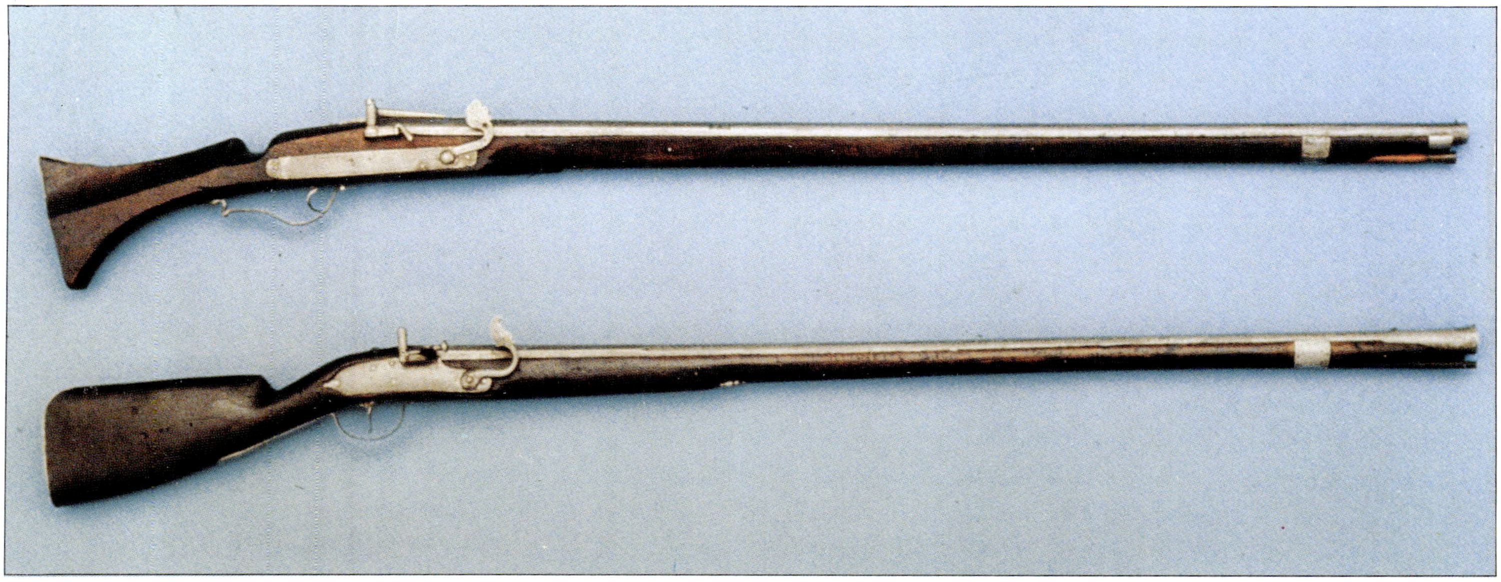

Two 17th century English matchlock muskets. The top one was made about 1630, the bottom one in 1690.

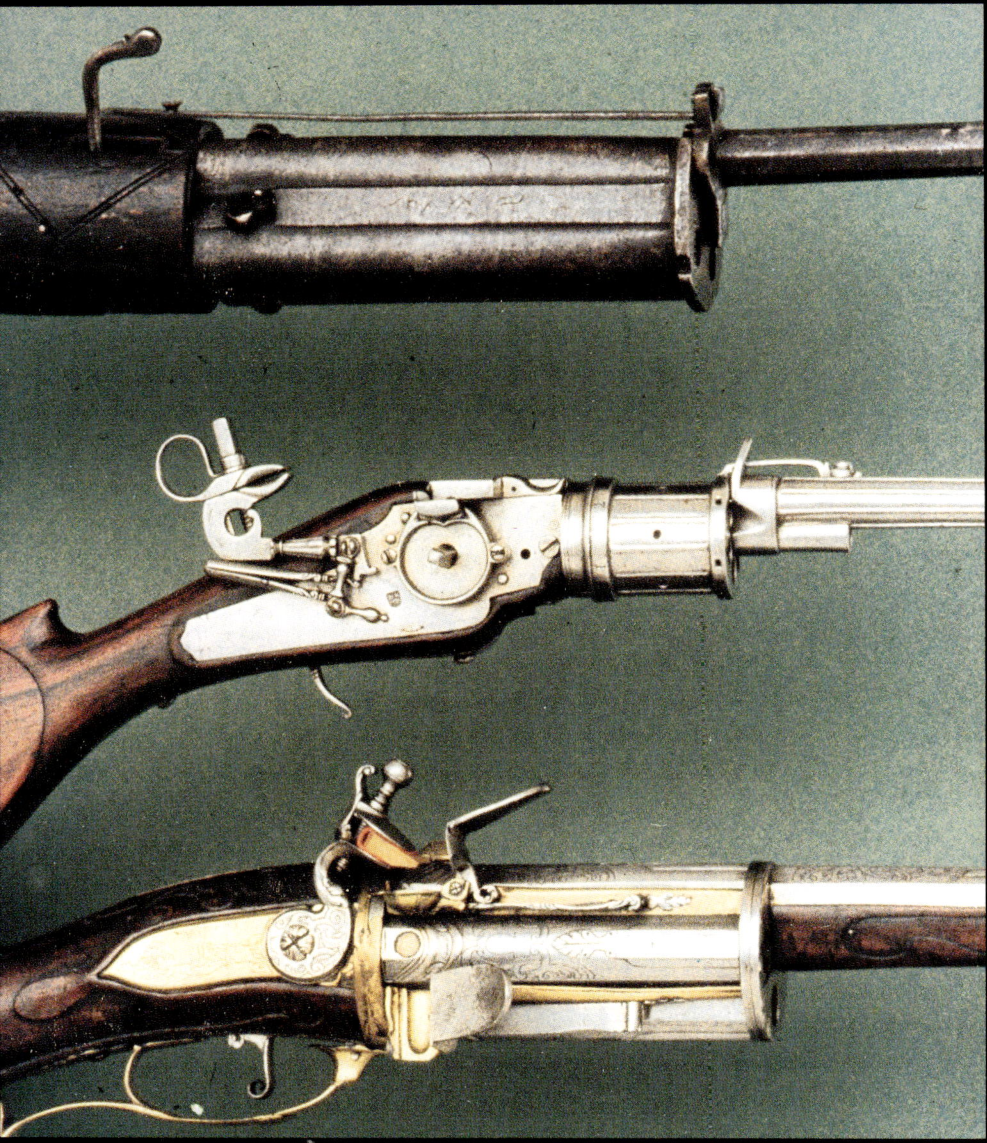

The guns that inspired Samuel Colt (from top to bottom): Indian matchlock, wheellock, flintlock, short musket of 1770.

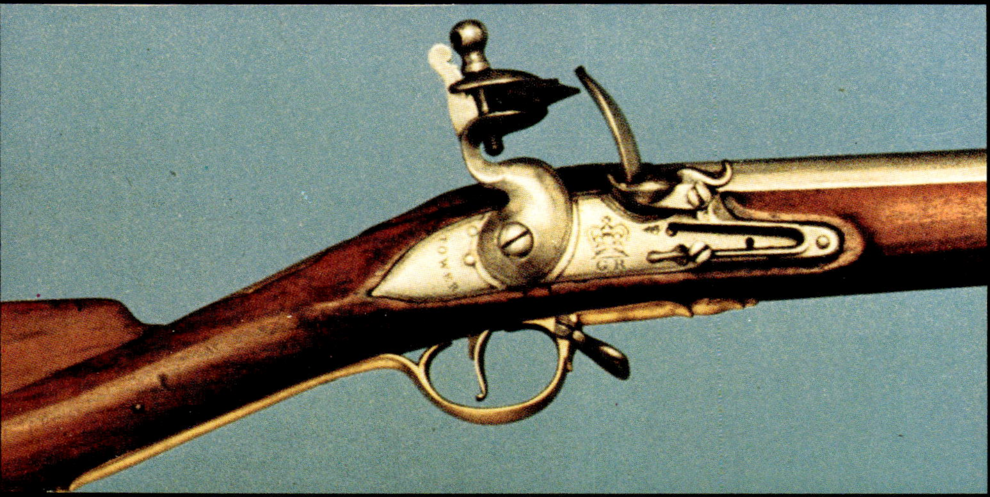

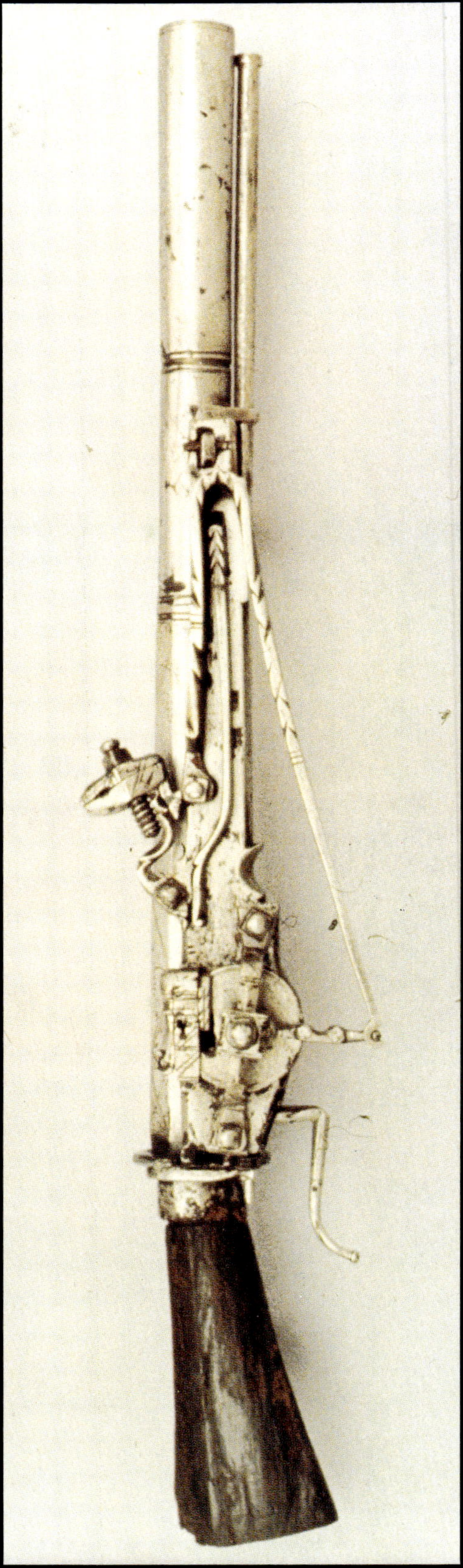

An Italian pistol originating from about 1520. It uses an early wheellock design. Note the unguarded trigger.

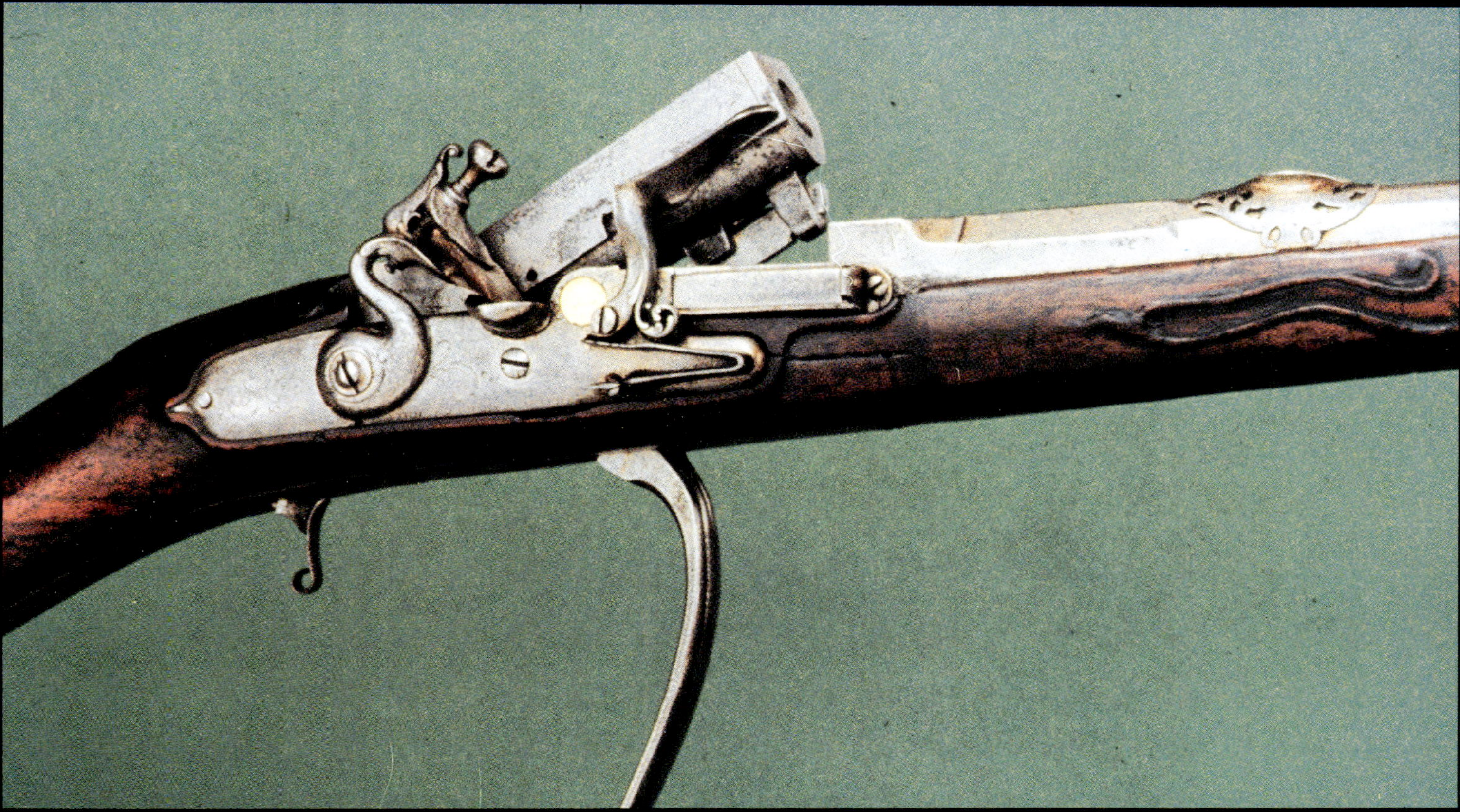

Above: A flintlock breech-loader by Birchell.

Below: Early repeating rifles with sliding locks.

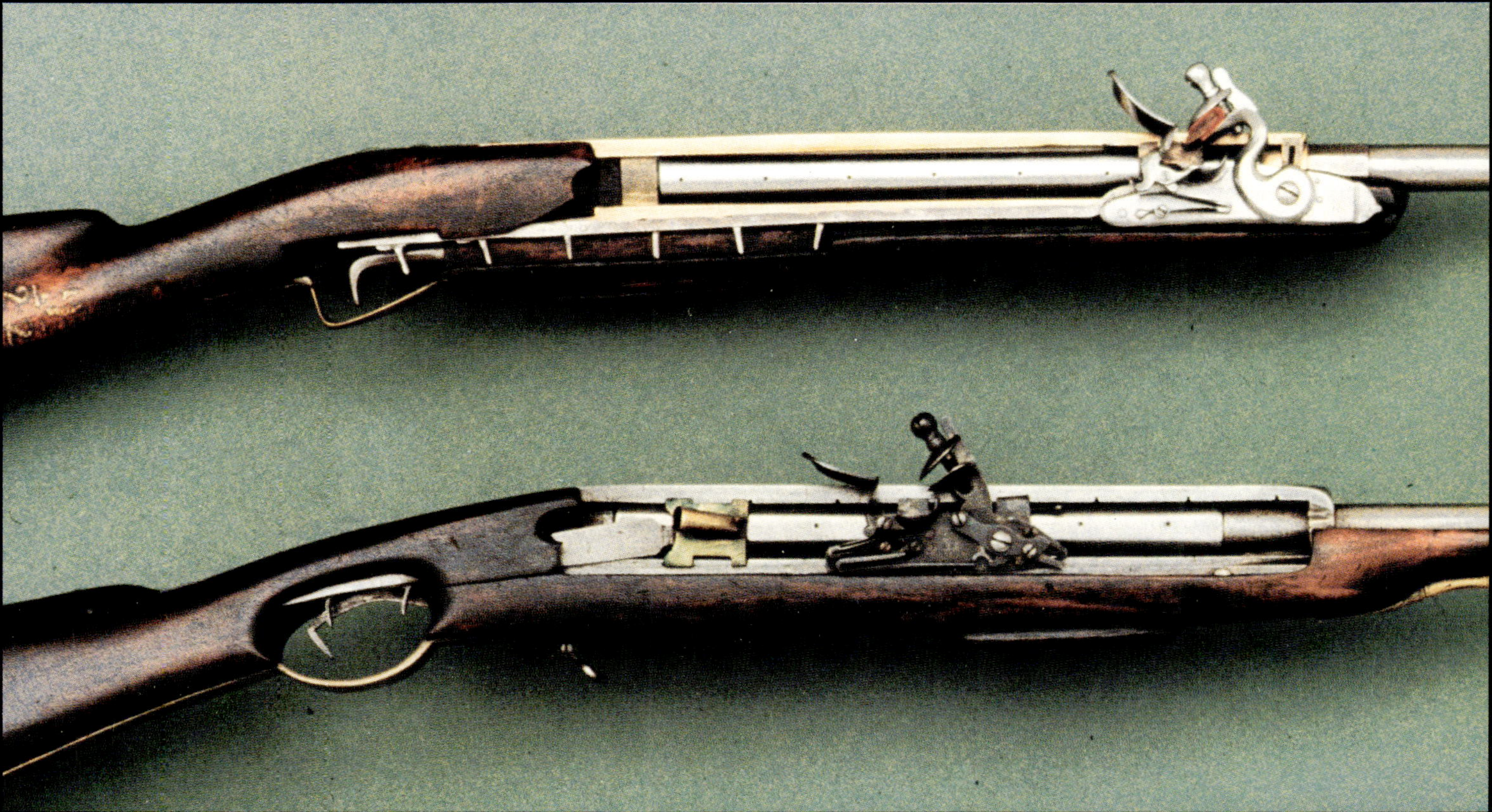

Reverend Alexander Forsyth who developed the first percussion system in 1805.

After the matchlock came the wheel-lock, invented in about 1500, which uses a roughened wheel, driven by a spring which had first to be wound up or 'spanned'. When the trigger was pulled the wheel spun, its surface in contact with a piece of flint, so that sparks were directed into the pan of powder.

About 50 years later the flintlock appeared (first called the 'snap-lock'). By this mechanism a flint was securely held in the jaws of the 'cock' and allowed to move forward, when released by the trigger, to strike a carefully positioned piece of steel and so produce the requisite sparks. With later minor refinements this was the height of technical development in small-arms ammunition systems, and it remained the only method of ignition until the 19th century.

All these systems were clumsy and prone to accident. The pan was fitted with a lid fairly early in its development — an endeavour to keep out damp and prevent the loose powder from being blown away or otherwise dislodged. The pan cover later formed the steel on which the flint struck, so that the blow performed two functions, generating a spark and throwing open the pan to expose the powder. Indeed, the whole business of loading was wide open to accidents: the powder had to be sifted down the barrel; the ball inserted and rammed; a wad rammed on top to prevent it from falling out again or rolling away from the powder; loose powder had to be sprinkled into the pan and vent; the flint had to be prepared correctly and properly set into the jaws of the cock ... It was remarkable that the firearms worked at all.

One constant factor in flintlock weapons was the appreciable pause between pulling the trigger and the gun actually going off. Watch a flintlock being fired and it is possible to distinguish the separate events; first the flint ignites the pan and then, a fraction of a second later, the discharge of the gun itself follows. If such a delay was inconvenient for soldiers, it was positively traumatic for sportsmen. At the first flash of the flint and pan many a quarry would take fright and flee often far enough during the pause to escape the line of fire entirely (particularly if the gunner's aim had been poor in the first place).

It was this irritating characteristic of the flintlock gun which led the Reverend Alexander Forsyth, Minister of Belhelvie in Scotland, to devote much of his life to developing a better method of igniting the powder charge. In 1805 he developed the first percussion system, using a mixture of mercury fulminate and potassium chlorate which, when struck with a falling hammer, detonates and gives off a strong flash. By replacing the cock of the flintlock with a hammer, and arranging for a small measure of powder to be dropped at the mouth of the 'vent' leading into the gun chamber, Forsyth produced a system that was practically instantaneous. He offered the idea to the Board of Ordnance, but in 1807 they turned it down. Forsyth duly patented it and, in conjunction with the famous gunsmith James Purdey, began to produce percussion sporting guns. In 1842, after much wrangling, he was awarded £200 by the Board of Ordnance, who had by now appropriated the idea for their own use. After his death in 1843 an additional sum of £1,000 was distributed among his dependants.

Various mechanisms for containing and placing the igniting powder were tried; the system evolved that was eventually regarded as the best was to place the powder in a tiny copper cap, rather like a miniature top hat. Pressed carefully into the 'crown' of the hat and secured there with a dab of varnish, it was waterproof and could not fall out. The vent of the gun was then finished off with a screwed-in hollow tube, called the 'nipple', which was in line with the hammer. The cap was placed, open end down, on the nipple so that when the hammer was released it fell, crushing the cap and the powder and thus forcing the flash to drive down the vent and into the gun chamber to ignite the powder charge.

A double-barrelled percussion shotgun made by le Page Moutier of Paris which was shown at the 1862 exhibition at South Kensington.

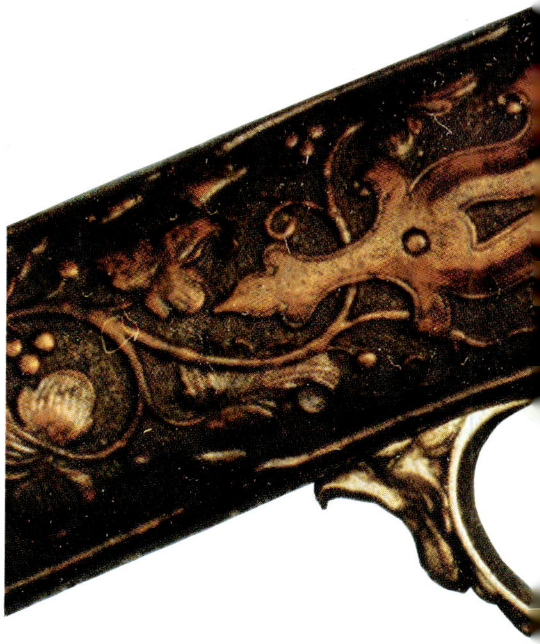

By 1840 the percussion system had almost entirely replaced the flintlock. The British Army adopted it in 1838, the Americans in 1842, and by that time almost every other army had also taken to it. In the sporting world its adoption was even faster, and it is safe to say that by 1845 the number of flintlocks still in use was negligible. Many gunsmiths made a good living by converting flintlocks to percussion, merely fitting a hammer and nipple in place of the cock and pan, since the rest of the gun remained the same.

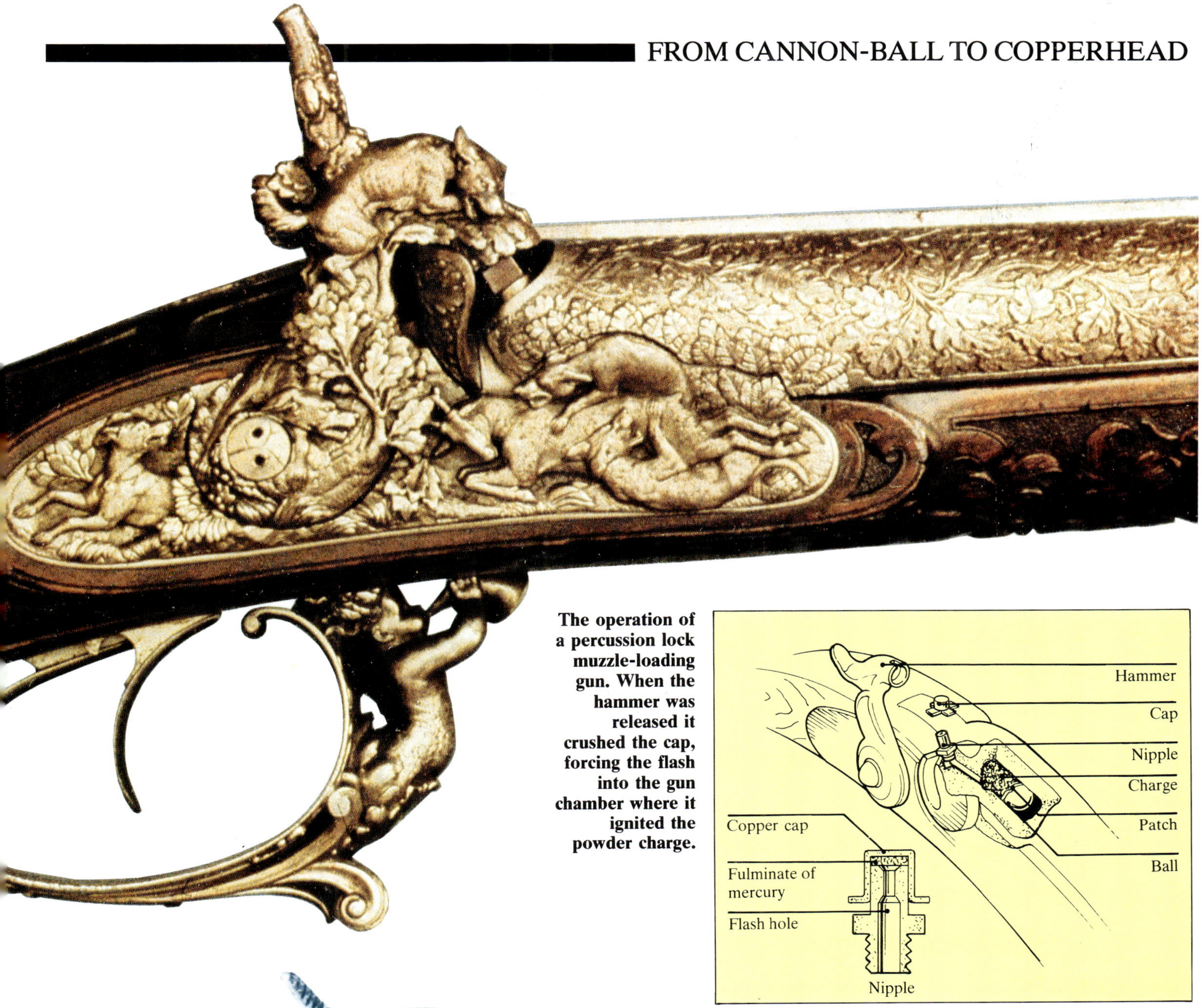

The operation of a percussion lock muzzle-loading gun. When the hammer was released it crushed the cap, forcing the flash into the gun chamber where it ignited the powder charge.

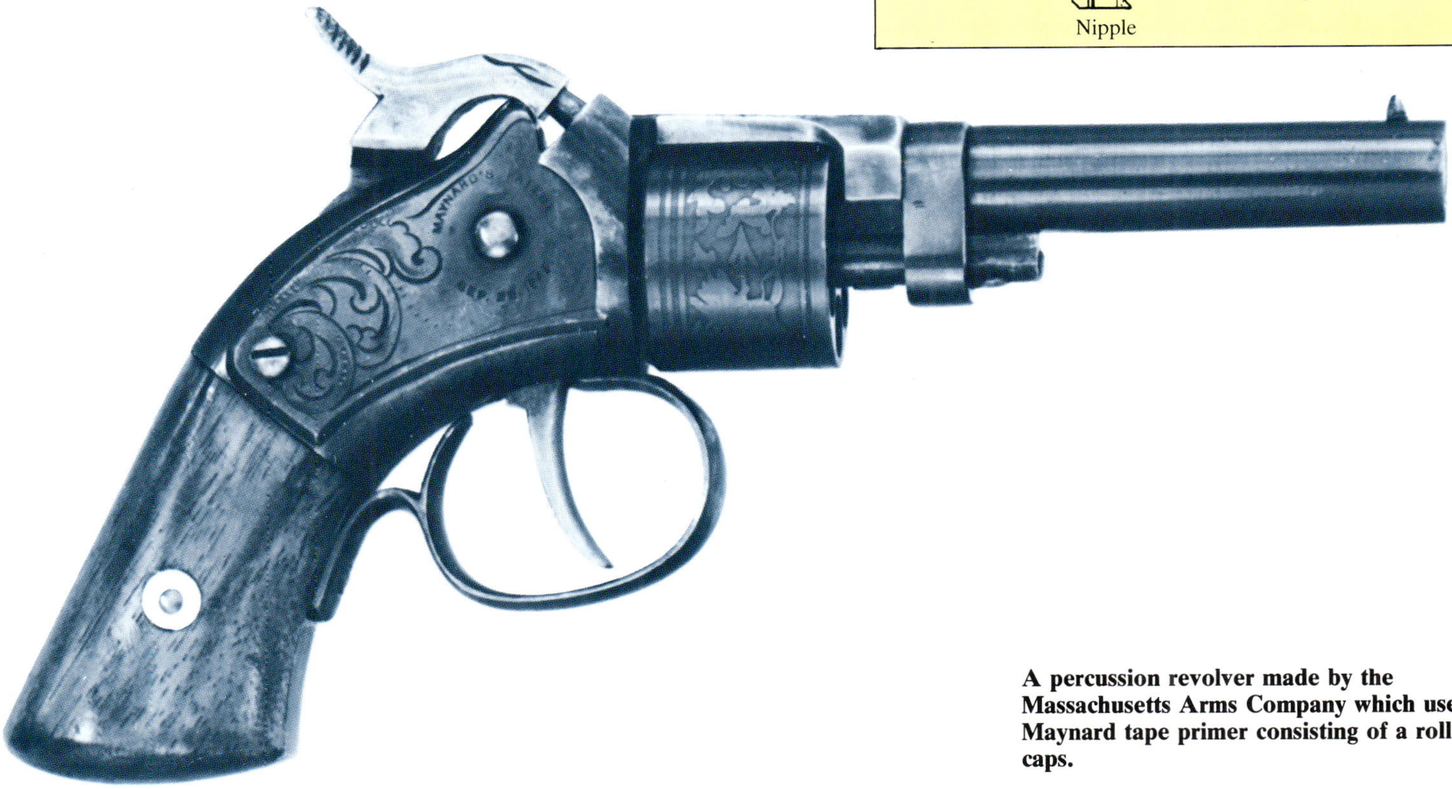

A percussion revolver made by the Massachusetts Arms Company which uses a Maynard tape primer consisting of a roll of caps.

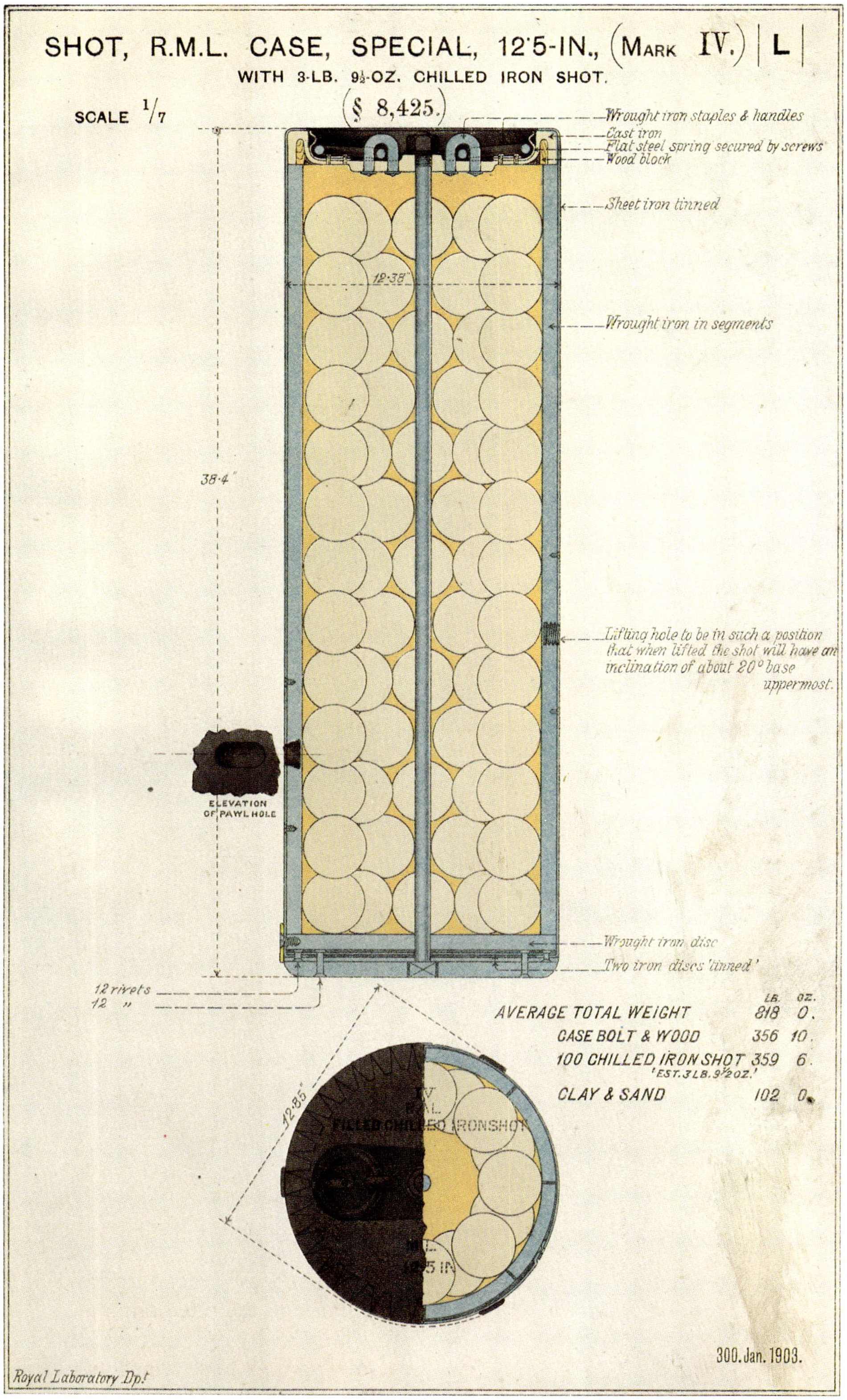

An early example of case shot. The diagram shows how the shot was packed.

So convenient a method of priming the weapon demanded an equally convenient method of loading, and by this time the casual sprinkling of powder down the muzzle had largely been overtaken by preformed cartridges, particularly for military use. The cartridge was a tube of paper containing the correct powder charge and the ball. To load, the paper was torn open and the powder emptied down the barrel; the paper was then crumpled and forced down on top of the powder as a retaining wad, making certain that no matter how the gun was manipulated the powder remained in the chamber and adjacent to the ignition system; the ball was thrust into the muzzle and rammed; the hammer cocked and a cap replaced on the nipple; and the weapon was ready to fire.

The spherical lead ball that had been the standard projectile for hand weapons since their introduction had certain defects. The biggest one was that it was directionally inaccurate. In order that the ball could be forced down the barrel it had to be of smaller diameter than the inside of the bore of the weapon. When fired, it thus tended to bounce from side to side as it went up the barrel, and its final direction depended largely upon which side of the barrel it bounced off last. In flight it was retarded by the air through which it passed, and because it was the same size viewed from any angle (it was not in any way aerodynamic), the effect of wind was totally random, there being no inbuilt stabilizing factor. The fault had been realized for many years, and there had been periodic attempts to improve matters.

The most usual method was to rifle the barrel of the gun; by cutting a series of spiral grooves in the bore it was possible to make the ball spin, and this gave it gyroscopic stability which tended to keep it on a constant path. In order to make the ball grip the rifling, however, it had now to be slightly bigger than the diameter across the raised part of the rifling (known as the 'lands'). To load the ball it had to be hammered down, engraving into the rifling as it went, so that it would ride in the grooves as it left. Such hammering usually deformed the soft lead ball, with the result that any accuracy gained from the rifling

Early spherical case shot or shrapnel. These were little more than hollow cannon balls filled with lead balls and a small central burster charge of gunpowder.

was generally lost due to the battered shape of the ball as it flew through the air.

The obvious solution was to arrange things so that the ball was below bore diameter as it went in but was somehow expanded to slightly greater diameter before it left. Achieving this aim brought forth some peculiar stratagems. One of the earliest systems was to have an upstanding pillar in the breech end of the barrel; when the ball was dropped in it stopped on the tip of this pillar, the powder of the charge being distributed around the pillar and under the ball. A few sharp blows with the ramrod then caused the ball to be driven against the pillar and expand outwards so that it bit into the rifling. When fired, the ball was engaged in the rifling and flew quite well.

It then occurred to some experimenters that the explosion of the charge might be used to do the expanding, if the shape of the bullet were altered. And so, for the first time, a cylindrical bullet with a pointed nose was employed. The base of this bullet was hollowed out to leave a relatively thin 'skirt' and the hollow filled with a hard metal cone. Loaded, the bullet slipped down inside the rifling and came to rest on the powder charge and wad in the usual way. When the charge was fired the sudden eruption of gas pressed on the metal cone, which in turn pressed on the soft lead skirt and forced it outwards into the rifling as the bullet gathered speed. The 'cylindro-conoidal' shape gave better flight characteristics, and this 'Minié Bullet' — named after its promoter rather than its inventor — was soon adopted for military use all over the world. The British Army tried it but objected to the metal cup: it was too expensive, too fiddling to manufacture and, if the truth were known, probably would have involved paying a patent fee. So the British adopted a bullet of their own which resembled the Minié in principle but used a far simpler method of expanding the base: a solid clay plug, which, impelled by the gases, achieved the expansion just as well as the expensive metal cone and did far less damage to the rifle barrel.

By the 1850s, therefore, the infantry of the major armies were armed with a rifled muzzle-loading musket which could deliver a heavy ball — usually of about .5-inch calibre — with considerable accuracy out to quite long ranges. This brought about something of a revolution on the bat-

Right: a spherical shrapnel shell. The bursting charge is separated from the bullets by a thin metal diaphragm. Note that the wooden base holds the boxer fuze at the top, away from the propelling charge.

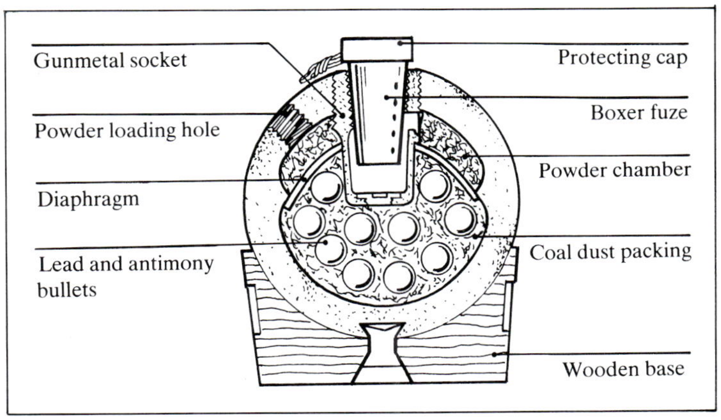

Below right: a cartridge for a muzzle-loading rifle using an expanding 'Minie' bullet.

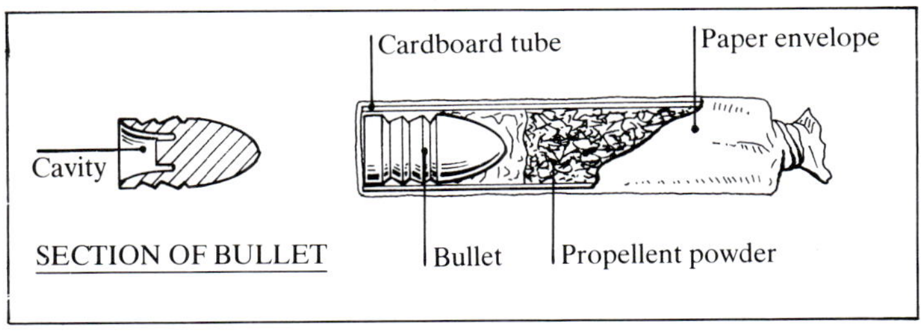

tlefield. For the first time it became possible for infantry to out-range artillery; the smoothbore guns of the day showed little improvement over those of two centuries before, but the rifle had made great strides, and the cannon was no longer the master of the battlefield. Guns brought into action in full view of the enemy perhaps 450 metres (500 yards) away, as had been the tactical principle for years, were at considerable risk, since the enemy infantry could shoot the gunners a good deal more easily than the gunners could deal with the mass of infantry. The time had come to make some fundamental changes in artillery.

There was just one item of ammunition which prevented complete disaster for the artillery of the time, and that was the shrapnel shell, named after its inventor Colonel Henry Shrapnel.

This projectile had been developed as a method of killing infantry at long range with more certainty than a solid cannonball could offer. In the Siege of Gibraltar towards the end of the 18th century a Captain Mercier had experimented with firing mortar shells from a suitably sized howitzer; fitted with a short fuze and fired with a reduced charge these shells burst in the air over the enemy infantry, showering them with splinters.

After the siege, Shrapnel (then a lieutenant) set about improving on the idea, and in 1784 he produced his 'spherical case shot'. This was a thin iron ball containing a filling of musket balls and gunpowder; a wooden fuze was inserted and when it burned through it ignited the gunpowder which then exploded to shatter the casing and blow the musket balls in all directions. Shrapnel's idea was to 'extend the range of case shot' by allowing the projectile to get to the enemy before it burst, so that he reproduced the effect of case shot but took it from the gun muzzle and placed it among the enemy.

The whole point about Shrapnel's development is that he did not simply cram as much gunpowder as possible into the shell in order to obtain a powerful blast. The amount of gunpowder was deliberately kept low so that all it did was open up the shell — the balls obtained their velocity from the forward motion of the shell at the time of the burst. After many tests and trials it was formally approved in 1792, but it remained known as 'spherical case shot' until it was officially renamed 'Shrapnel shell' in 1852 in memory of its inventor.

A typical example of case shot, sectioned to show its constituent parts.

Apart, though, from shrapnel the artillery had few means of overcoming the new superiority of the infantry. The gunmakers therefore urgently began to look at methods of rifling a cannon and, more important, of developing suitable projectiles. There was no possibility of using the expansive properties of lead, as did the shoulder arms. Ideas had been put forward from time to time in the past: in 1821 Lieutenant Croly of the 1st Regiment of Foot had suggested rifling the gun barrel and using a lead-coated ball; a similar idea was later put forward by the Swedish Baron Wahrendorff in 1846; in 1842 a Frenchman, Cavalier Treulle de Beaulieu, had suggested a cannon with deep grooves and with studs on the shell to ride in the grooves; and in 1845 a Major Cavalli of the Sardinian Army had proposed a gun rifled with two grooves and a shell having two ribs to match. Although some desultory tests were made of these various ideas, nobody took them particularly seriously until the Crimean War in 1854. Then money suddenly became available and soldiers began to show some interest, especially when confronted with long-range infantry fire. The British Government took the decision to provide a number of guns rifled according to an idea by Henry Lancaster, an English gunmaker. Lancaster's idea was to make a gun with an oval bore which was twisted; the projectile was oval in section and had opposite sides planed on the skew so as to match the twist of the bore.

The guns were not a success. The shot showed a tendency to jam in the bore, either on loading or firing, damaging the bore as a result, and the already poor accuracy rapidly worsened. They were soon withdrawn from service and the Army cast around for a better idea.

One was provided by a Mr William Armstrong, a solicitor from Newcastle-on-Tyne, who had a controlling interest in an engineering company. Appalled by reports from the Crimea which highlighted the great difficulties involved in moving an 18-pounder (weighing three tons, or 3,000kg) through the mud at of Inkerman, Armstrong sat down and designed a totally new type of gun, his object being primarily to save weight and distribute what weight there was in a more scientific manner. At the same time, he introduced the concept of loading the gun from the breech end rather than from the muzzle, rifled it, and developed a new type of ammunition.

The details of Armstrong's system of gun manufacture is less significant to this historical outline than the manner of loading, which was both revolutionary and very important for future developments. The rear end of the barrel terminated in a sort of box, open at top and bottom and with a threaded hole in the rear end opposite the entrance to the barrel — or the 'chamber' as the rear end was now called. The projectile was loaded in through the hole, followed by the propelling charge — a bag of gunpowder — and then a slab of steel called the 'vent piece' was dropped into the top of the box to close the rear of the chamber. It was then jammed tightly against the chamber, so as to make a gas-tight fit, by putting a large screw through the hole in the 'box' and winding it tight. The face of the vent piece had a copper plate which helped to make an effective seal to prevent the gas leaking out when the cartridge was fired.

Firing was carried out by having a hole in the vent piece and using a 'friction igniter', a goose-quill filled with powder and with a topping of phosphorus compound. A jagged piece of metal passed through the compound and when this was jerked free the roughness ignited the phosphorus — a system analogous to a common match being struck on a rough surface — and the powder was fired, shot flame down the vent, and fired the cartridge.

The gun was rifled with several shallow grooves. The projectile was coated with a layer of lead; moreover, it was more or less shaped as shells are today — a cylinder with a rounded nose. The shell sat in the smoothbored chamber and when it was exploded into movement by the powder charge the lead coating bit into the rifling, which began just in front of the rest position, and the shell left the barrel

Above: Sir William Armstrong, the inventor of the breech-loading gun.

Below: The mechanism of the Armstrong rifled breech-loading cannon.

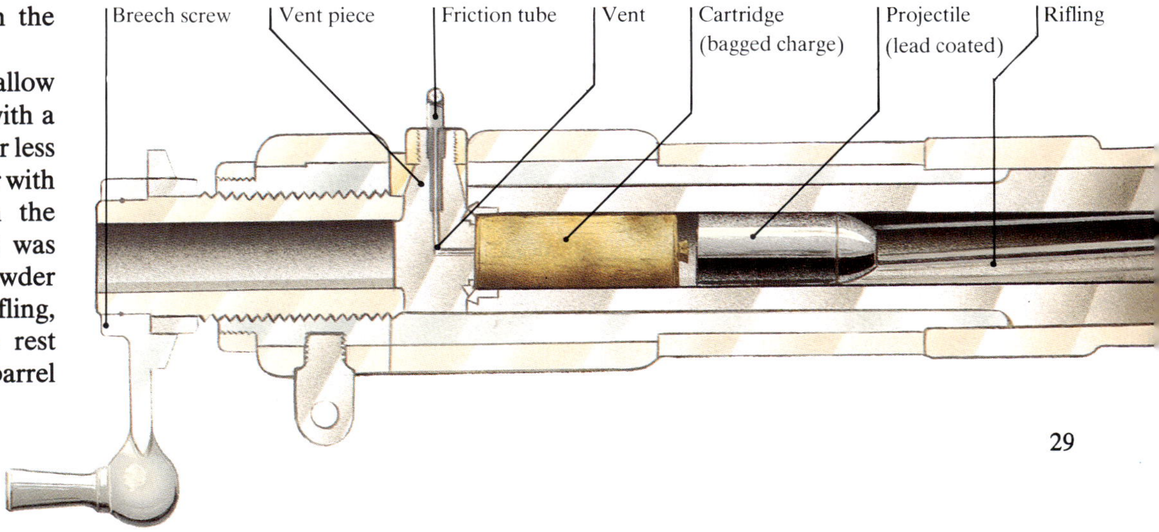

Left: *Gloire*, the first French ironclad whose construction forced the development of heavy artillery in Britain.

Right: A friction igniter — an improvement on the early goose-quill type. Here a copper tube is used instead.

Below: The British response to the *Gloire* was the *Warrior*. This was the first British ironclad.

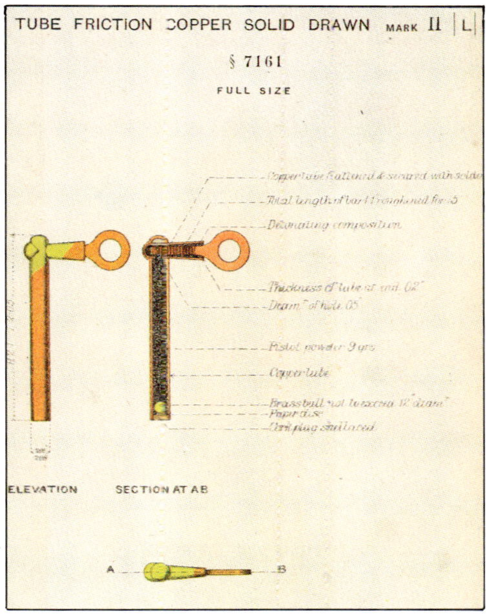

spinning. This gave the Armstrong gun remarkable accuracy in comparison with any other cannon, and it also meant that the shell arrived at the target point-first, unlike a cannon-ball which might be in any attitude when it arrived. Armstrong rounded off this technical *tour de force* by inventing a time fuze, one which relied on a length of gunpowder and could be adjusted by simply twisting a brass cap to set an arrow against various figures engraved on the body of the fuze.

Several thousand Armstrong guns were built for the British Army and Navy between 1855 and 1865 — but then came a sudden reversion to muzzle-loading, which has frequently been held up as an example of British resistance to modern innovation. In fact, it was a reasoned response to a difficult problem the British Navy now faced.

The French Navy had caused the problem by launching the world's first ironclad warship and compounded it by uttering belligerent noises even as Napoleon III began talking of enlarging the French empire. The British Navy immediately replied by setting about building ironclads of their own, but the pressing problem was to find guns capable of defeating these armoured ships. A 68-pounder smoothbore, provided with a special hardened steel shot, could just about manage to penetrate the iron armour at short range. But such a tactic was no solution because the French ships, which were largely armed with shell-firing guns, would reduce the wooden British warships to shattered wrecks before much impression had been made on their own armour. The Armstrong gun, efficient as it was, was weak in its breech closing and thus unable to withstand a powerful enough charge to send a penetrating shot with sufficient velocity to smash the armour. The only solution was to go back to massive muzzle-loading guns that could take heavy charges and fire potent projectiles at high velocity.

Armstrong lead-coated shells — on the right the lead has been removed to show the retaining notches.

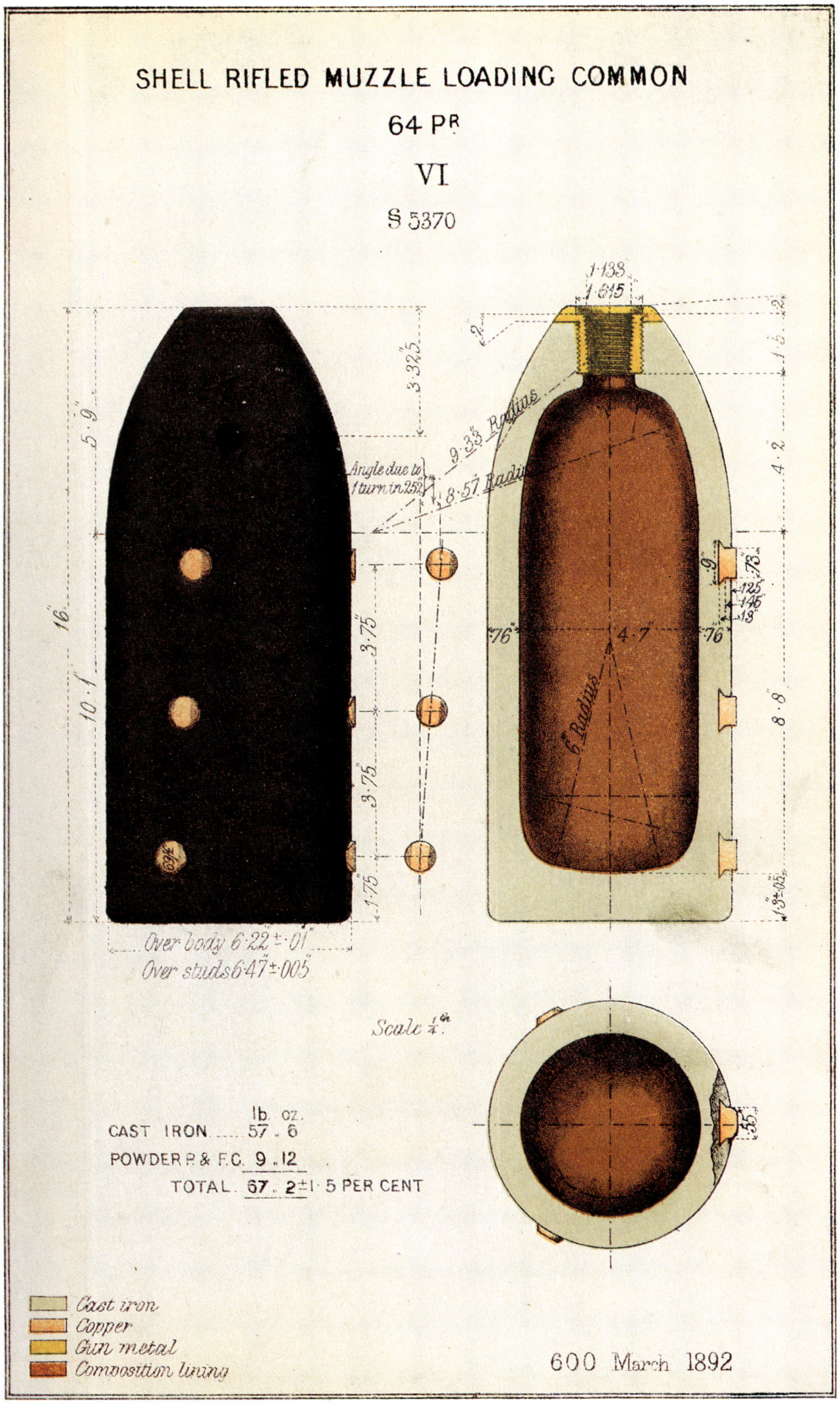

A 64-pounder studded shell — the copper studs engage the rifling and spin the shell as it goes up the barrel.

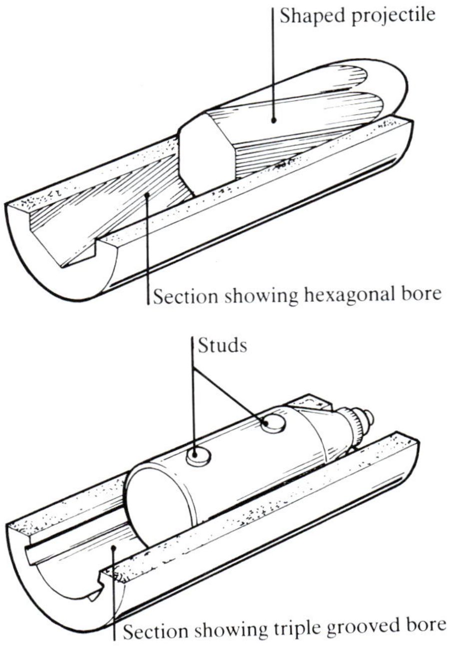

The rival rifling systems to impart spin to muzzle-loading projectiles. Above: The Armstrong. Top: The Whitworth.

It so happened that a suitable design was ready to hand. In 1863 another British manufacturer, Joseph Whitworth, had offered a new system of rifling in which, again, the bore was actually smooth but was hexagonal and with a twist in it. The projectile was likewise hexagonal, the sides being angled so as to match the angle of twist. He objected to the Armstrong monopoly of providing British guns and demanded a comparative trial; a Whitworth muzzle-loader, an Armstrong breech-loader and an Armstrong muzzle-loader were selected for competition, and the Armstrong muzzle-loading system was adjudged best. The Armstrong muzzle-loader used three deep grooves in the barrel and three rows of soft metal studs on the shell; the studs were introduced into the grooves as the shell was entered into the muzzle, rode down as the shell was rammed, and then rode up the grooves again, spinning the shell as it was fired. It was simple and reliable, whereas Whitworth's system, like the similar Lancaster, jammed its shot too often for comfort. And so the Armstrong breech-loading gun was quietly retired, and the RML (rifled muzzle-loader) came in its place.

Whatever the system, one thing was certain; the day of the cannon-ball was over, and the elongated shot or shell was now the rule. But actually perfecting elongated shot and shell to perform in the way the soldiers wanted was a slow business and led to some long and expensive trials. And yet, as is often the case, the best ideas came from outside the mainstream of ammunition development; an example is the piercing shot adopted in Britain.

The prime demand in the late 1860s was for something to go through 60 centimetres (24 inches) or so of wrought iron armour supported by a foot or more of solid teak wood, which was the accepted method of construction of the early ironclad ships. Cast iron had been the normal material for making cannon-balls, and it had carried over into Armstrong's shells, but cast iron flung against a wrought iron shield simply shattered. Attempts were made to construct steel shells, but steel-making was still something of an infant art in the 1860s, not entirely understood and not capable of regularly producing flawless billets from which shot could be made. Moreover, of course, steel was expensive.

What complicated the issue was that there were two schools of thought on the question of how to defeat armour: the 'racking' and the 'punching' schools. The 'rackers' wanted to fire a big heavy projectile at low velocity and thus 'rack' or strain the entire structure of the armour so as to bring about its collapse and expose the interior of the ship. The 'punchers' wanted high-velocity light shot which could punch through the plate so that both it and the fragments of armour it displaced would act as projectiles to do damage behind the armour. In short, the rackers were out to destroy the armour and the punchers were after whatever the armour was protecting. To a great extent the punchers were proved right and the rackers wrong, but their argument led in the meanwhile to a lot of time-wasting on the theory of flat-headed and round-headed projectiles.

As all this was going on, a Captain Palliser of the 18th Hussars — a regiment not noted at that time for scientific application — devised a pointed projectile which he cast with its nose down in a water-cooled

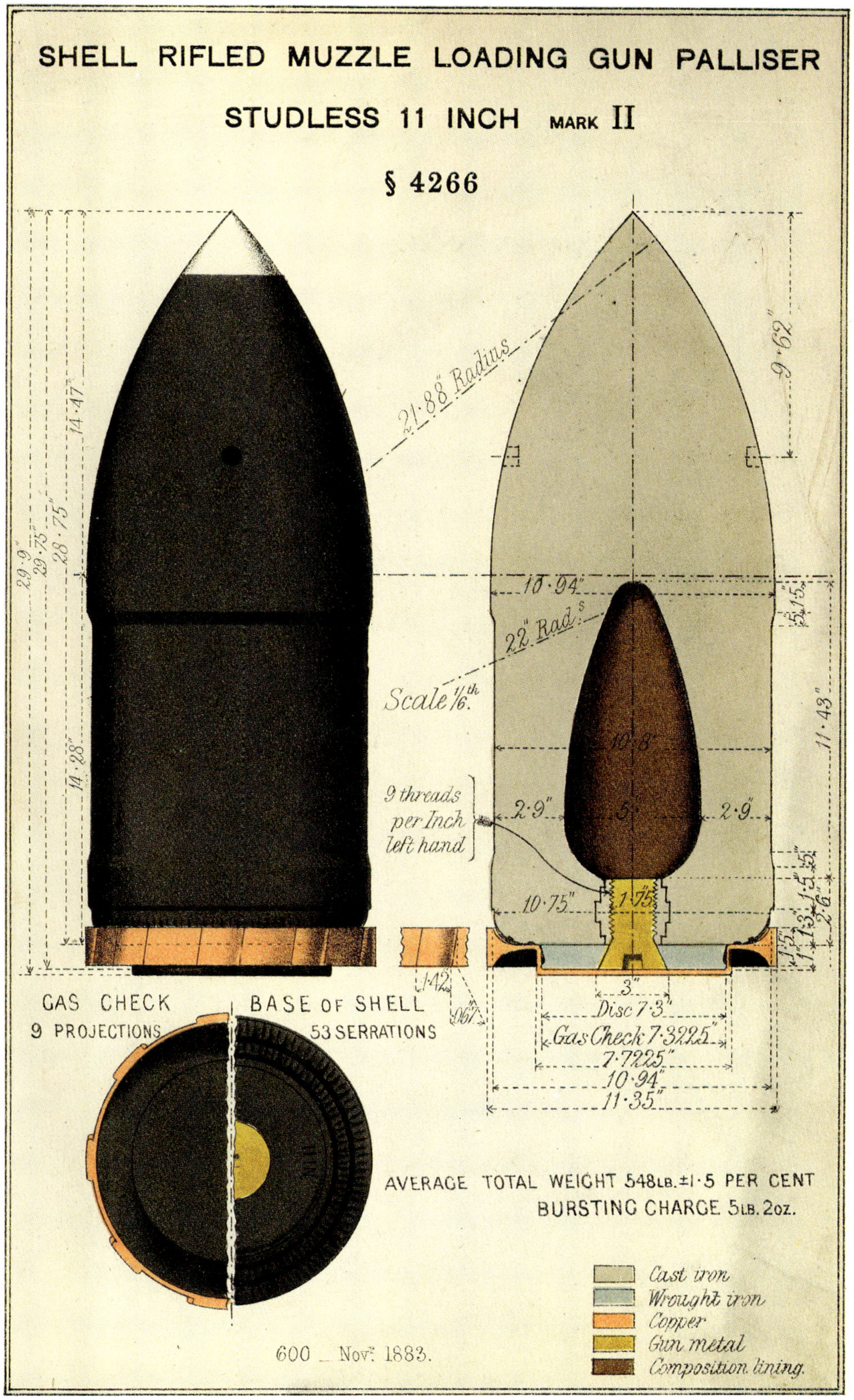

An 11in Palliser shell. The tip was cast nose-down in a water-cooled iron mould which made the point extremely hard.

iron mould, with a normal sand mould for the rest of the body. Giving a rapid chill to the nose made the point of the shell extremely hard, and it proved to be a most efficient projectile, one that was rapidly adopted for the RML guns. It is on record that in 1879, during the short war between Chile and Peru, the Chilean *Almirante Cochrane* fired a 9-inch Palliser shell which penetrated the gun turret of the Peruvian *Huascar*, going through 12 centimetres (5.5 inches) of iron plate, 32 centimetres (13 inches) of teak and a further centimetre (half an inch) of steel to explode inside the turret, kill most of the crew, and completely destroy the gun.

For those targets of lesser thickness the 'common shell' was devised; the word 'common' in this case indicates its application to 'common' targets (ie, anything other than thick armour). The common shell was pointed, like a piercing shell, but of plain steel or wrought iron, and was filled with a cotton bag containing coarse gunpowder. After the filling, the bottom of the shell was closed with a simple plug. When the shell struck the target, the sudden check to its flight caused the gunpowder to be thrown violently forward, and friction between the individual grains of powder was sufficient to cause it to explode and burst the shell. It was true that when the shell was fired there was a certain liability for the filling to set back violently, but careful insertion of the powder bag, making sure it was safely lodged at the rear end of the cavity, generally took care that the shell did not explode prematurely.

The greatest problem was adapting the shrapnel shell to the new rifling guns.

Right: A sectioned 18-pounder shrapnel shell dating from World War I.
Centre: The 18-pounder shrapnel shell complete with fuse.
Far right: A sectioned 64-pounder shrapnel shell dating from the 1870s.

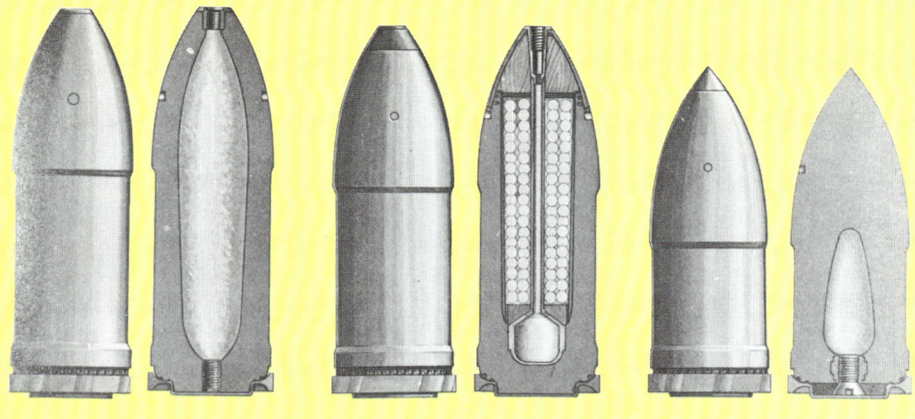
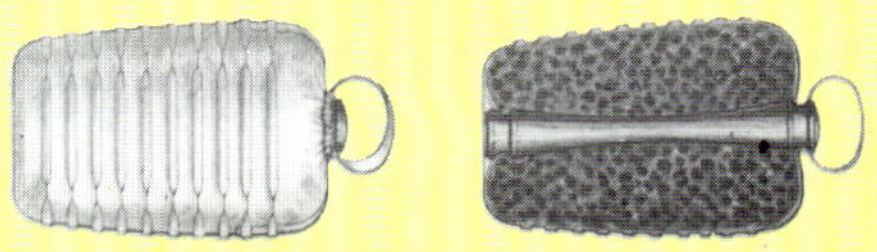

Common, shrapnel and palliser shells for muzzling-loading guns.

Eventually, Colonel Boxer, Superintendent of the Royal Laboratory at Woolwich Arsenal (which was the official ammunition factory) developed a design in which the nose of the shell was lightly pinned to the body and held a socket for a fuze. Beneath the fuze a tube ran down to a chamber in the base of the shell which was filled with gunpowder. Above this chamber was a loose round plate, the 'pusher plate', and above this, taking up the space in the body of the shell around the central tube, was the filling of musket balls packed in resin and coal dust. The shell was fired and the fuze lit, a flash being sent down the central tube to explode the chamber of powder after the fuze had burned through. This explosion forced the pusher plate up, carrying the balls with it, and in turn this force sheared the pins and threw the head of the shell off the body. That then fell to one side and the force of the gunpowder blast ejected all the musket balls in a spreading cone in front of the shell, following the trajectory which the shell was taking. So that if the fuze were correctly set, the shell could be burst in the air some 13 metres (40 feet) above and before the enemy and could then blast him with several score (or hundred, depending upon the size of the shell) musket balls at high velocity. As a man-killer, shrapnel was unrivalled.

Below left: Three wooden time fuzes dating from the mid 19th century.

Below: A diagram showing the construction of a Boxer time fuse.

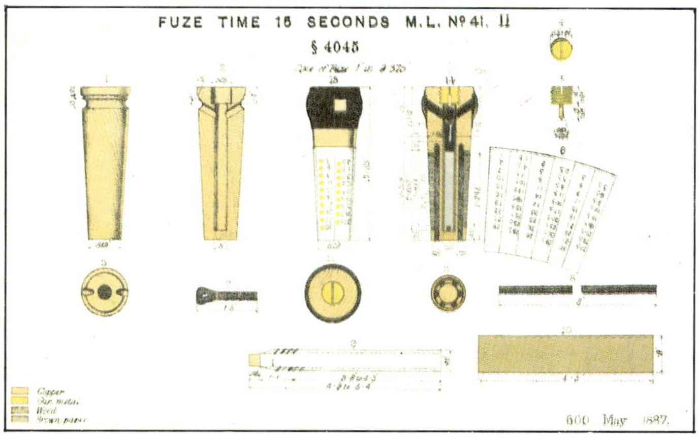

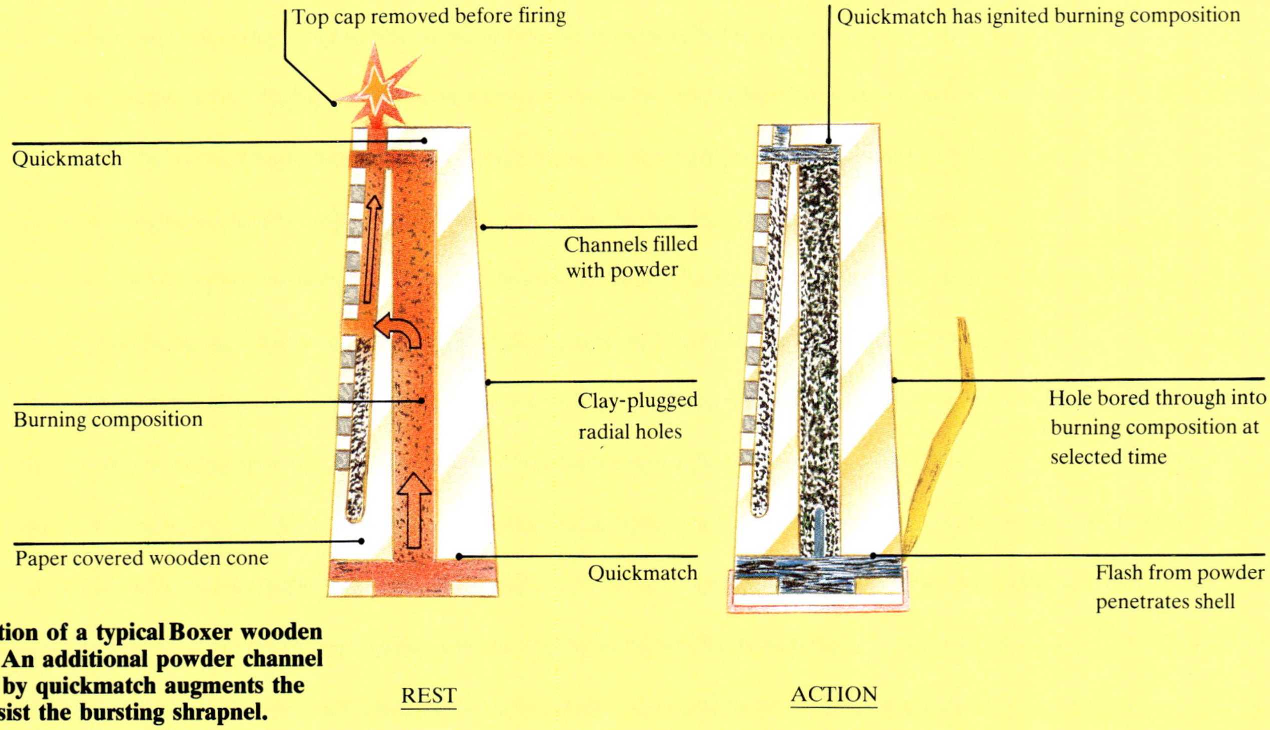

The operation of a typical Boxer wooden time fuze. An additional powder channel connected by quickmatch augments the flash to assist the bursting shrapnel.

It will be appreciated that with a shell as efficient as this, and with guns improving in their range and power, the old-style fuze (which had to be cut to set a time of burning) was somewhat inefficient. Boxer therefore designed a new pattern; it resembled the old insofar as it was a tapering plug of beech wood, but it had some considerable improvements. The central hole was bored in from the top, stopped before reaching the bottom, and was filled with slow-burning powder; above it was a length of quickmatch, coiled up in a recess in the top of the fuze. Parallel with the central hole, and off to the sides of the fuze, another six holes were bored in from the bottom, not reaching the top. More holes were bored in from the sides to connect with these six channels, and the whole system was then filled with gunpowder and the fuze covered in varnished paper. The location of the holes in the sides was indicated by spots of ink, and each hole was given a figure corresponding to the length of powder at that point in the system, each one-tenth of an inch being worth half a second of flight.

The fuze was 'bored' by a gimlet prior to firing. The desired time having been selected, the gimlet was forced into the hole corresponding to the required length, and bored through the interior wood until it entered the central channel. The fuze was then put in the shell and the shell fired. The flash-over around the shell ignited the quickmatch in the head; this lit the slow-burning composition in the central hole; in turn this burned steadily until it came to the point to which the gimlet had been bored in, whereupon the flame passed through the hole and ignited the gunpowder in one of the outer channels. The gunpowder immediately flashed down its length to the bottom of the fuze and ignited the contents of the shell.

The flash-over which lit the fuze was a mixed blessing. Whereas it was useful in that respect, it caused problems in the gun itself, washing away the steel of the bore as it forced its way past the shell and wasting energy which should have been applied behind the shell to give it more velocity. After experiments with rope wads and other ineffective methods, a 'gas-check' was devised in the form of a saucer-shaped copper plate bolted to the bottom of the shell. Its diameter was slightly less than that of the shell body, so that on being rammed down the muzzle it passed within the rifling. When the charge exploded behind it, the soft copper was forced outwards by the gases and bit into the rifling, sealing off the escape of gas and allowing all the gas to be devoted to propelling the shell. This considerably improved the wear and tear on the gun — but it now failed to ignite the fuze, and so Boxer had to re-design his fuze by putting a small firing pin and cap in the nose. As the shell was suddenly accelerated, so the firing pin was forced back to hit the cap and this ignited the central powder channel.

The most acceptable thing about the rifled gun, so far as ammunition designers were concerned, was that the shell now arrived point-first with regularity. This meant that for the first time it became possible to fit a percussion, or impact, fuze which would operate when the shell struck the target. There was nothing complex about the design: it was merely the usual sort of plug with a firing pin and cap, so that on impact the pin was driven into the cap to fire a small charge of powder which, in turn, flashed back into the shell. The shells used with this type of fuze were for anti-personnel effect in the field and either

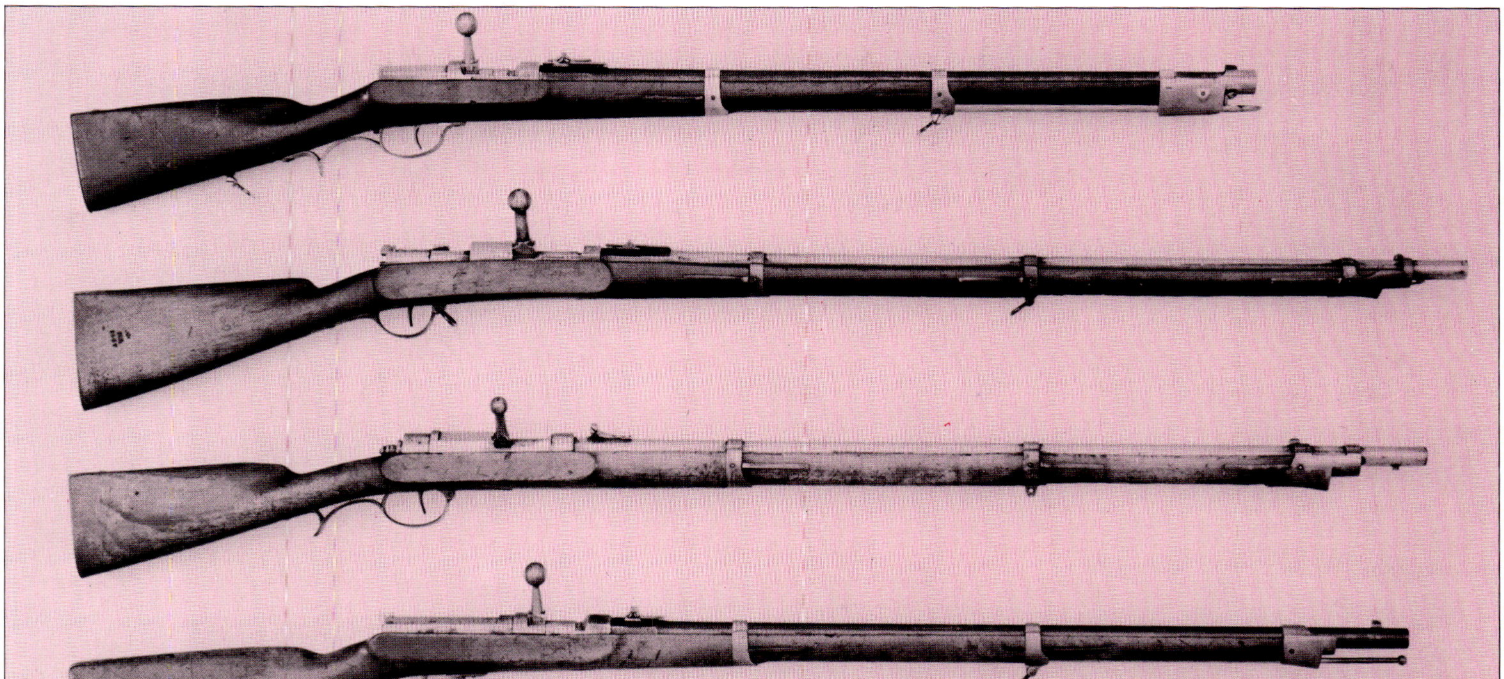

An assortment of Prussian Dreyse needle rifles (from top to bottom): model 1854, used by the navy and marines; two experimental model 1841s and a model 1862 which was used by the infantry.

relied upon the shattering of the wrought-iron casing by the gunpowder inside or, in a few cases, used artificial methods of producing the requisite fragments. One common pattern was the 'ring shell' in which the body was built up of a series of segments of iron encased in lead or zinc, so that when the central bursting charge exploded the segments were blasted separately in all directions as they burst the outer casing.

At this point the artillerymen may be seen to have regained the position they had earlier lost to the infantry, and we may go back to see what had been happening in the small-arms world.

Breech-loading had, in fact, made an impact on small arms earlier than it had on artillery. The German gunsmith von Dreyse invented the 'Needle Gun' in 1838, basing his invention on an idea first put

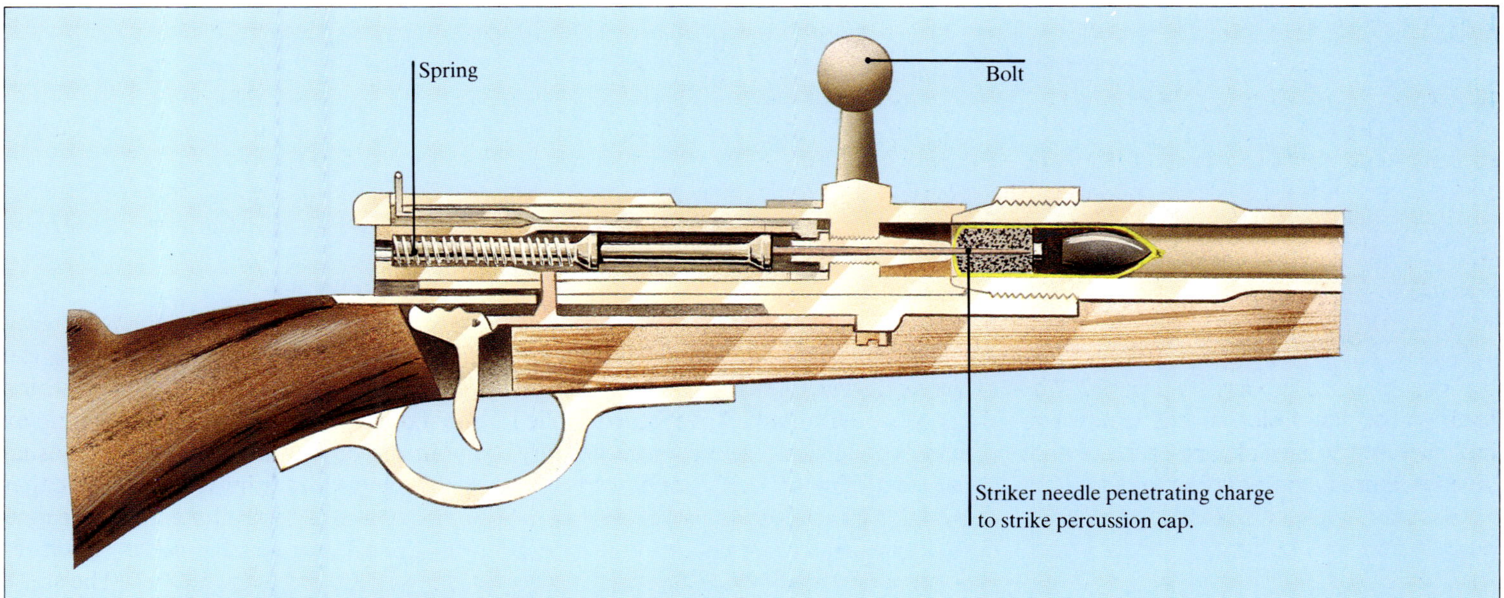

A section through a Prussian needle gun showing how it operates.

forward by Pauly, a Frenchman, some years before. In brief, the rear of the gun was closed by a bolt which worked like a common door-bolt; when thrust forward it forced its tapered end into the rear of the gun chamber, and the handle, turned down in front of a lug on the gun body, prevented the explosion blowing the bolt open. A long spring-propelled needle ran down the centre of the bolt. The Dreyse cartridge consisted of a paper package comprising a conoidal bullet with a percussion cap in its base, and a paper cylinder of gunpowder. This was placed in the chamber and the bolt closed. When the trigger was pulled the needle flew forward and pierced all the way through the cartridge until it struck the cap and fired it. This ignited the powder charge and the bullet was propelled up the rifled barrel. The bolt was then re-opened, the firing pin cocked, a new cartridge inserted, and the gun was ready to fire again, far faster than a muzzle-loader could be prepared.

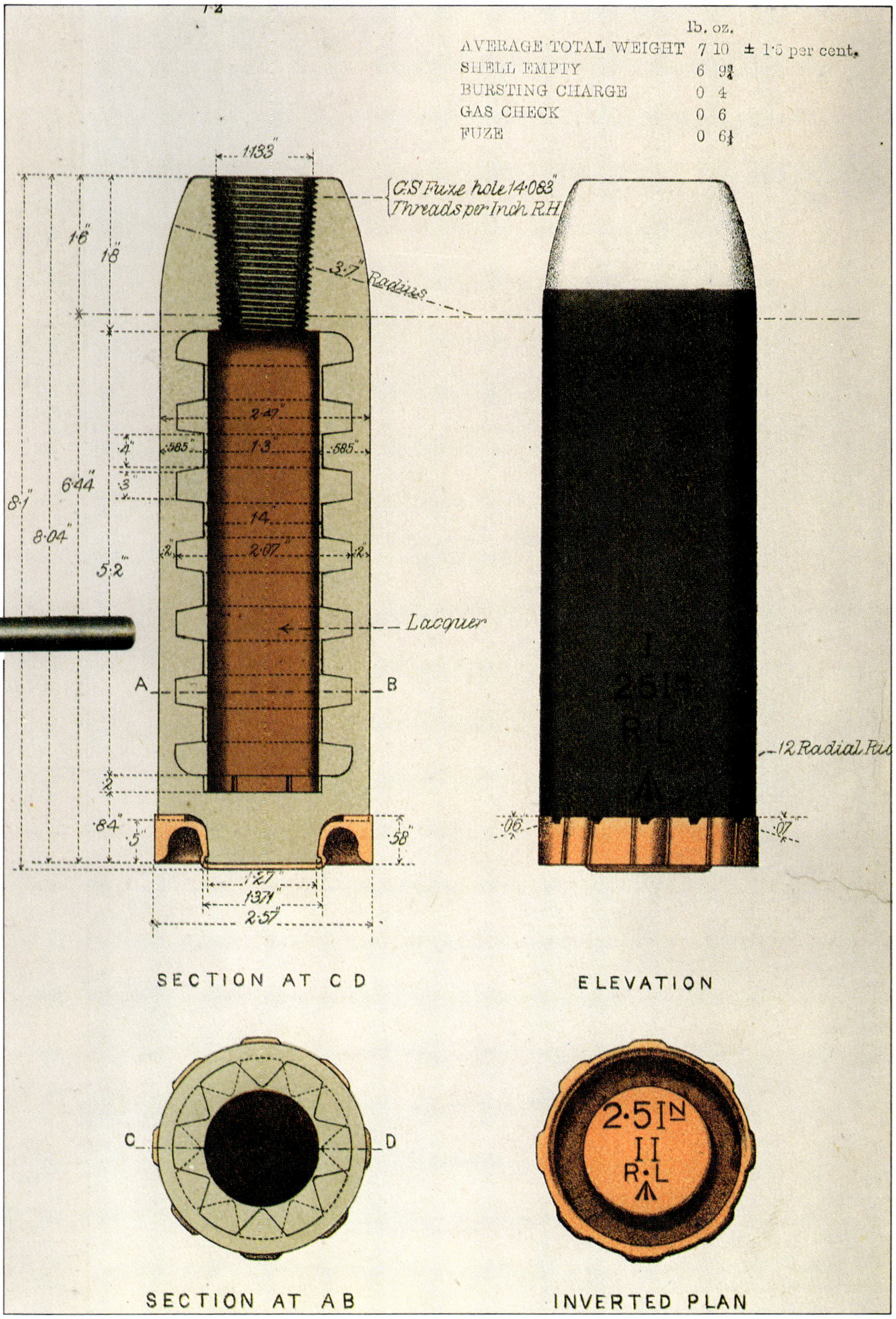

Above: A selection of American Sharps carbines. The paper cartridge was sliced open by the closing breech block.

Above right: A 2.5in ring shell which would be muzzle loaded. The ridges around the base would engage the rifling inside the barrel.

The French Army, ever fearful of what the Germans might be up to, were alarmed at the thought of this advanced weapon on the other side of their frontier and set about developing something similar: the 'Chassepot', named after its inventor. The Chassepot also used a bolt system, and the bolt carried a thick rubber sealing ring at its front end. The cartridge was similar to that of the Needle Gun, except that the cap was at the back end of the cartridge so the firing pin did not need to pass entirely through the cartridge. Both of these weapons had their drawbacks, but for their time they were efficient and formidable. The performance of both guns aroused a good

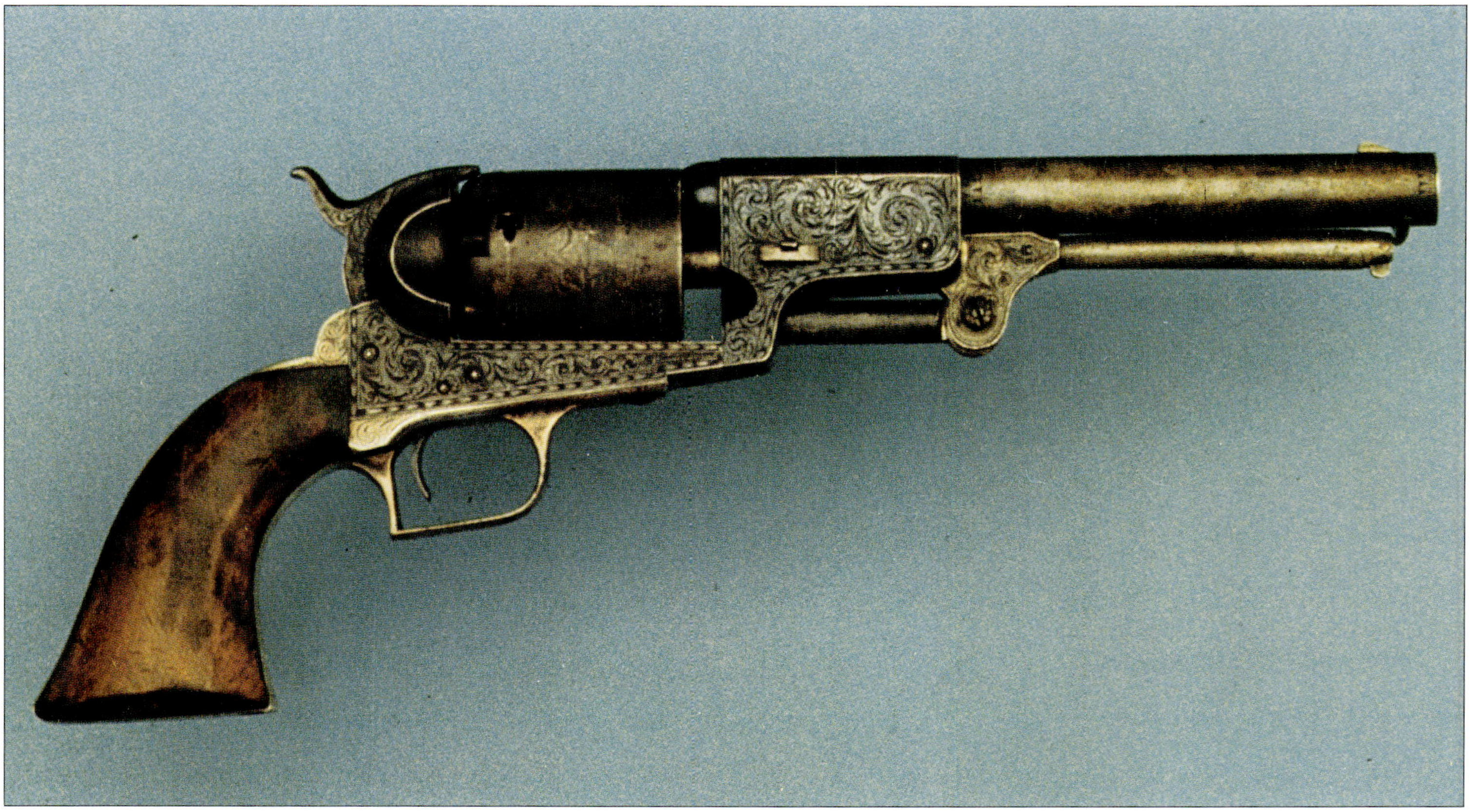

The Colt revolver. It owed its pre-eminence to Samuel Colt's master patent on the revolving cylinder.

deal of interest when France and Germany met in the war of 1870.

The American Sharps carbine showed a different approach to breech closure, using a metal block that was lifted and lowered by a lever underneath the butt. A self-contained paper cartridge was again used, which was loaded into the chamber and the breech closed; as the breech block came up so it sliced off the rear end of the paper cartridge, leaving the powder loosely exposed inside the chamber. A percussion cap was then placed on a nipple and hammer-fired (in other words, the Sharps cartridge did not have self-contained ignition as did the Dreyse and Chassepot designs).

Another interesting idea from about this period (the 1850s) was the Volcanic rifle, developed by — among others — two Americans called Smith and Wesson who were to go on to greater fame. The Volcanic used a bullet that had a hollow rear section filled with powder and closed by a cardboard wad with a central hole; it was inserted into the breech and the breech block raised to close. A cap was placed on a nipple on the breech and hammer-fired. The flash passed through the hole in the wad, ignited the powder, and the explosion blew the bullet from the gun. The idea was sound, but the breech sealing was practically non-existent, and the quantity of powder in the bullet was so small that the Volcanics could not achieve much velocity. After about six years of endeavour to solve the problem, Smith & Wesson gave up the struggle and turned their attention to a more promising project.

What they turned to was a revolver. At that time Samuel Colt had the revolver business thoroughly sewn up, thanks to a master patent covering mechanical methods of rotating the revolver cylinder. But the patent was due to expire in 1857, and Smith & Wesson intended to step in with a new design. Casting about for some novelty, they chanced on a new device that had been invented in France — the metallic rimfire cartridge.

The idea of putting everything — bullet, charge and cap — into a single unit had, as is demonstrated by the existence of the Needle Gun, attracted many experiments, but the failure rate was high, largely because of the problems involved in sealing the breech. Eventually, in 1836, a French gunmaker called Lefaucheaux devised a weapon in which the barrel was hinged and could be 'broken' away from the breech — the pattern which is today familiar in shotguns. He made use of a self-contained cartridge, designed by a man called Houiller, in which most of the case was of paper but with a short brass cap at the rear end, the bullet at the front, and the powder charge inside the paper section. Inside the brass end-piece was a percussion cap, with a short pin above it, aligned so as to protrude through the side of the brass section. The cartridge was loaded into the breech, the pin passing down a slot in the end of the barrel as the gun was closed. The pin now stood proud of the barrel where it could be struck by a hammer; striking fired the cap, the charge exploded and expelled the bullet, and the brass section was expanded tightly against the walls of the chambers by the explosion, so sealing the breech against any escape of gas to the rear.

The 'pin-fire' cartridge revolutionized gun design and shooting overnight — but another design that was already being drawn up would eventually replace it. Another Frenchman, Flobert, took the percussion cap, removed the rim from the front and inserted a tiny lead bullet. He then adapted the Lefaucheaux gun and made the hammer strike a short pin which, in turn, struck the percussion cap to fire the bullet. Such a weapon was obviously small and weak, but it proved popular for indoor shooting at targets. Flobert then improved the idea by changing the shape of the cap slightly, giving it a protruding rim which butted against the rear of the barrel and kept the loaded cartridge in the proper place. The rim was folded and hollow, and filled with percussion mixture. Flobert filled the rest of the tube with powder and placed a bullet in the fore end. His rifle was so designed that the hammer struck against the rim of the cartridge, crushing the composition inside against the edge of the barrel and so igniting the powder. The rimfire cartridge was born.

Smith and Wesson now adopted this rimfire for their revolver. The cartridge required a cylinder with holes bored all the way through, so that it could be loaded from the rear. When Smith & Wesson discovered that a patent existed for this arrangement, they bought up the patent and by so doing put themselves in a position to control revolver manufacture as soon as Colt's patent expired.

The success of the Smith & Wesson revolver with the rimfire cartridge stimulated interest, and many alternative designs of rimfire appeared, particularly in the United States of America, during the following years.

But the rimfire, as it was built in those days, had one defect — it tended to be relatively low-powered. If the rim was to be crushed, the case metal had to be fairly soft, and this argued against high-powered cartridges, since they might well blow out through the soft rim.

Above: A cased Smith and Wesson revolver. They exploited the expiry of Colt's patent in 1857 with the introduction of the metallic rimfire cartridge in the US. This had been invented in France by a man named Houiller, while the breaking action had been devised by Lefaucheaux.

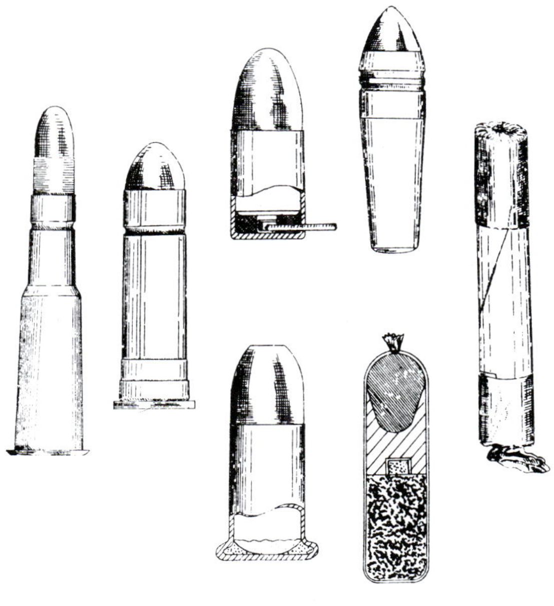

Right: A selection of early small arms cartridges, including a dumdum-style bullet with a flat head and an explosive shell.

The next move was to take the existing percussion cap and the metallic cartridge and marry the two together, placing the cap in the centre of the case base and striking it with a hammer or a firing pin. There were several separate attempts at this before the final successful designs appeared, but by about 1875 the centre-fire metallic cartridge was in being and fast becoming the standard method of charging a hand or shoulder firearm. It became possible to make cartridges of any desired power in the centre-fire pattern, because the cap was supported from behind by the breech closing system.

By the late 1870s the breech-loading hand arm was appearing in all sorts of guises and with all sorts of mechanisms, and the artillery designers began to take a second look at the possibility of breech-loading for cannon. Engineering had made some advances since Armstrong's day and there now seemed to be a possibility to make gun breeches of sufficient strength to withstand heavy charges. Two systems appeared to hold particular promise: the first was the sliding block breech developed by Krupp of Germany, and the second was the interrupted screw breech which appeared in France.

The Krupp system resembled the Armstrong insofar as the gun ended in an open-sided box; but instead of Armstrong's 'vent piece' the Krupp gun used a slab of steel which slid sideways to close the chamber. After experimenting with various ideas for sealing the breech against the escape of gas, Krupp adopted the metallic cartridge case which had proved so successful in small

The breech of a French gun, circa 1800, showing the de Bange breech screw. The breech is opened by rotating the handle to the vertical to unlock the thread.

Left: The shell and cartridge case — standing upright behind — for Krupp's largest gun, the 80cm railway gun.

Below: A drawing from 1870 showing the Krupp coastal defence gun. A crane was needed to load the ammunition.

arms. Despite later development, he never changed from this system in subsequent years, even for his largest gun, the 80cm (31.5-inch) calibre railway gun of World War II, a monster which fired a 7,100kg (7-ton) shell. The metal cartridge case for this weapon was 1.3m long and 96cm in diameter, and carried a charge weighing 2,240kg (just over two tons).

The interrupted screw breech constituted a more difficult engineering problem (which is one reason Krupp stayed clear of it once he had a working system). The rear end of the gun had an enlarged diameter behind the chamber which was cut with a very coarse screw-thread; this then had longitudinal segments of the thread milled away. The breech block was a cylinder which was made with a matching screw-thread and then, again, had sections milled away, so that both chamber and block had three plain and three threaded sections along their lengths. If the block were now aligned so that the threaded sections slid into the plain sections in the chamber, the block could be inserted into the gun. Once inserted to its full depth, close against the end of the chamber, it could be given a one-third turn so as to bring all the threaded sections into engagement. This would then resist any pressure that might tend to thrust the block out again.

The problem now lay with sealing the breech block and chamber joint. The earliest guns did it by fitting a short steel cup on the front of the block; this entered the chamber behind the cartridge and was expanded tightly against the chamber walls by the explosion, in just the same manner as

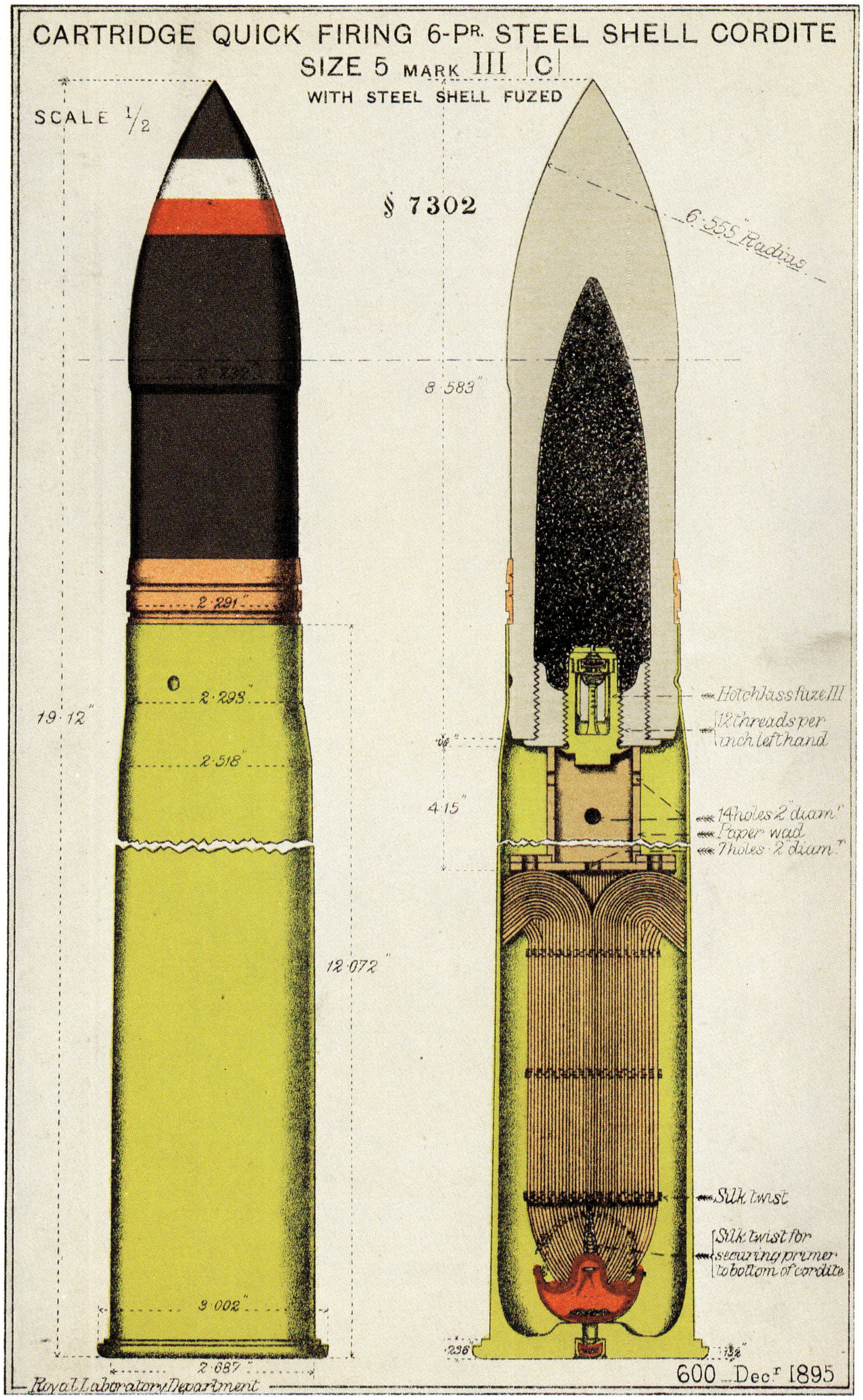

A fixed round of heavy ammunition — a quick-firing 6-pounder shell, sectioned to show its construction.

Above: A contemporary drawing of a battery of French 75mm quick-firing guns at practice. Fixed rounds considerably speeded the process of reloading.

Right: A fixed round of shrapnel for the 13-pounder horse artillery gun widely used in World War I.

a cartridge case was expanded inside a small arm. However, experience soon showed that the steel cup not only soon began to erode from the hot gases but was often battered and bent as the breech block was opened and closed. A better idea came from another French designer, de Bangé, who fitted a resilient pad of asbestos and oil to the front end of the block and then secured it in place with a 'mushroom head' of steel. The stem of the mushroom passed through the breech block and was secured by a nut at the rear; it also had a vent bored through to accommodate the ignition flame. The sealing pad was trapped beneath the mushroom head, and when the gun fired the pressure pushed back on the head and squeezed the pad outwards to seal against the side of the chamber. Even without going into abstruse mathematics it can be shown that the outward pressure on the pad is always in excess of whatever the interior pressure inside the gun may be, so that sealing is always assured. The de Bangé system, with periodic refinements, has been used ever since on major-calibre artillery.

The next step in artillery ammunition was to marry the shell and cartridge together into a single unit, as was common in small arms ammunition. This would allow a gun to be loaded much more quickly by simply inserting the complete one-piece round, instead of loading the shell, ramming it, then loading the cartridge behind it. The first guns to adopt a method using 'fixed ammunition' were three- and six-pounder guns (1.85- and 2.224-inch calibre) for naval use. These became known as 'quick-firers' because the speed of loading allowed a very high rate of fire to be achieved, up to perhaps 20 shots per minute with a skilled crew. It was also made possible by the fact that each gun was anchored securely on a steel mount to a steel deck on a warship, so that the force of recoil was absorbed by the ship's structure.

When the same principle was attempted in a field-gun, problems arose: such a high rate of fire was impossible because the gun on firing rolled back under the force of recoil and the gunners had to get it back into place and re-lay before another shot could be fired.

The French Army solved this problem with their famous 75mm gun of 1897. It used a hydraulic cylinder beneath the barrel which carried a piston rod connected to the gun in such a way that at the recoil it was dragged through the cylinder, its movement resisted by oil. As it displaced oil, so it built up pressure in a compartment of nitrogen gas, which also acted to brake the movement of the gun. Once the gun's recoil stopped, the gas pressure forced the piston head back inside the cylinder and pulled the gun barrel back into the firing position. The gun carriage remained stationary — only the barrel moved — and the gunners could thus remain clustered around the breech. A fixed round of ammunition was used and the '75' could achieve 20 shots a minute with ease. It became the pattern for the field-guns of the 20th century.

During the time that breech-loading was making its return, rifling was also being improved. The universal method now evolved was to cut a large number of spiral grooves into the bore of the gun and fit the shell with a 'driving band' or 'rotating band' of soft copper. This was sunk into the steel wall of the shell and anchored securely; as the shell entered the bore of the gun the copper bit into the rifling and the shell took up spin.

During the last 20 years of the 19th century a major revolution in ammunition technology took place when gunpowder was at last toppled from its monopoly position as the universal explosive. As a propelling charge gunpowder had its faults. It was susceptible to damp, and deposited dirt and fouling in the gun barrel, necessitating frequent cleaning. It also gave out a cloud of white smoke when it fired, which immediately gave away the position of the weapon and, when hundreds of weapons were in use, contributed to covering the battlefield in an impenetrable 'fog of war' which made command and control virtually impossible. As a bursting charge for putting inside a shell it was, again, prone to damp, gave off white smoke when it burst (which was acceptable since it at least showed where the shell had landed), broke the shell up into a few large fragments, and was always liable to premature explosion due to friction if the filling had been carried out carelessly.

Throughout the 19th century there had

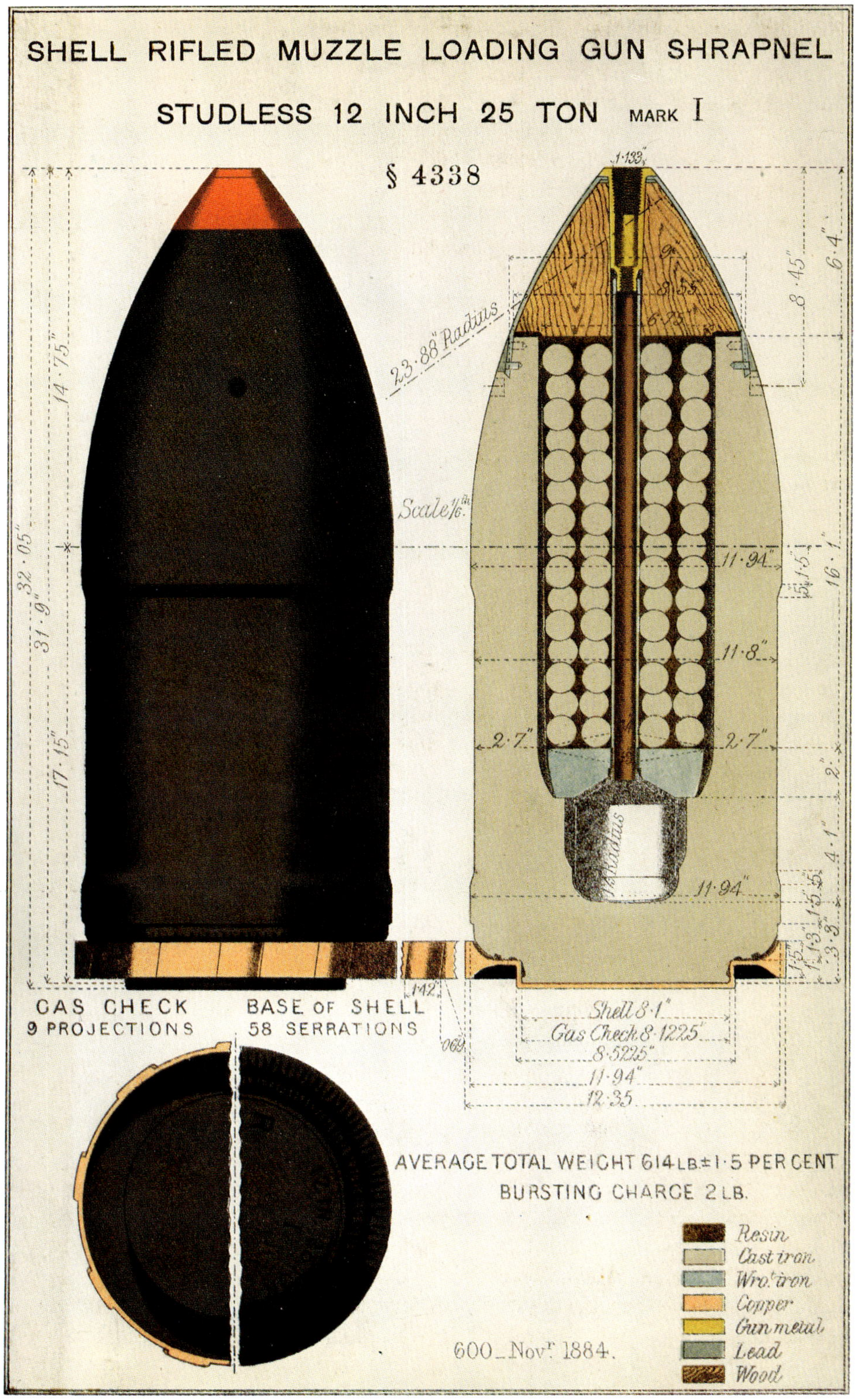

A shrapnel shell for the British 12in rifled muzzle-loading coastal gun. This ammunition provided an effective defence against landing parties.

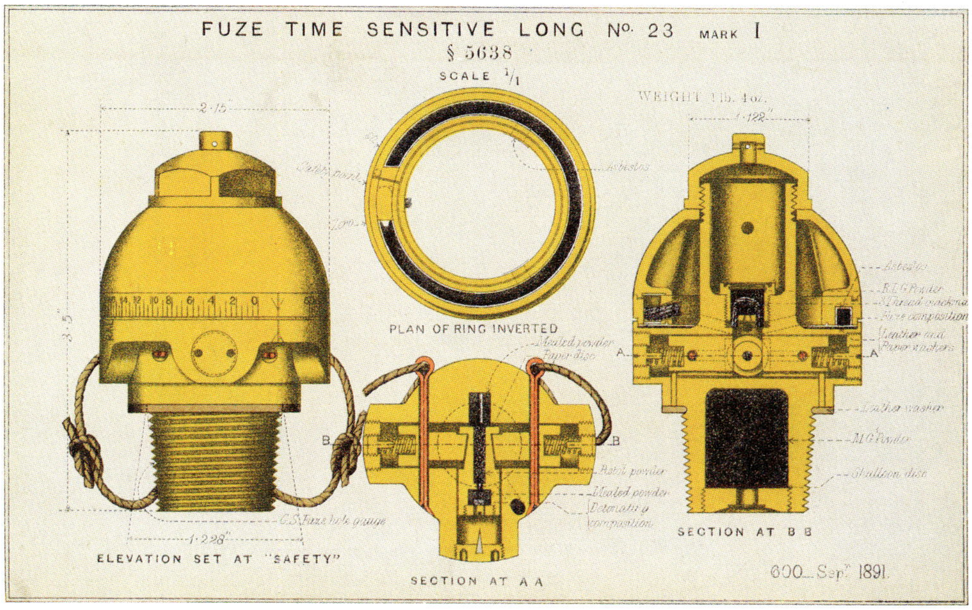

A metal time fuze — the length of fuze is adjusted by turning the top.

been innumerable advances in chemistry. In 1846 Schönbein had discovered 'gun-cotton', made by the action of nitric and sulphuric acids on cotton. It was a devastating explosive, although it proved difficult to manufacture (several factories were blown to pieces before the process was mastered). It eventually found use as a demolition explosive and as a filling for mines and torpedoes — but as a propellant it was entirely useless: the principal difficulty was that when ignited gun-cotton would detonate, rather than explode. An explosion is a rapid burning process and is, to some degree, capable of being controlled; a detonation is the molecular disruption of a substance — much faster and more violent than an explosion, and quite uncontrollable. So making a propelling charge of gun-cotton simply blew the gun to pieces as it detonated.

Nitroglycerine was discovered in 1846, but it was not until the 1870s that Nobel managed to produce it in quantity without blowing up the factory and, again, it was so violent and sensitive that nobody could see any application for it in ammunition.

The first successful substitute for gunpowder was discovered by Major Schultze of the Prussian Artillery in 1865; it was made of nitrated wood impregnated with potassium salts. It was followed by 'EC Powder', made in England by The Explosives Company, which was nitrated cotton impregnated with potassium nitrate. These powders exploded, and they were widely adopted for sporting use in shotguns, but they proved to be too violent for rifled weapons, and it was not until 1886 that Vielle in France developed the first 'smokeless powder' which could be used in a military rifle. The French Army had selected a small-calibre rifle, the 8mm Lebel, as their future weapon and the Vieille powder — named *Poudre B* after General Boulanger — became the standard propellant.

The British 'War Department Chemist', Professor Abel, spent several years working on various compounds before he perfected Cordite in 1889. This was a mixture of nitroglycerine and nitro cellulose made in the form of cords or sticks (and hence its name); because it was possible to make it in any desired size it rapidly became the standard military propellant for both small arms and artillery.

Compared with gunpowder, smokeless powders had great advantages: they were virtually smokeless, they were much less susceptible to damp, they would last longer in store and, weight for weight, they were much more powerful. Moreover, the size and shape of the individual sticks or grains could be made exactly as required so that the burning rate could be controlled to give the desired ballistics. Against all this was the drawback that the flame temperature of these powders was much higher than that of gunpowder, and they therefore had a severely erosive effect on the interior of the gun barrel. A weapon which might fire 10,000 shots with gunpowder could well be worn out after 5,000 with smokeless.

By 1914 the general design of ammunition had reached a fairly uniform standard throughout the world. Revolvers used brass-cased cartridges with lead bullets. The few automatic pistols used similar cartridges but usually with bullets having a lead core and a covering of copper, because they were usually higher-velocity weapons and a lead bullet tended to leave a deposit in the barrel when fired at these velocities. Rifle ammunition was similarly brass-cased, using a bullet with lead core and copper or steel jacket. Artillery ammunition used either a brass case to contain the charge, or a silk bag, depending upon whether the gun used a sliding block breech or an interrupted screw system. Shells were either 'common', 'piercing' or 'shrapnel'.

The common shell was still filled with gunpowder, although some armies had developed more violent fillings — the British used 'Lyddite', which was their name for solid picric acid, and many other countries used it also under different names. The Germans were beginning to use TNT (tri-nitro-toluene) which was more powerful than picric acid but was much more difficult to detonate.

Piercing shells were by now made of specially hardened steel, and had a very small filling of explosive in the base, together with a delayed-action fuze which caused the explosive to detonate after the shell had pierced the armour of the target.

To make the shrapnel shell work, the time fuze was no longer a bored-out piece of wood but a complex metal device which screwed into the head of the shell. Usually of brass, it consisted of a core containing a firing pin and cap which ignited a train of gunpowder set into a brass ring surrounding the core. Beneath this ring the body of

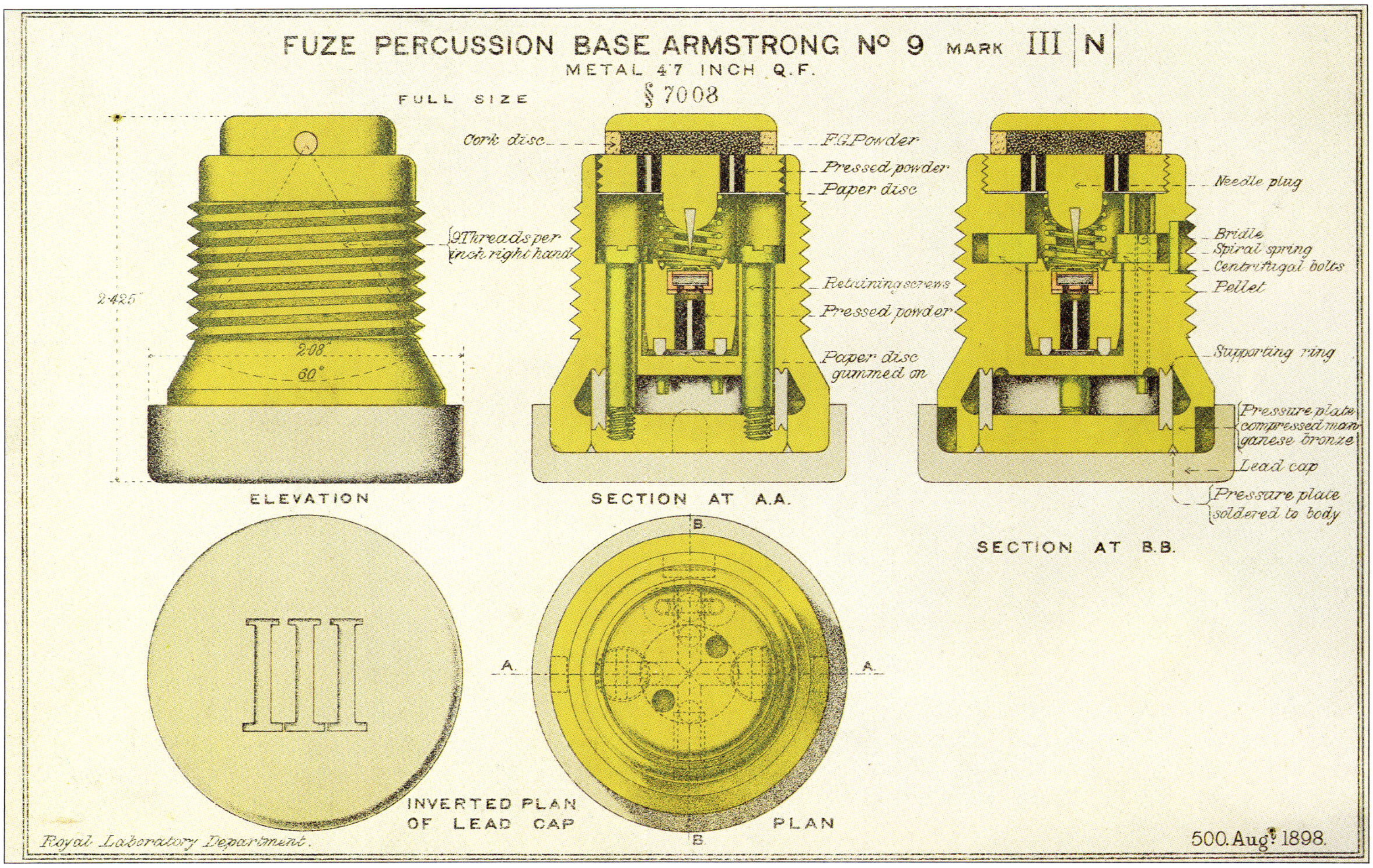

Armstrong percussion base fuze.

the fuze carried another, similar, train of powder which led to a small compartment at the bottom of the fuze, also powder-filled. Before loading, the fuze was set to the desired time by turning the lower, moveable, ring until an arrow was correctly adjusted against an engraved scale. On firing, the pin hit the cap, giving a flash which ignited the powder in the top ring at the specified point along the train. This burned, at a carefully regulated speed, until it came to the end of the train whereupon it flashed down through a hole in the bottom of the ring and lit the train in the fuze body. This, in turn, burned until it ignited the charge of powder which then made the shell function. Turning the loose ring varied the length of the powder train which had to burn before the various transfers of flame took place, and so regulated the operating time of the fuze.

World War I brought several new items into the ammunition world because new targets made their appearance and new tactical ideas demanded specialized ammunition to make them work. One of the first innovations was the hand-grenade. Although a grenade had been seen in the past, it had generally been a small cast-iron mortar shell with a simple length of slow match fuze; the fuze was lit and the shell tossed (for example) over the ramparts of a fort to upset any attacking parties in the ditch. When that sort of warfare faded, so did the use of the grenade, and it was not until the Russo-Japanese War of 1904 that it reappeared in small numbers. This led to some research on the idea, so that when grenades were demanded in 1914 for trench warfare, there were a few designs in existence and more were soon developed. As is often the case, it took time to get these designs into production and for the first few months the soldiers were filling jam tins and empty containers of all sorts with gunpowder and stones, fitting a short length of fuze, and throwing them at the enemy.

The rifle-grenade was developed for those occasions when the hand-grenade was wanted but for long distances. In the first designs the projecting was done by fitting the grenade with a long thin rod which was inserted into the barrel of a rifle; a blank cartridge (one without a bullet) was inserted into the breech, the weapon elevated to about 45°, and the cartridge fired. This blew the rod and grenade out, the rod acted as a stabilizing tail, and the grenade would go to 100 or 150 metres range. Unfortunately, the sudden check to the gas pressure as it endeavoured to lift almost one kilogram (2lb) of grenade out of the barrel often caused local bulging of the barrel, and rifles used for firing rod-grenades were soon useless.

West's spring gun is one method of catapulting a grenade silently. A No 23 hand grenade sits on the throwing arm and the 'gun' is ready to fire.

The next move was to clamp a cup on the muzzle of the rifle and put the grenade inside, with a round plate behind it to act as a gas check. The blank cartridge now developed sufficient gas to fill the cup and blow the grenade out without doing any damage to the rifle barrel, so it could still be used to shoot bullets. But the strain on the weapon of firing heavy grenades often loosened the woodwork and made the rifle inaccurate.

Next came the trench mortar, a short-barrelled gun for pitching shells into the air and dropping them into the opposing trenches. The German Army had a design ready when the war began, but it was an expensive and complicated breech-loading weapon. After several home-made designs had been tried, an English engineer called William Stokes came forward with a design he had perfected in his own factory. It was the essence of simplicity: a smooth-bored tube, closed at the rear end and with a fixed firing pin, supported on a baseplate and a simple two-legged stand. The 'bomb' was a

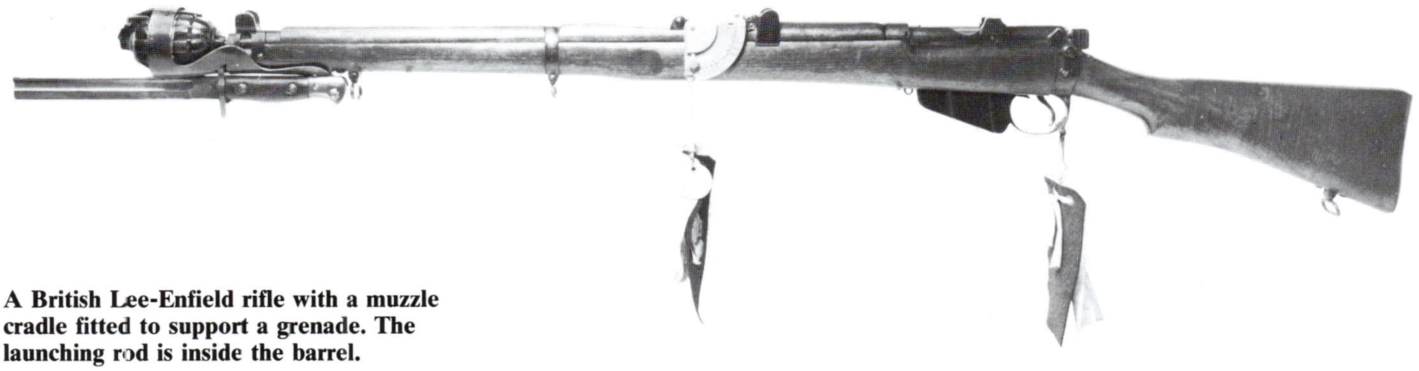

A British Lee-Enfield rifle with a muzzle cradle fitted to support a grenade. The launching rod is inside the barrel.

cylinder of cast iron with a modified hand-grenade fuze on the front end and with a perforated tube carrying a shotgun cartridge full of smokeless powder at the rear end. When this was dropped down the tube the shotgun cartridge cap struck the fixed firing pin, and the subsequent explosion lifted the bomb out of the tube and propelled it for about 700 metres on a high trajectory so that it dropped almost vertically into the enemy trench.

The development of aircraft for warfare set off a demand for guns for air defence, which posed considerable problems. A normal shrapnel or explosive shell could deal with aircraft if the gunner could get it close enough, but the major problem was that he had little or no idea where his shell had gone in relation to the target. The first requirement was to mark the flight of the shell so that corrections could be made — and this led to the development of the tracer shell.

The first types used a liquid, carried in a compartment below the main explosive filling of the shell; this had a small port leading to the outside, plugged with solder. As the shell was fired, the friction as it went up the bore melted the solder and as the shell spun through the air the black liquid was thrown out through the port by centrifugal force, leaving a dark line in the air behind it. As might be imagined, this was a good enough system in broad daylight but of limited value when the light was poor, and it was soon replaced by pyrotechnic tracers. These used magnesium compounds pressed into a tube screwed into the base of the shell. The flash from the charge lit the magnesium and this burned as the shell went through the air, giving the impression of a streak of light marking out the trajectory.

The next problem was that if the shell missed the target — as most of them did — it was liable to fall back to earth and then go off, to the discomfort of anyone it happened to land on. Understandably, impact fuzes had to be abandoned and only time fuzes used, so that even if the shell missed it would detonate in the sky at the end of the set time. Fragments would indeed still rain down on the people beneath, but they were less liable to do harm than a primed and

Above: A group of German staff officers inspect captured equipment, including a British 9.45in trench mortar.

Above left: These aerial torpedoes — so called because of their long tail-fins — were fired out of trench mortars and delivered a large charge at close range.

Right: An early British mortar, the Vickers 1.75in 'toffee apple'. The round bomb has a stick tail inside the mortar barrel.

filled shell. The discovery then followed that time fuzes fired into the upper reaches of the atmosphere performed differently from those fired at ground-level targets, and it took the better part of three years' hard work and testing before all the problems were solved.

One thing that caught all contemporary belligerents unawares was the immense appetite for ammunition that modern warfare was consumed by. The replacement programme for field-gun shells for the French Army, for example, called for the manufacture of 3,600 shells per day; this may sound a lot — but the French had 1,200 guns in the front line, so the programme really meant three shots per gun per day. Factories had to be built to manufacture the shells and the explosives, the fuzes and the cartridges, and assembly plants had to fit them all together . . . and it all took a year or more to reach the point at which the supply could keep up with the demand. The German Army were caught as badly as anyone else, and even before the end of 1914 they were looking for some substitute for explosive which they could put into shells. The answer was to use gas.

It is often said that the Germans first used gas against the British at Ypres, Belgium, in April 1915, but this is not true; the first use of gas was in shells fired against the Russians on the Eastern Front in January 1915. These shells were filled with xylyl bromide, a tear gas, but because of a slight oversight on the part of the Germans they failed to have any effect. What they had overlooked was the severity of the Russian winter: the gas froze solid and failed to disperse. The next use, chlorine gas released from cylinders against the British at Ypres, was far more effective, which is why it is often regarded as the first application of the gas weapon.

Left: The King examines a gas bomb at the Gun School Helfaut, 7 July 1917. Below: Section view of captured German gas shells. Right: Loading a smoothbore 18cm German projector. Far right: Phosphorus bullets being used to shoot down a Zeppelin during World War I.

But from this beginning the gas war proliferated. All three contesting armies used gas shells fired from a variety of guns and howitzers. Trench mortars were also used since they were able to carry a greater percentage of gas in their bombs. The most effective device was the British 'Livens Projector', a wide-barrelled mortar which pitched a drum containing 13.6kg (30lbs) of gas into the enemy trenches. These projectors were strung out in long ranks behind the British trenches and fired simultaneously so as to deliver the maximum amount of gas in the shortest possible time. They proved to be extremely effective.

Small-arms ammunition also developed in response to particular problems. Again, the first demand was for a bullet that would allow machine-gunners to fire at aircraft, and this led to the development of tracer bullets. The principal aerial target in the early part of World War I was the Zeppelin airship, and demand rose for an incendiary bullet capable of igniting the hydrogen gas that kept the Zeppelin airborne. This development was closely linked with the tracer development — since one bullet, properly designed, could do both jobs — although some specialized incendiary bullets appeared. The small size of rifle and machine-gun bullets ruled out the use of complicated fuzes or fillings, and the eventual result was a hollow bullet filled with phosphorus and with one or two holes drilled into it. These holes were plugged with solder and, as with the early tracer shell, friction in the bore melted the solder and spin caused the phosphorus to trickle out of the bullet in flight. Phosphorus has the useful property of being self-igniting when exposed to the air, so there was no requirement for caps for ignition systems. The burning material itself would ignite the hydrogen as it pierced the gas-bags.

When the first tanks arrived on the battlefield there was a demand for bullets which would penetrate the armour. The answer came in the form of a bullet which carried a hard steel core inside the conventional copper jacket, so that when it struck, the core passed through the copper and penetrated the armour plate.

Phosphorus was also used on the ground to provide smoke so that an obscuring cloud could be laid on the ground to conceal the movement of troops behind it. All that was necessary was to put sufficient explosive into the shell to crack it open and allow the phosphorus to reach the air, whereupon it caught fire and produced a dense white smoke. It was also found that the flying phosphorus made an excellent incendiary material when it landed on buildings or equipment, and it turned out additionally to be an extremely painful substance when it landed on the body. The biggest drawback was that phosphorus-filled ammunition had to be very carefully manufactured so that it did not leak in storage — more than one ammunition dump was destroyed by a fire begun by a leaking smoke-shell.

The star shell was perfected in order to see what the enemy was up to at night. It had existed in one form or another for many years, usually as a simple shell which blew open and ejected one or two firework stars which lay on the ground and burned. But a better result was obtained by bursting the shell in the air and suspending the star from a parachute, so that it floated slowly down and illuminated a large area beneath it as it did so.

At the end of the war, in 1918, ammunition development virtually stopped. The survivors had, after all, lived through 'the war to end wars', so there was a general slackening of all military development, and the principal activity was to take the wartime designs, many of which had been hurriedly put together, and consider refinements and improvements. One of a few significant developments of the period was the British invention of a new type of smoke-shell. The white phosphorus shell was simple to make but had the technical drawback that the heat of burning warmed up the air around the smoke; the hot air

Above: White phosphorus smoke shells contain enough explosive to crack on impact, exposing the phosphorus to the air.

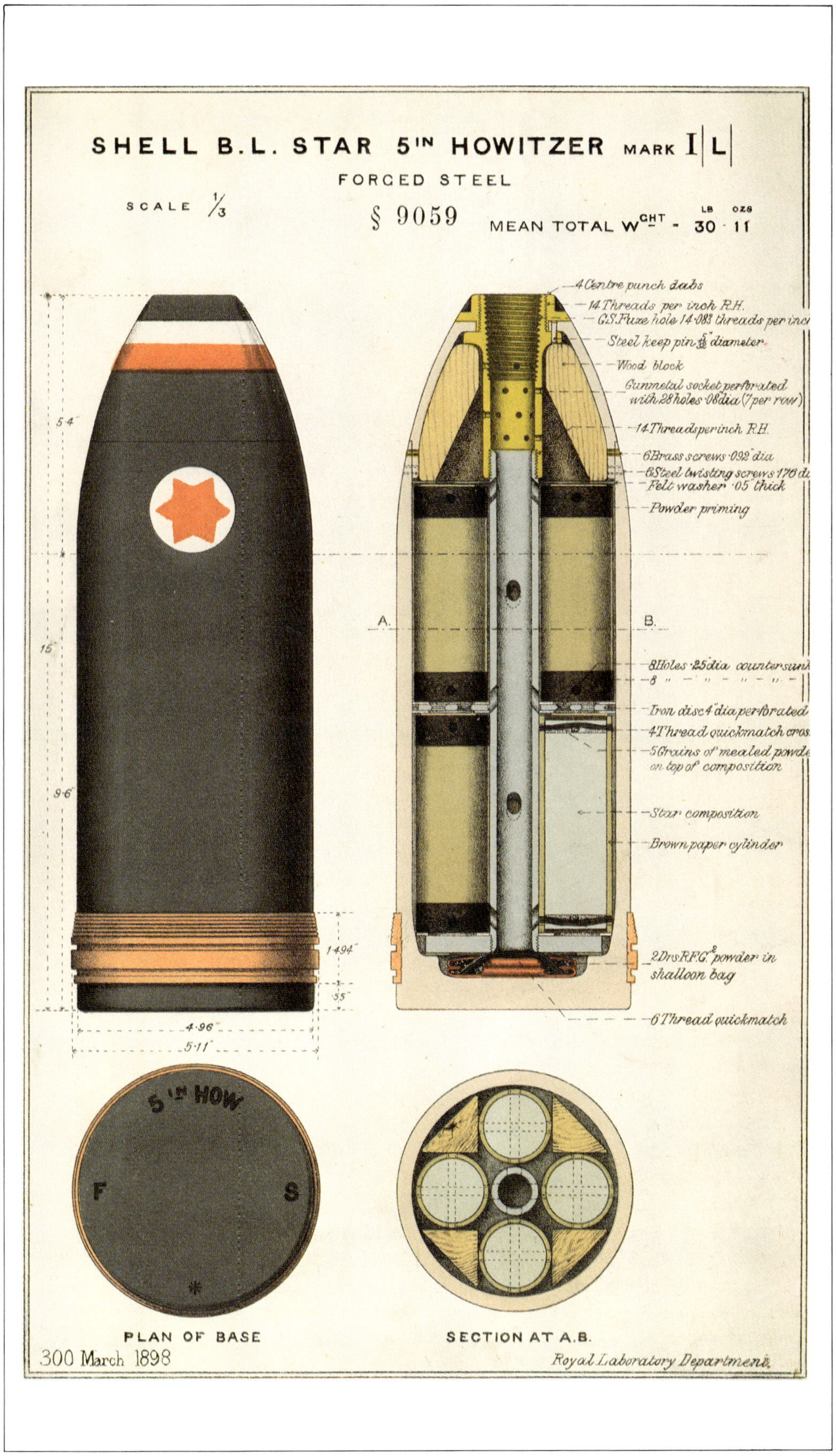

Right: The construction of an early version of the star shell which did not have a parachute to slow its descent.

rose and took the smoke with it, so that instead of a low-lying cloud to obscure the view, there was a series of pillars of smoke with frequent gaps. The new idea was the 'base ejection' shell in which the smoke composition, based on hexachlorethane and powdered zinc, was packed into canisters each of which had a perforated tube running through its centre. Three or four of these canisters could be packed into a shell body, topped by a pusher plate and a charge of gunpowder, and the base of the shell was screwed or pinned into place. A time fuze at the nose ignited the gunpowder expelling charge at the proper time, the charge exploded, and the flash passed down the central tubes of the canisters and lit the smoke composition. The pressure due to the explosion then forced down on the canisters, burst open the base of the shell and allowed the canisters to fall to the ground where they continued to emit smoke. The empty shell went on to fall harmlessly, while the smoke developed by the mixture was cool and stuck close to the ground to make a thick and continuous screen.

The other major change was the final abandonment of the shrapnel shell. Shrapnel was at its best against masses of troops advancing in the open, but the 1914-18 war had been conducted largely in trenches and under cover, and shrapnel had had little effect in these conditions. The high-explosive shell was far more useful, being able to wreck the cover and kill by fragmentation and blast, and so the high-explosive shell became the standard artillery projectile in the 1930s.

There was also a change in the operation of time fuzes. In 1916 the British had picked up fragments of German shrapnel shells which had landed more or less undamaged and had discovered that the fuze contained a clockwork mechanism. It was a fully-wound clock which ran for up to 60 seconds, and at the end of the set time a catch was released which fired a detonator to operate the shell. Work began in Britain on developing a similar fuze, but the difficulties were enormous and it was not until the middle 1920s that a successful design was achieved. Unfortunately the mass-production of such a device demanded a watch-making industry — which Britain did not have — and so the mechanical time fuze was produced only in very small numbers for demonstration purposes. It was to take the outbreak of another war, and the loosening of the public purse-strings, to get such a device into production.

In the 1930s a number of German officers began to question the form of the infantryman's rifle and its ammunition. The standard type of rifle, which had evolved around the turn of the century, was a weapon firing a powerful cartridge, capable of accurate shooting out to 900 metres (1,000 yards) range and of being used in a machine-gun to as far as 2,700 metres (3,000 yards). But careful examination of soldiers who fought in the war, and study of records, seemed to indicate that few soldiers ever saw an enemy farther away than about 350 metres (400 yards), and very few ever attempted to fire a rifle at ranges over 450 metres (500 yards). It seemed illogical to burden the man with a heavy rifle and a powerful cartridge when he did not need the power they had been designed to deliver. It would be more sensible, said the Germans, to make a cartridge with a shorter case and lighter bullet; this would give less recoil, the rifle could be shorter and lighter, and the man could carry more cartridges.

The reasoning was perfectly sound, but the objection was simply that Germany had several hundreds of millions of cartridges of the regulation size, and changing the rifle and all that ammunition would be prohibitively expensive. And so, although some design work was done on a 'short cartridge', nothing else happened.

When war broke out again in 1939, armies went to war with much the same weapons and ammunition that they had ended with in 1918. Behind the scenes, however, intensive research was in progress which would bear fruit later in the war.

The first area of concern for the soldier in 1939 was how to stop a tank. Most countries had small anti-tank guns — about 37–40mm calibre, firing solid steel shot — and the infantry were generally provided with high-powered rifles firing armour-

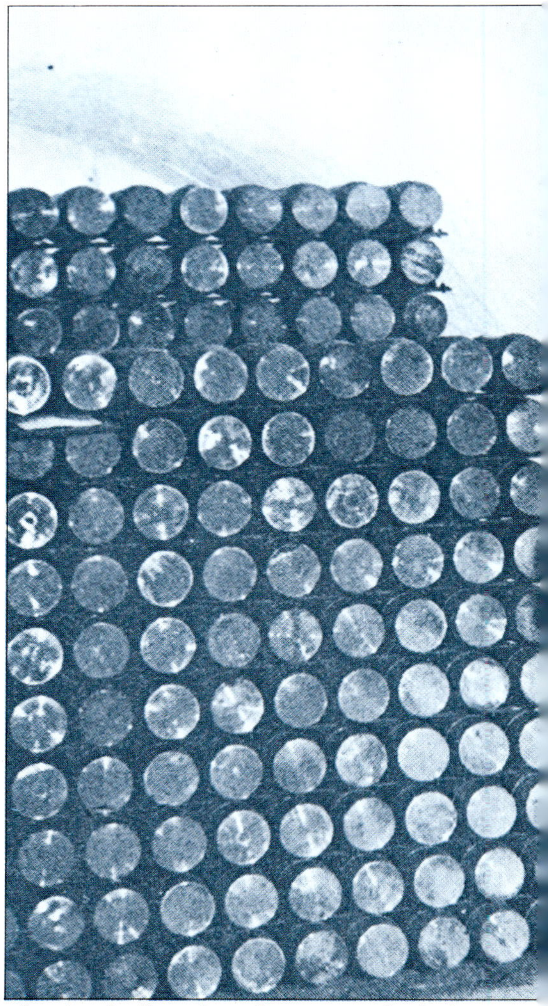

piercing bullets, but the way tank design was progressing it was only a matter of time before these would be incapable of dealing with the increased thickness of armour. It seemed that the only way to defeat armour was to throw a very hard projectile at very high speed, so several designers simply made guns with more powerful cartridges. But they then ran up against a new problem: when a steel shot hits a hard target at more than about 800 metres per second (2,600fps) velocity, the shock of impact is so great that the steel shot disintegrates without making much impression. To prevent this, one solution was to protect the tip of the shot with a soft steel cap which, more or less, acted as a 'shock absorber' and allowed the tip of the shot to penetrate. Unfortunately, the best shape for this 'penetrative cap' was not the best flight shape, so a second 'ballistic cap' of thin steel had to be put in front. But these caps merely lifted the 'shatter velocity'

An underground American shell store where hundreds of 155mm shells were being prepared for issue.

slightly, and the problem soon came back again.

In 1938 two mysterious Swiss gentlemen announced to the various military attachés in Switzerland that they had a 'new and powerful explosive' which could penetrate armour. They announced a demonstration, and the military attachés reported back to their masters and arranged for various ammunition experts to attend.

At the demonstration, the Swiss fired rifle-grenades at a sheet of armour and, sure enough, blew holes right through it. They then mentioned large sums of money, and the experts went away to think about it. More than one of the experts realized that what they had been watching was not a new explosive at all but a fairly old idea that had been revived and apparently been made to work by the two Swiss. The idea was called the 'Monroe Effect', because in the 1880s an American experimenter called Monroe had been doing some work on gun-cotton and found that if he laid a slab of gun-cotton on a sheet of steel and fired it nothing much happened to the steel. But if he turned the slab so that the incised letters 'USN' (for the US Navy, who made the gun-cotton) were against the steel, then the letters were reproduced deeply in the surface of the steel. This became something of a parlour trick among explosives scientists, but in spite of some work on it during the 1914-18 war nobody managed to put it to practical use. Now it seemed that the Swiss had done so, and the various experts went back to their own laboratories determined not to be outdone by a pair of amateurs.

The result of their work began to appear late in 1940 as the first 'hollow charge' munitions went into service. The principle of these projectiles — rifle-grenades and artillery shells — was that the nose was hollow, and the body held a charge of high explosive inside a cavity hollowed out in the form of a cone or hemisphere. This cavity was then lined with a thin copper plate of appropriate shape, and the fuze was arranged to detonate the explosive at the end farthest from the cavity and liner. The detonation wave, sweeping forward through the explosive, collapsed the liner, and the cavity shape 'focused' the blast into a very fast moving jet of hot gas and molten metal which punched through steel armour as if it were cardboard. A 75mm shell could easily defeat 100mm of armour plate. Although the hole was small — no more than 1.2 centimetres (half an inch) across — the hot jet passed through into the tank and did immense damage if it struck something vital, such as a fuel tank, ammunition or, of course, a member of the crew.

The hollow-charge principle could also be applied to demolition weapons. Its first use during the war came when German glider-borne troops used such bombs to punch through the armoured turrets of Fort Eben Emael during the attack on Belgium in 1940.

While the hollow charge was being developed, work had gone forward on trying to solve the shot-shatter problem, and the Germans decided to adopt tungsten as the material to attack tanks. But tungsten is about twice as heavy as steel, and making a full-sized shot of tungsten inevitably meant low velocities which would prevent penetration. The taper-bored gun developed to circumvent this shortcoming, fired a shot that had a tungsten core and a light alloy body capable of being compressed. When the shot was loaded it was of large calibre and light in weight; as it passed down the tapering barrel the alloy was squeezed down until it emerged from the muzzle small in diameter and heavy in relationship to its diameter, which gave it good carrying power and allowed it to keep up its high velocity. The tungsten did not shatter when it struck the tank, but pierced cleanly and then broke up inside to ricochet around and do damage.

One of the steel turrets of the fortress of Eben Emael in Belgium which was penetrated with shaped charges.

The British, confronted with the same problem, designed a projectile in pieces, one that had a 'sub-projectile' of tungsten surrounded by a full-bore 'sabot' or support of light alloy. When loaded, the projectile was much lighter than a standard steel shot and so moved off at very high velocity; the construction was such that as it left the muzzle the light alloy sabot would split in pieces and fall away, leaving the sub-projectile to fly to the target. This 'Armour Piercing Discarding Sabot' projectile appeared in 1944 and was to prove the most potent anti-tank weapon of the war. The German design, good as it was, did not survive the grave shortage of tungsten which affected Germany from 1943 onwards; they had the choice: they could use tungsten for machine tools or fire it off in ammunition — and the machine tools won the day.

The search for a suitable anti-tank weapon for infantry was now becoming desperate. The answer seemed to lie with the hollow-charge principle, which relied entirely upon the explosive in the projectile and was independent of range or velocity. The first weapon of this type to appear was the British 'PIAT' or 'Projector, Infantry, Anti-Tank', a weapon that used a novel method of discharging the projectile. Instead of using a barrel, it used a thick steel rod or 'spigot'; the bomb had a long hollow tail, with fins, and the propelling cartridge was inside this tail. The bomb was laid in a trough in front of the spigot; when the trigger was pulled the spigot rod shot into the tail of the bomb and fired the cartridge, and the explosion blew the bomb off the spigot and through the air. The hollow-charge warhead could pierce three inches of armour, enough to deal with most tanks of the time, and the weapon had a range of about 115 metres (125 yards).

By now the Americans had entered the

Above: An Israeli 90mm APFSDS shot.

Two shaped charge projectiles. On the left is the British 3.7in pack howitzer, developed during World War II for use against Japanese tanks. On the right is the PIAT shoulder-fired bomb, the standard infantry anti-tank weapon of the British Army from 1942 to 1950.

The original American 2.36in bazooka, complete with its shaped-charge rocket.

The 2.36in bazooka in action with the US infantry in Normandy in 1944.

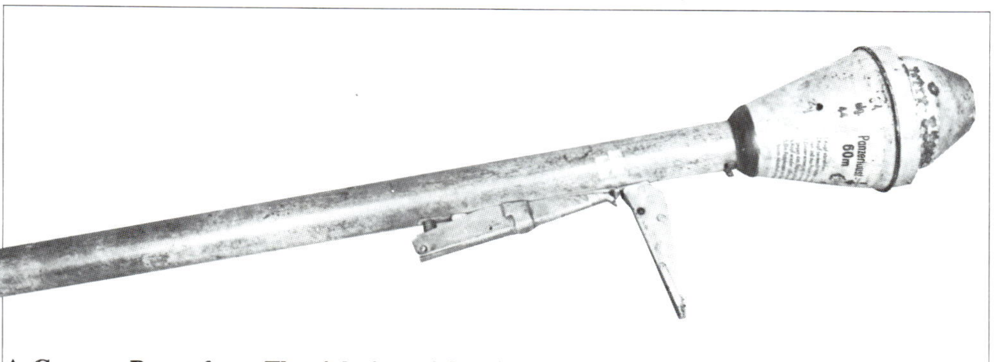

A German *Panzerfaust*. The sight is upright with the trigger behind it. Pressing the rear end down fired the propelling charge and launched the bomb.

war. Their contribution to the infantry's problem took the form of a light tube which a man could place on his shoulder and from it launch a short rocket with a hollow-charge warhead. This was the famous 'Bazooka', which became the principal US anti-tank infantry weapon. Many were given to the Russians, who lost some to the Germans. In no time at all the Germans had their own copy in action, the *Panzerschreck* ('Tank Terror'). But the Germans were not entirely happy with this rather cumbersome weapon, and they devised the *Panzerfaust*, a much smaller tube which could be tucked under the firer's arm and which launched a grenade with a massive hollow-charge warhead.

The *Panzerfaust* was not a rocket; it was among the last of a line of German developments of the recoilless principle. The recoil of a gun had always represented a problem to be overcome, and in major calibres it had meant heavy and complicated hydraulic braking systems on top of heavy and complicated carriages. Dispensing with recoil would remove a great deal of the weight from guns, and this had been the aim of a number of inventors.

The first successful recoilless gun was made by Commander Davis of the US Navy during World War I, and consisted simply of two guns back-to-back. In fact it was a single chamber with two barrels; one held the service projectile, and the other had a similar weight of bird-shot and grease. When the charge was fired in the chamber these two, the shot and the 'counter-shot', went off in opposite directions; both barrels therefore recoiled equally, cancelling out the recoil of each other. The Davis gun was mounted on British Royal Naval Air Service seaplanes as an anti-submarine weapon in 1918, but there is no record of its use in combat.

The Germans applied this countershot idea in a scientific manner. They reasoned that, provided the product of mass and velocity for the shot and countershot were the same, the countershot could be quite light as long as it moved fast enough. The end result was a gun which fired a jet of gas at extremely high velocity through a vent behind the breech block, thus balancing the recoil due to the shell being fired from the barrel. These 'Light Guns' offered 105mm firepower for very little weight and were used to arm German paratroop units. The *Panzerfaust* used the same principle, a small black-powder charge firing the bomb from the front of the tube and exhausting a blast to the rear which counterbalanced the recoil.

Above: A Soviet 82mm recoilless rifle on display at the Warsaw Military Museum.

Below left: Ammunition for a recoilless gun. Note the characteristic perforated case.

Below: A British 105mm light gun firing from a camouflaged position.

The other target worrying soldiers was aircraft. The aeroplane had made great strides since the days of the Zeppelin and was now capable of flying at great speed and altitude. A wide variety of anti-aircraft guns were produced by all the combatants — but the gun was only part of the solution. Finding the target was the major part, solved by the development of radar. The final part was getting the shell to burst as close as possible to the aircraft. The mechanical time fuze had become the standard type of fuze for anti-aircraft fire. It was far more accurate and reliable at high altitudes than the powder-burning fuze but, even so, setting the correct time on the fuze was a matter of educated guesswork. It occurred to some British scientists involved in the development of radar that it might be possible to make a fuze that would pick up the radar energy reflected from the target and react to this when the signal strength indicated that the shell was within striking distance. Tests showed the theory to be correct, but the circuitry and components needed to make it work were far too big to fit inside any projectile. They therefore turned to a new idea, putting a tiny radio transmitter in the fuze, together with a receiver which picked up the reflection of the radio signal from the target and fired the shell when the signal was strong enough. No suitable manufacturing facilities existed in Britain so the idea was passed to the USA in late 1941, together with several other technical secrets. The 'Proximity Fuze' was developed, largely by the Eastman Kodak Company, and first saw service in the Pacific in 1943. It proved invaluable in defending US ships against *kamikaze* suicide aircraft, and it was instrumental in defending Britain against the German flying bomb attacks in 1944.

The Germans, having exhausted most of the possibilities offered by conventional ammunition, moved into a new area when they began to develop guided missiles. As is well known, the 'V-Weapons' did immense damage in Britain and Belgium. The V-1 flying bomb and the V-2 rocket were, however, virtually unguided; they were merely given an initial direction and left to their own devices. But other weapons were further advanced technically. A gliding bomb, steered by radio from the aircraft that dropped it, was used in the Mediterranean in 1943 and sank the Italian battleship *Roma.* Beam-riding anti-aircraft missiles were under development as the war ended, as were tactical ground-to-ground missiles and wire-guided anti-tank missiles.

The lines of development were taken up after the war by the victor nations, and the atomic bomb, developed in USA, was allied with them to produce the fearsome armoury of missiles which now exists. The development of these weapons is largely a matter of electronics and physics, and the ammunition side of the story is relatively conventional; high-explosive warheads for missiles resemble shells in many respects, but their size allows exceptionally complex fuzing and fragmentation control to be achieved.

A modern rifle grenade in position on an assault rifle, ready to be fired.

Left: Proximity fuzes since World War II have shrunk due to modern electronics.

Below: Two modern mechanical time fuzes for field artillery shells. The one on the right has a percussion element in the nose.

The British 120mm Wombat recoilless gun fitted with an aiming rifle and image-intensifying night sight.

German ideas for a short cartridge for infantry rifles came to fruition in 1943 with the development of a special 7.92mm cartridge and a new weapon, eventually called an 'assault rifle', to go with it. This development, too was taken up by other countries, and today most major armies are using assault rifles with short cartridges in calibres that would have been thought ridiculous fifty years ago. But, again, this story is really the story of firearms development. The 5.56mm cartridge of today differs only in size from the 7.92mm or .30 cartridge of sixty years ago, and the essential principles remain the same.

In the last five or so years, however, a new dimension has arrived in ammunition, and that is the application of some of the principles of guided weapons to artillery projectiles. In the past the artillery shell was entirely uncontrollable once it left the gun muzzle; today, the use of 'remotely guided munitions' promises some changes in artillery employment. It is now feasible to fire a hollow-charge shell from a howitzer and control the last few seconds of flight so that it strikes a specific target. It is also possible to fire a howitzer shell which bursts above a target area and releases a number of individual 'sub-munitions' which, using radar techniques, can detect targets and home in on them. Similarly, long-range rockets can be fired to discharge a load of mines in front of an advancing tank force, and once the force has thus been stopped, to bombard them with lethal sub-munitions.

This brief overview of the development of ammunition has of necessity been confined to some of the most important developments. It cannot go into the level of detail on individual items necessary to tell the whole story, but it does fulfil its main objective by showing the general trend of ammunition development.

A Drill version of the Copperhead terminally guided artillery shell — fins spring out after it has left the gun.

A-Z
OF AMMUNITION

A

Note: Words in *italics* in the descriptions indicate that the word in question forms an entry elsewhere, and reference should be made to it.

ACCELERATOR Trade name for *discarding sabot* small arms ammunition manufactured by the Remington Arms Company. Calibres .308 Winchester and .30-06 are available.

ANTIMONY Metal which is mixed in small quantities with lead to make bullets; the proportion varies from 98 per cent lead, 2 per cent antimony to 90 per cent lead and 10 per cent antimony. The addition of antimony makes the lead hard enough to resist deformation during loading and firing, while remaining soft enough to be engraved by the rifling.

ANVIL A component of a *primer cap* or cartridge case; the *cap composition* is nipped between the firing pin or hammer and the anvil so as to explode and generate flash to ignite the propelling charge.

A cartridge case cut in half to show the anvil and one flash hole.

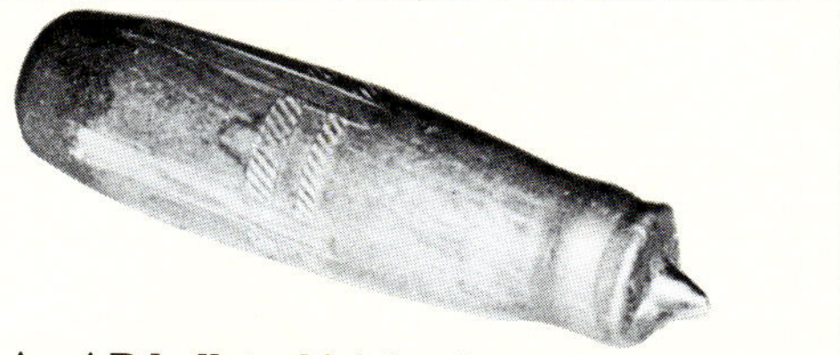

An AP bullet which has been defeated by armour, showing the flattened jacket tip and the armour-piercing core driven forward through the tip.

ARMOUR-PIERCING Type of military bullet designed to penetrate light steel armour. It is formed of a hard steel core surrounded by a lead sleeve, both carried in the usual type of jacket. This method of construction allows the core to be given the optimum point for piercing while the entire bullet has the optimum shape for accurate flight. The purpose of the lead sleeve is partly to add mass to the bullet, so that it will sustain its velocity, and partly to provide a yielding substance beneath the jacket which will give way when the rifling engraves the jacket material. If there were no lead sleeve the engraving would press directly on the steel core and would be ineffectual. On impact the lead sleeve and jacket are arrested while the piercing core continues through the target, the collapsed lead acting as a support to hold the bullet in alignment during the penetration. The AP bullet was first patented by Roth of Vienna in about 1903 but, there being little use for it at that time, it was neglected until the 1914-18 war when it was introduced for snipers in order to counter the small iron shields which were being used as loopholes in entrenchments. When the tank appeared on the battlefield it became the principal target for AP bullets. By 1939 tanks had become more heavily armoured and were largely impervious to small-calibre weapons, but the AP bullet found a fresh application in aerial combat, being used to counter attempts to protect aircrew and parts of the aircraft. Today its principal use is against light armoured vehicles such as armoured personnel carriers.

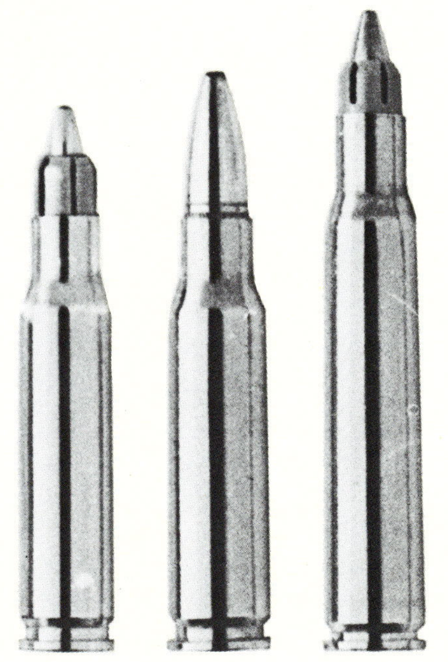

Remington 'Accelerator' cartridges flank a conventional rifle cartridge.

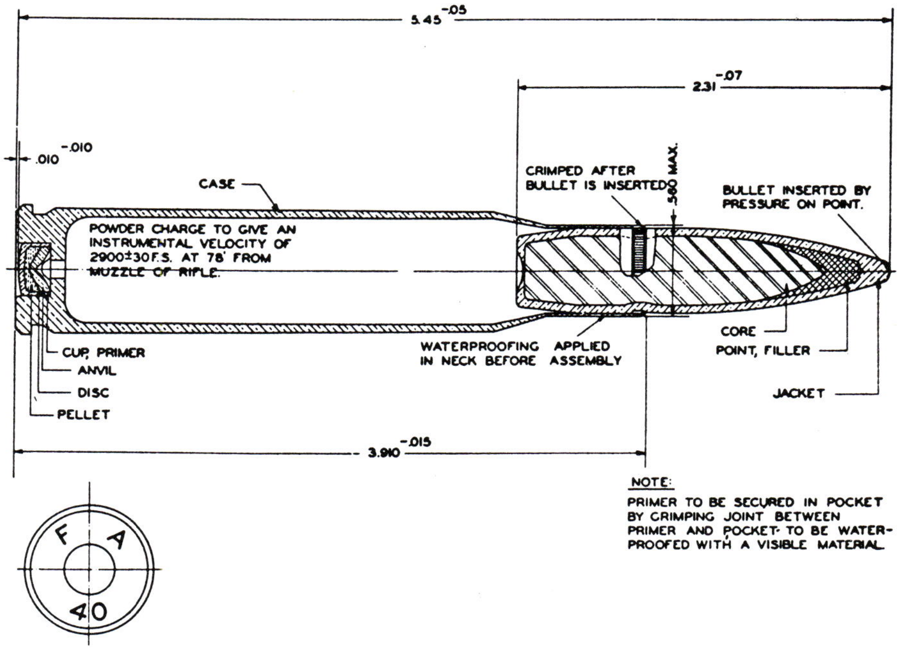

An American assembly drawing for a typical ball cartridge.

B

Unusual ball bullets: left to right, Two French THV, metal-piercing police bullet, standard 9mm Parabellum round.

The plastic baton from a standard 37mm round (centre), with a 12-bore shotgun cartridge (left) and a 9mm pistol cartridge for comparison.

BALL Term used to describe military bullets which are entirely inert and intended for anti-personnel and general use, in order to distinguish them from specialized bullets, eg, armour-piercing or tracer; so called because the original bullets for muskets were spherical.

BASE DRAG In flight the bullet cleaves the air and the air passes down its sides until it reaches the base. Behind the bullet is an area of low pressure caused by the displacement of air, and the airstream swirls into this area, setting up turbulence. This retards the bullet's speed, and the effect is most significant at subsonic velocities. Base drag can be reduced by tapering of the bullet towards its base; see *Streamlined*.

BATON ROUND The correct name for what is popularly known as a 'rubber bullet'. It is a riot control projectile, generally a plain cylinder of rubber or plastic of a size to suit a standard 12-bore, 26mm or 37mm riot gun and fired by a low-powered charge to attain a muzzle velocity of about 60 metres per second and a range of about 100 metres. Fired in accordance with its design intention, ie at ranges in excess of 25 metres and at the ground, so as to bounce up and strike rioters on their lower limbs, the baton round will cause a painful blow but no injury. The first baton rounds were developed for the Hong Kong Police and were of wood, but these proved to be liable to splinter on impact with the ground, and the splinters caused wounding. The wooden baton was then replaced by a rubber one, and the bullet's use spread throughout the world. Rubber batons were found to bounce indiscriminately and were superseded by PVC-type plastic batons which are more predictable in their behaviour and less likely to ricochet to long distances. In the most recent designs, the projectile is mushroom-shaped which gives better accuracy and less liability of injury if used improperly. A design used in Germany has four segments, joined at the front end. In flight the segments stream in the wind and the projectile is almost cylindrical; on impact the four segments fly forward, spreading out into a cross which spreads the force of the blow and makes the baton less likely to inflict severe injury.

Spark photograph of a rifle bullet in flight, showing the turbulent wake which causes base drag.

The three phases in a bullet's trajectory.

BEATEN ZONE That part of a bullet's trajectory between the *first catch* and the *first graze*; ie, that portion of its flight when it is close enough to the ground to strike a human target.

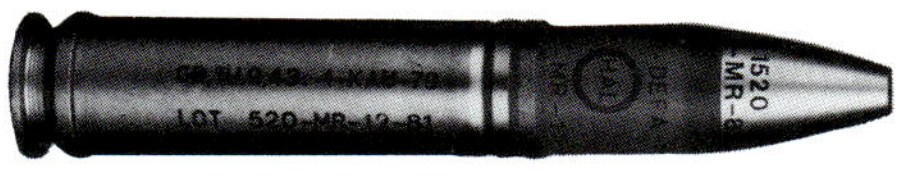

A cartridge, with belted case, for the French 30mm Aden/DEFA cannon.

BELTED CASE A cartridge case having a prominent raised belt around its body just in front of the extraction groove. The purpose is twofold: to position the case and projectile accurately in the chamber of the weapon, and to strengthen the rear of the case for use with high-powered propelling charges. Usually to be found on powerful sporting rifle cartridges and on some military cannon cartridges, it has also been adopted for the 12-bore cartridge used by the US Army Close Assault Weapon in order to prevent this high-powered cartridge from being used in a conventional 12-bore shotgun.

BERDAN PRIMER Primer cap for a small arms cartridge, designed in the 1860s by Colonel Hiram S. Berdan of the US Army Ordnance Department. The distinctive feature of the Berdan primer system is that the anvil forms part of the cartridge case, there being a number of flash holes alongside it to permit the passage of the ignition flame from the cap chamber to the propelling charge inside the body of the case. The Berdan primer is used in almost all military ammunition and in a large proportion of sporting ammunition. See also *Boxer Primer.*

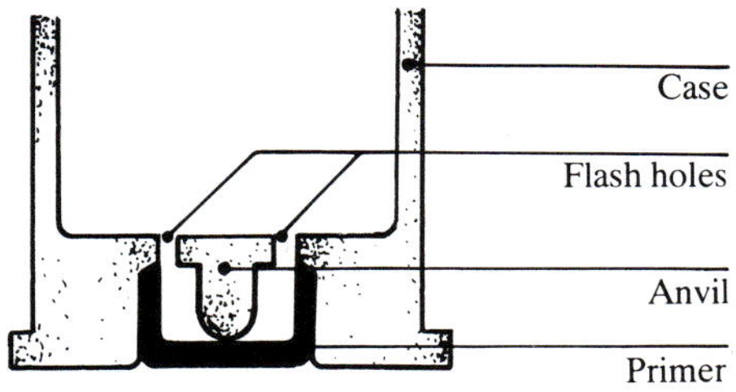

The standard Berdan Primer.

BLACK POWDER Compound, also known as gunpowder. It is a mechanical mixture of potassium nitrate, charcoal and sulphur in the proportions 70:15:10, though these proportions have varied during its history. The materials are ground separately into fine powder, then mixed in the wet state; the resulting 'cake' is dried and then crushed, and the grains are run through sieves to classify the powder according to size. The size of the grain effects the burning speed, the smallest grains producing the fastest burning. The finest grade, mealed powder, is used in the manufacture of pyrotechnics, in fuze compositions and ignition devices. The granular powder is classed from FFFFg (grains measuring about 0.017in across) to Fg (about 0.069in across), the smallest grains being used in pistol ammunition, the larger in rifle, shotgun and cannon ammunition. Although black powder has several applications in other classes of ammunition, principally as a component of ignition systems, it is used today as a small arms propellant only by those hobbyists who fire old or replica guns. Its principal defect, as a propellent for firearms, is that it generates a large volume of white smoke and the solid combustion products are deposited in the barrel of the weapon as a sticky fouling.

BLANK CARTRIDGE A cartridge without a bullet, containing a charge designed to generate a loud report on firing. Used for theatrical performances, for training game dogs, etc, and for military training. The cartridge often contains black powder, since this burns very rapidly and thus produces gas at an extremely rapid rate so as to generate the noise. Military blank uses a special grading of smokeless powder in order to produce the same effect. The powder is retained in the cartridge case by wadding, the mouth of the case being crimped or folded over to retain the wad; as a result, wadding is frequently ejected from the muzzle of the weapon and can cause injuries up to as much as five metres from the gun if carelessly handled. Military blank ammunition often had either a paper or wooden mock bullet in order to facilitate loading through the rifle magazine, but these were a source of danger and the practice has almost entirely ceased. Today's military blank cartridges often have the mouth of the case extended and folded so as to simulate a bullet, but so designed as to split open on firing without discharging anything from the gun. More recently, blank cartridges have been made of plastic material with the nose weakened so that it splits under pressure but does not eject any solid material.

Varieties of blank cartridge: left to right, plastic, plastic and metal (with a wooden bullet), and two examples of all-metal blanks with the case mouth rose-crimped.

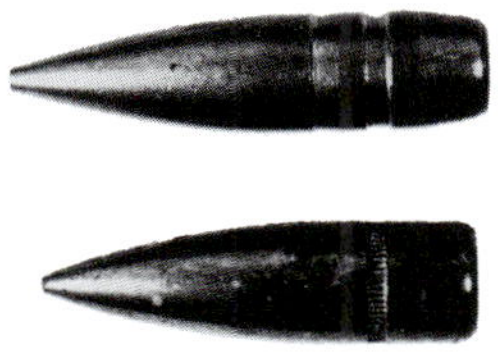

Comparison between a 7.62mm Spitzer bullet (top) and a 7.62mm boat-tailed bullet.

BOAT-TAILED Term used in the USA to describe a bullet in which the rear section is tapered to reduce base drag. Known in Britain as 'streamlined'.

BOTTLE-NECKED Description of a cartridge case shape in which the diameter of the case is considerably greater than that of the bullet, the front end of the case being sharply reduced in diameter to retain the bullet. This form is used where it is desired to have a large internal volume (to hold the propelling charge) in a relatively short cartridge. If the case were straight-sided and of about the same diameter as the bullet, it would need to be longer in order to have the same volume, and this length might not be compatible with the weapon's mechanism. The form was first used extensively in military cartridges during the period 1870-80, particularly for weapons using the Martini breech mechanism which precluded a long, thin cartridge shape.

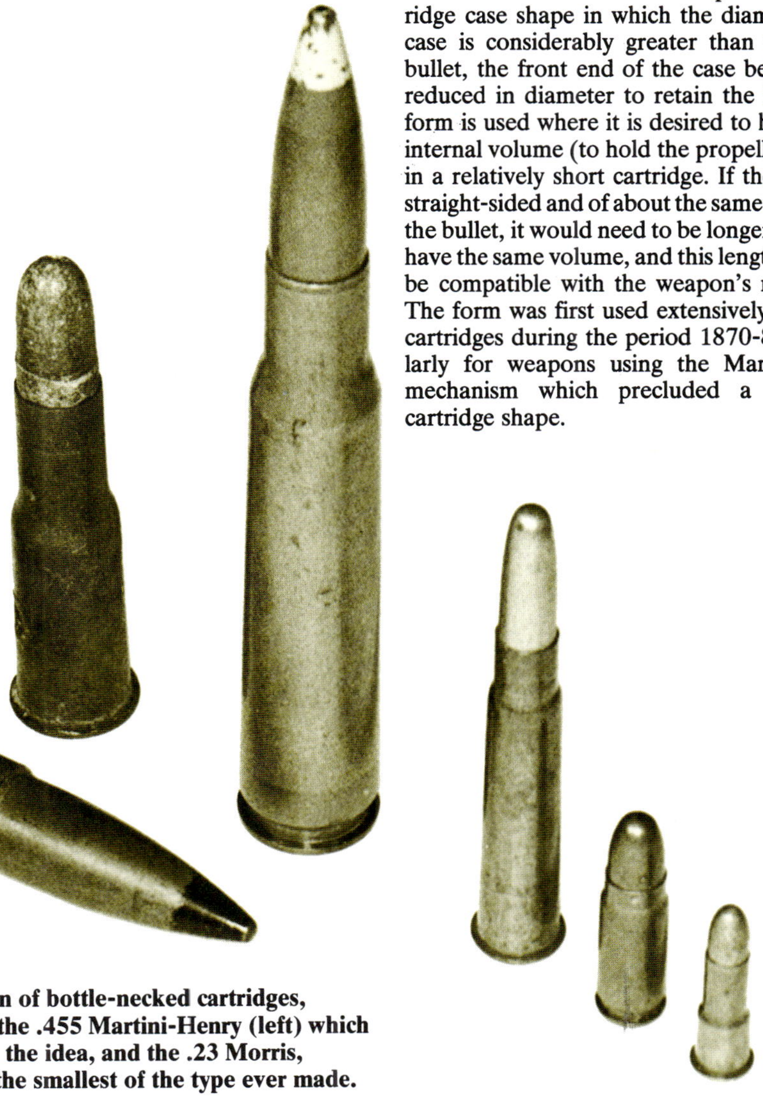

A selection of bottle-necked cartridges, including the .455 Martini-Henry (left) which pioneered the idea, and the .23 Morris, probably the smallest of the type ever made.

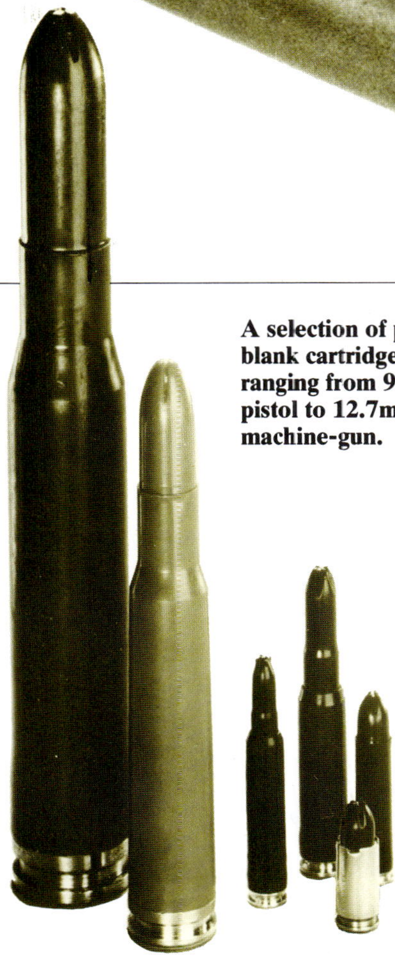

A selection of plastic blank cartridges, ranging from 9mm pistol to 12.7mm machine-gun.

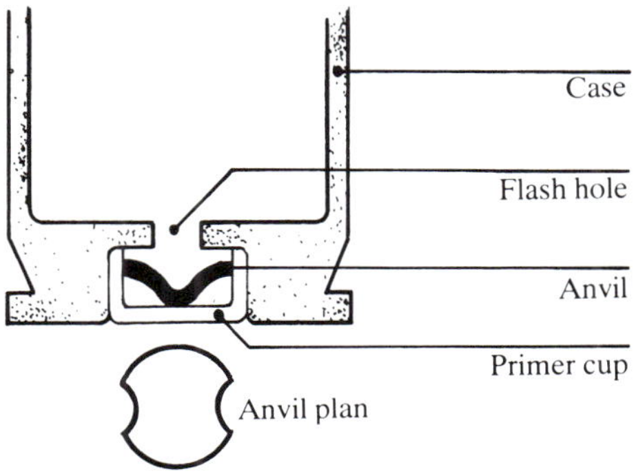

A Boxer primer. Compare the primer assembly on this type with that found on the two-hole Berdan opposite.

BOXER PRIMER A primer cap for a small arms cartridge, developed in 1866 by Colonel Edward M. Boxer, Superintendent of the Royal Laboratory at Woolwich Arsenal. The significant feature of the design is that the anvil is a separate component forming part of the cap assembly, and the case has a single, central flash hole in the cap chamber. Although rather more difficult to manufacture, caps of this type are much easier to use when reloading fired cases, the complete cap unit being easily expelled from the case (by a thin rod pushed through the flash hole) and replaced by a complete new unit. They are therefore found in commercial ammunition, which can be reloaded, but rarely in military ammunition, though in the past they have been used in military ammunition issued to isolated stations where reloading was practised.

BUCKSHOT Lead balls used as projectiles in shotgun ammunition. Derives its name from its original use against large game such as deer. Buckshot ranges in size from '000' (0.36in diameter, 98 balls to the pound weight) to 'No. 4 Buck' (0.24in diameter, 344 to the pound). Terminology differs according to the country of origin: '00 Buck' (in the USA) is known as 'SG' in Britain, 'SSG' in Canada, 'B8' in the Low Countries and 'Posten III' in Germany, and other sizes have similar differences. '000' is used as a military loading in combat shotgun ammunition.

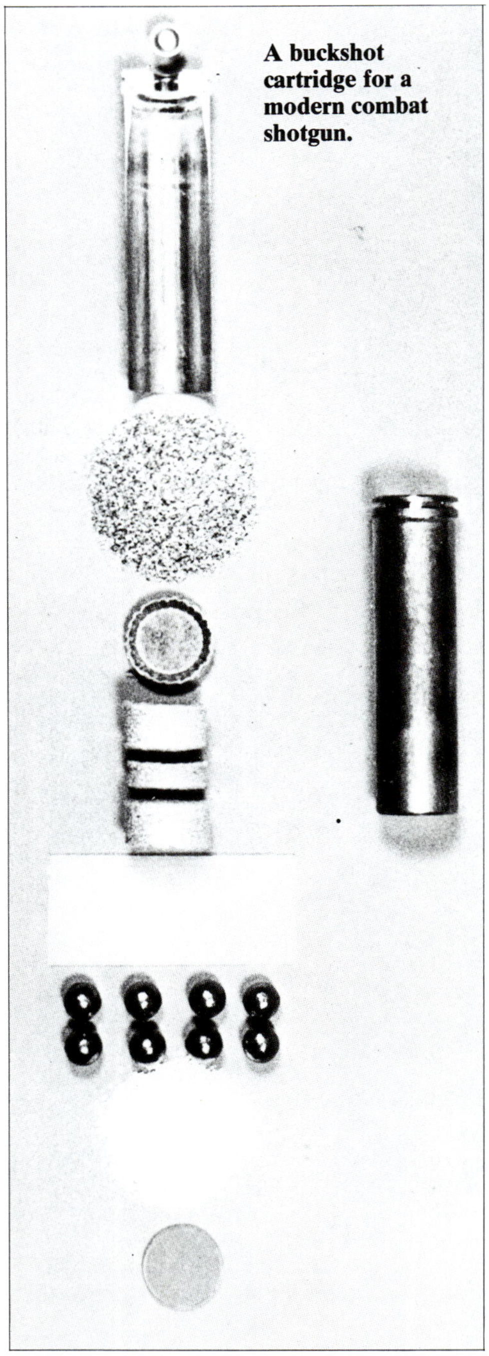

A buckshot cartridge for a modern combat shotgun.

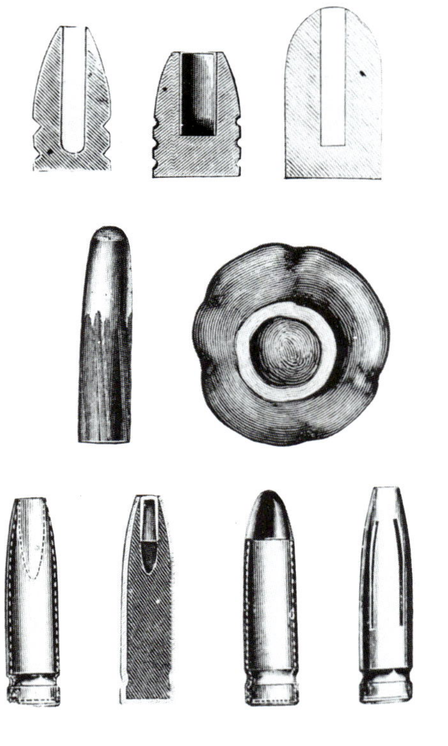

A selection of sporting bullets. The main concern of the designers was to produce a shattering or mushrooming effect so as to kill game as quickly and surely as possible.

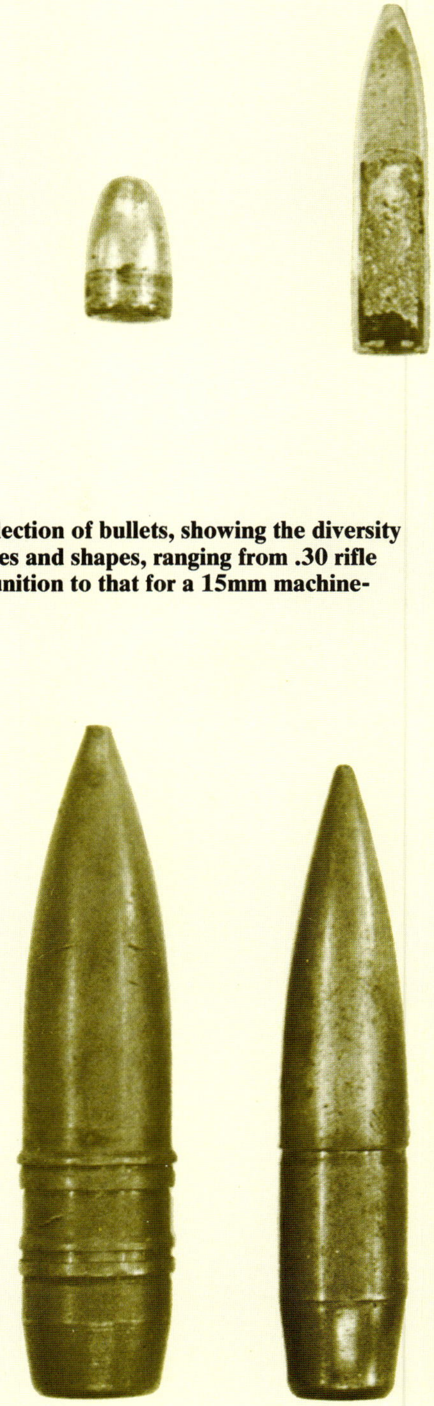

A collection of bullets, showing the diversity of types and shapes, ranging from .30 rifle ammunition to that for a 15mm machine-gun.

BULLET General term for the projectile fired by any small arm (ie, of a calibre up to and including 15mm).

Revolver bullets are usually round-nosed and made of lead, though for target shooting a flat-headed form is preferred. A more modern construction is to have the rear section of the bullet jacketed, leaving the lead core exposed; this gives better deformation on impact, as does a hollow-nosed bullet, both of which are used for hunting.

Bullets for automatic pistols are generally ogival, sometimes round-nosed and usually jacketed, because soft-nosed bullets tend to deform during the loading process, particularly in the more powerful types of pistol, and many fail to feed properly into the breech. Moreover, the velocities in automatic pistols are generally higher than those found in revolvers and thus there would be a considerably greater likelihood of lead fouling in the bore. Where full-jacketed bullets are not used, semi-jacketed, with an exposed lead point, will be found.

Rifle bullets are ogival and may be streamlined or flat-based. Military ball bullets are invariably jacketed because the Hague Convention prohibits any bullet which deforms on impact. The usual form of construction for military ball bullets is a lead/antimony core enclosed in a steel and gilding metal jacket, though in recent years bullets in the 5.56mm calibre have been made with cores partly of steel and partly of lead. This compound construction improves penetration, due to the steel forward portion of the core, without seriously impairing flight characteristics. The Soviet 5.45mm rifle bullet has a short, empty space inside the tip of

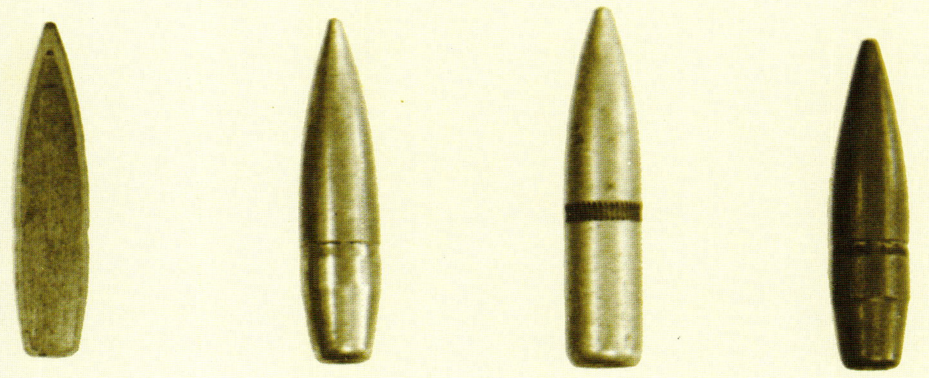

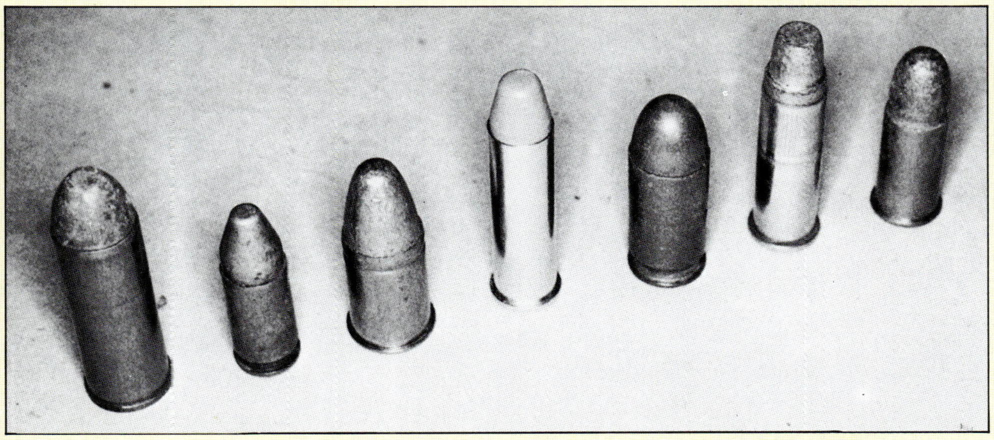

Above: A variety of pistol bullets: left to right, .45 Colt solid lead; 9mm Luger Conical; .455 Webley lead ogival; .357 KTW Teflon-coated metal-piercing; .45 Auto jacketed, .38 Special semi-wadcutter and .38 Smith & Wesson lead round-nosed.

the jacket, before the steel core begins, and this appears to promote a folding of the tip on impact, so that the bullet develops a curved path inside the target and has serious wounding potential.

Military bullets for other tactical applications — eg, tracer, armour-piercing, etc — are dealt with under their individual headings.

Rifle bullets for sporting use may be jacketed or may have the core tip exposed for better deformation on impact. Every commercial manufacturer has his own ideas about bullet construction, usually developed in the endeavour to control deformation, improve accuracy or consistency, or to reduce wear, and most of them appear to achieve their objectives.

C

CANNELURE A groove formed in the outer surface of a bullet or cartridge case. In a lead bullet, cannelures are intended to be filled with lubricant, which helps reduce lead fouling in the barrel of the weapon; they are usually concealed inside the cartridge case. In jacketed bullets, cannelures are used variously to press in the mouth of the cartridge case to achieve a secure joint, to allow identification of a bullet which, while similar in appearance, differs from some other bullet, or to lock the jacket and core together.

A cannelure may be pressed into a cartridge case for identification purposes or to suit some particular characteristic of the weapon's feed system; 25mm cartridges, for example, have a cannelure near the shoulder into which a rib on the feed belt link fits, positively locating the case so that it does not slip backwards or forwards in the link and thus jam in the gun's feedway. When the gun fires the internal pressure invariably blows the cannelure out flat so that it can be difficult to see on a fired case.

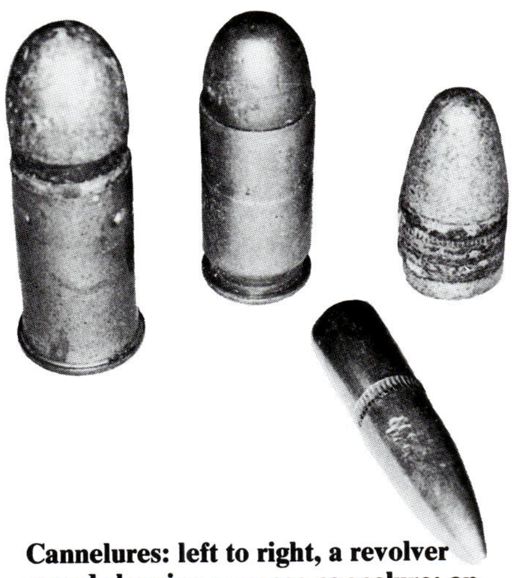

Cannelures: left to right, a revolver round showing a grease cannelure; an automatic pistol round; with a cannelure in the case to locate the bullet correctly; a revolver bullet with grease grooves; and a Tracer rifle bullet.

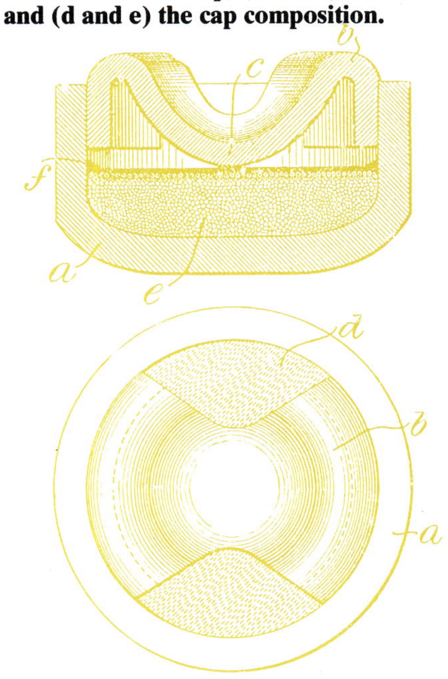

A typical primer cap, showing (a) the soft metal of the cap, (b and c) the anvil, and (d and e) the cap composition.

CAP (also called Primer) The ignition system used with a centre-fire cartridge; also the ignition unit used with a percussion muzzle-loading weapon. Present-day caps come in various sizes to suit the varying size of cartridge cases, and may be either of *Berdan* or *Boxer* pattern. Caps are formed of either zinc or copper, or a compound of both, so as to be malleable and deform easily under the blow of the firing pin. Beneath the metal is a layer of sensitive cap composition (see below) which is sealed in place by a layer of varnish which also waterproofs the composition. Impact of the firing pin nips the composition against the anvil and the resulting flash passes into the body of the cartridge case and ignites the propelling charge. The cap is retained in the cartridge case either by being a tight fit (usually when it is desirable to be able to remove the cap for reloading the case) or by the upsetting of the case metal around the cap so as to lock it into place; 'staking' or 'ringing' are the usual methods, both of which cause a small amount of case metal to overlap the cap and hold it in position. This is only required for the purpose of transportation; once inside the gun the breech mechanism fits closely behind the cap and supports it against major rearward movement, although excess pressure in a case can sometimes cause the cap to be blown out as the breech is opened. The joint between case and cap is referred to as the 'cap annulus' and is usually sealed with waterpoof varnish; in military ammunition this varnish is often coloured to indicate the nature of the bullet with which the case is loaded.

CAP COMPOSITION Sensitive composition filled into a cap or primer and which bursts into flame when struck a violent blow by the firing pin.

Early compositions were based on fulminate of mercury, potassium chlorate and antimony sulphide; the first was sensitive to the blow, and the other two increased the heat and flame of the explosion. However this composition had drawbacks: the mercuric salts from the explosion attacked the brass of the cartridge case, weakening it and making it unsuitable for reloading, and the potassium salts were deposited on the bore of the gun and promoted rusting. In the days when gunpowder was the principal propellant this mattered little, because the washing necessary to remove powder fouling also removed the salts. But when smokeless powders came into use, and cleaning was less rigorous, the salts began to make their presence known. The military continued to use the fulminate mixture because they rarely bothered with reloading, strict discipline ensured ample washing of the rifle bore, and the power and reliability of the mixture was considered more important. But commercial manufacturers began to abandon fulminate mixtures during the early years of the 20th century, changing to compositions based on lead styphnate and tetrazene which did not affect cartridge brass or promote rusting. Military applications for non-corrosive compositions came late and not until the late 1950s could one be certain that military caps were non-corrosive.

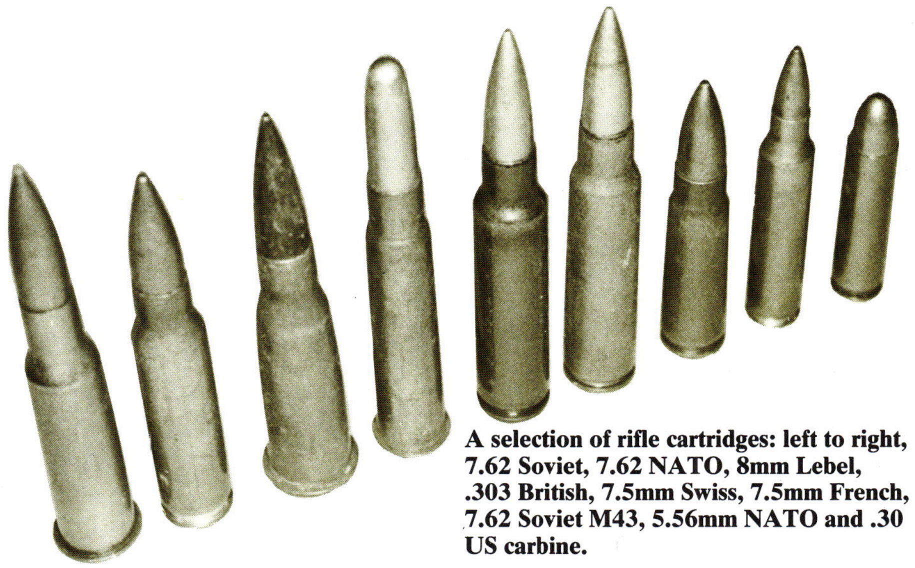

A selection of rifle cartridges: left to right, 7.62 Soviet, 7.62 NATO, 8mm Lebel, .303 British, 7.5mm Swiss, 7.5mm French, 7.62 Soviet M43, 5.56mm NATO and .30 US carbine.

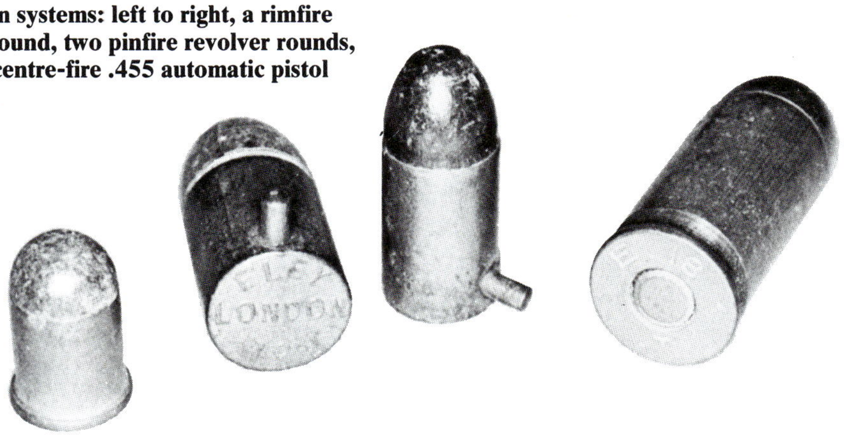

Ignition systems: left to right, a rimfire 9mm round, two pinfire revolver rounds, and a centre-fire .455 automatic pistol round.

CARTRIDGE That part of a complete round of ammunition which forms the propelling charge; also, and particularly in small arms ammunition, an alternative term for the complete round — cap, case, charge and bullet.

CARTRIDGE BRASS Mixture of 70 per cent copper and 30 per cent zinc from which cartridge cases are traditionally made. Selected because of its ease of manufacture and because it has the property of elasticity. This allows the

Above: Examples of rimless and semi-rimless cases; the bottle-necked cases are rimless, the straight cases semi-rimless. The case at the extreme right is a Bergmann 6.5mm of 1894 with no extraction groove or rim of any sort.

Below: Stages in the manufacture of a .303 rifle cartridge case. The brass case must be annealed (heat-treated) between each process.

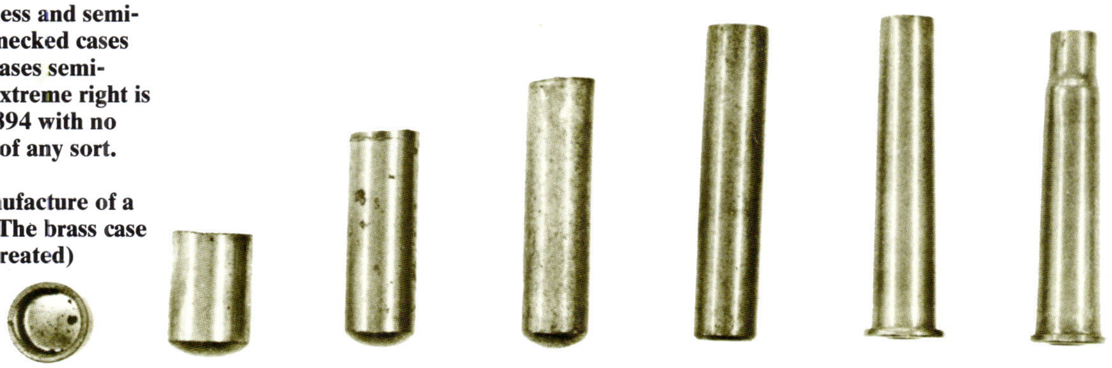

case to expand under the pressure of the exploding charge and adhere tightly to the chamber walls, thereby sealing the breech against any escape of gas, and then allows it to contract to its former dimensions so that it can be easily extracted. It also has the useful property of oxidizing immediately after manufacture so forming a protective coating over the metal which inhibits further corrosion or oxidization.

CARTRIDGE CASE Metallic case, usually of brass, forming the propellant container in a round of small arms ammunition. The case performs other vital functions: it locates the bullet correctly relative to the bore of the weapon; it supports the bullet; it supports or carries the means of ignition; it seals the breech against the unwanted escape of propellant gas when the gun fires; it protects the propellant from damp; and it acts as an insulator between the propellant powder and the hot walls of the chamber in a rapid-firing gun.

Although brass is the traditional metal, steel is often used. The steel must be coated to prevent it from rusting. Plastic varnish is the most usual method today, though in the past steel cases have been brass- or zinc-coated or given chemical treatment to form a protective coat. Brass is preferred for commercial ammunition, particularly when reloading is likely, and steel is almost entirely confined to military ammunition. Plastic materials have also been tried, as have various light alloys, but these have rarely been successful in conventional cartridges, though plastic is extensively used for training and blank ammunition and light alloy cases have been used in some aircraft cannon ammunition in the 20-30mm calibre range.

The metal case is made by a series of stamping and drawing movements. A thick disc is stamped from a sheet of brass; this is then forced into a cup shape and gradually, by several draws, extended to more than the desired length. The mouth of this cup, which contains any impurities present in the metal, is cut off and discarded, and the base is formed with its extraction rim or groove and cap chamber. The cap anvil and fire holes are formed, and the case is necked down to the correct size, if necessary. It is finally trimmed to the correct length and polished. Between most of these stages it is necessary to heat-treat the cases to a specified temperature and allow them to cool at a controlled rate, since brass has the tendency to 'work-harden' as it is forced through dies, becoming brittle and liable to crack or tear in subsequent processes.

Steel cases are made in a similar way, but the different metallurgical properties of steel mean that the thickness of the case and the distribution of metal is not the same as in a brass case. It is not practical to make a steel case from the drawings of a brass one.

Plastic cases may be entirely of plastic, though this is confined to low-pressure training ammunition for revolvers. The more usual method is to make the base, with the extraction rim or groove, from metal and then mould the plastic to it.

CASELESS CARTRIDGE A cartridge in which the conventional metal case is not employed. Instead, the propellant is mixed with a binder so as to become a hard mass which can be shaped as required. A cap, made entirely of combustible material, is fitted into the base of this block of propellant and the bullet is recessed into the forward end. The object is to eliminate the extraction and ejection of the empty case after firing, and so speed up the action of a weapon. It also has the advantage of lightening the round of ammunition (brass or steel cases are heavy) and of reducing the demand for copper and zinc which are both critical materials, particularly in wartime.

It has its drawbacks, however: it does not seal the breech, so this has to be done by some mechanism in the weapon, and it does not provide an insulating 'heat sink' between the propellant and the hot chamber of the gun, so conventional propellants are unsuitable. The

An experimental caseless cartridge (right) alongside a conventional round of comparable ballistics.

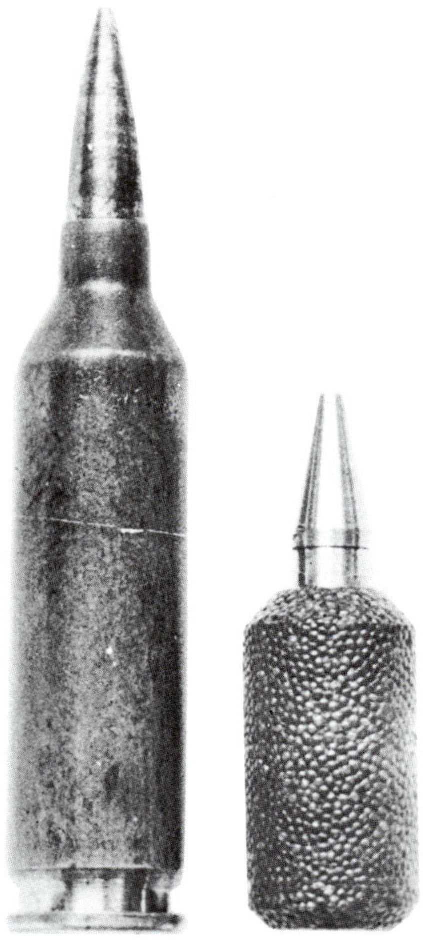

most advanced caseless cartridge at present is the 4.7mm pattern developed in Germany by Heckler & Koch and Dynamit Nobel for the former company's G-11 rifle, scheduled to enter service with the Federal German Army in the early 1990s. This cartridge uses a moderated high explosive as the propellant rather than a conventional nitro-compound.

CENTRE-FIRE Term used to describe cartridges in which the ignition cap is centrally placed in the base, distinguishing them from *rimfire* or *pinfire* cartridges. In early designs the cap was sometimes concealed within the case, and the firing pin of the weapon had to deform the case metal to operate the cap. This type can easily be confused with contemporary rimfire cartridges.

CHARGER Device for loading cartridges into a weapon, often erroneously called a 'clip'. A charger is a spring steel clip which grips the rims of a number of cartridges, usually a number which will fill the magazine of the weapon. The action of the weapon is opened and the charger is inserted into a support so that the cartridges are above the mouth of the magazine. By thumb pressure, the cartridges are then swept out of the charger and into the magazine, and the empty charger is removed, usually by the thumb, sometimes by the mechanism of the weapon as the bolt closes to load the first round.

Chargers and a clip: left, the ten-round charger for the Soviet Simonov rifle; and right, the five-round Lee-Enfield charger. Between them is the eight-round clip for the US Garand rifle.

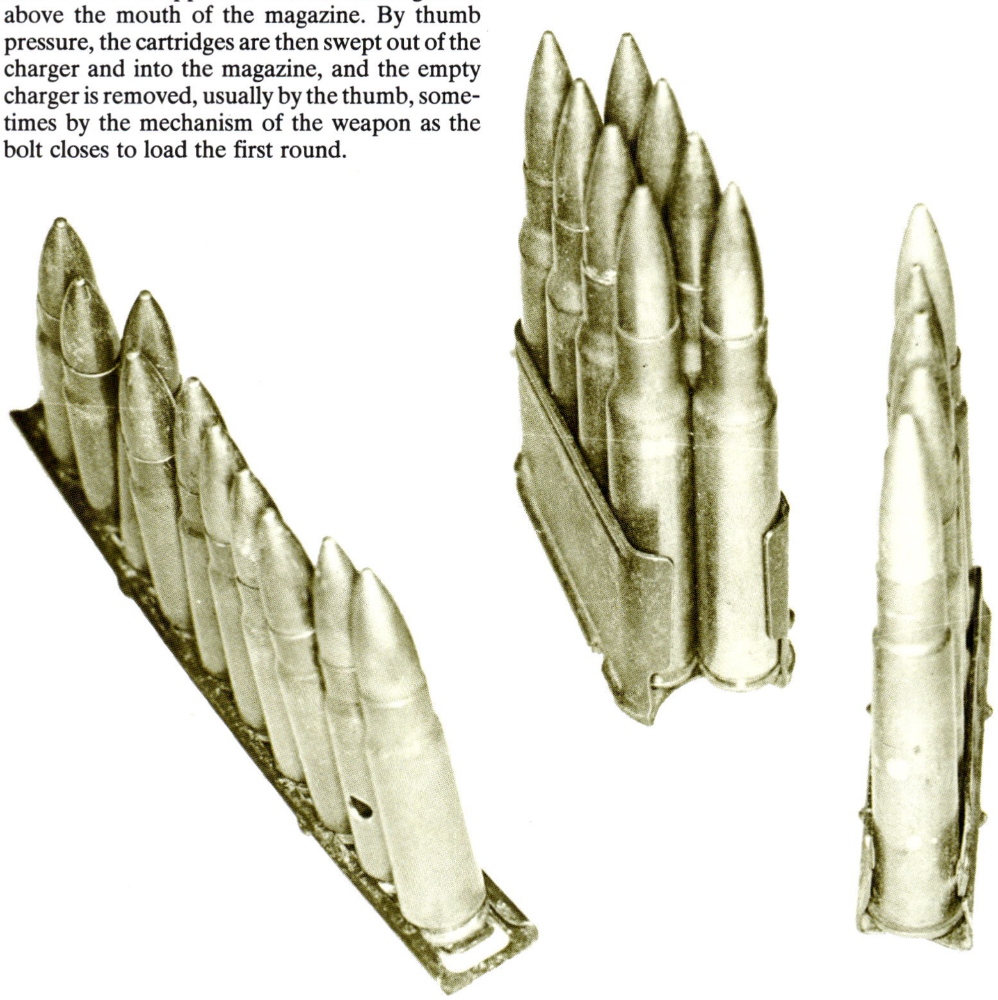

A centre-fire cartridge (right) and a rimfire cartridge.

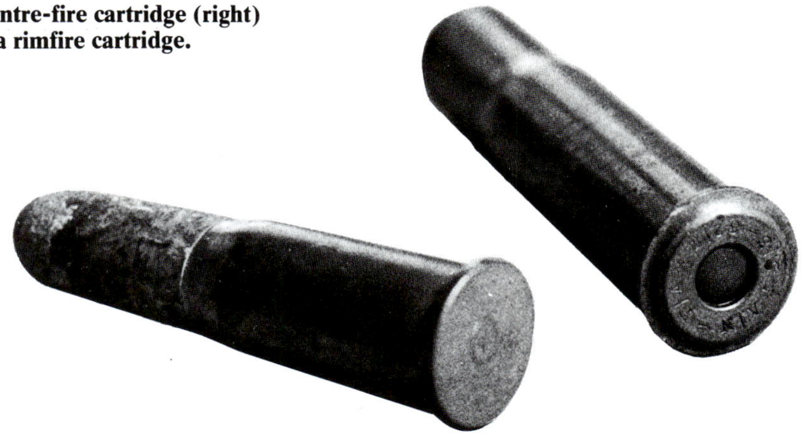

Clips are not confined to small arms. Here, an American gunner loads a clip of 37mm rounds into a light anti-aircraft gun.

CLIP Device for loading cartridges into the magazine of a weapon and frequently confused with a *charger*. The clip, like the charger, is a spring steel frame holding a number of cartridges, always a number which fills the magazine. The action of the weapon is opened and the complete unit (clip and cartridge) is forced into the magazine. The action of the weapon is then closed, and the first cartridge is loaded into the chamber. As the weapon is operated, a pusher arm forces the cartridges up in the clip until the last one has been loaded and fired, whereupon the clip is ejected from the magazine. The significant feature of clip-loading is that without the clip the weapon's magazine will not function, and it is impossible to 'top up' the magazine with loose rounds. Probably the best-known clip-loading weapon of recent years was the US Army's Garand rifle.

COMPOUND BULLET A bullet constructed by surrounding a core of dense metal with a jacket of lighter metal, the object being to obtain a small-calibre projectile of suitable weight, but which does not bring lead into contact with the rifling of the weapon. The idea was pioneered by Major Rubin of the Swiss Army in the 1880s and was quickly taken up by military designers who were seeking suitable bullets for the new generation of small-calibre rifles; it then moved across to the commercial field.

COPPER A constituent of cartridge brass; also, used alone as a material for caps; also, used in the form of small slugs in a device for measuring the internal chamber or bore pressure in firearms, in which case the pressure is given as so many 'CuP' or 'Copper units of Pressure'. This system is particularly popular in the United States; in Europe it is more usual to quote pressure in direct units such as lb/in^2 or kg/cm^2.

D

DARK IGNITION Method of igniting military *tracer* bullets in which the propelling charge flame first ignites a delay composition, and then ignites the tracer, so that the tracer does not become visible until the bullet is some distance from the weapon, usually 50-100 metres. The object is first, to prevent the emerging tracer bullet from dazzling the firer; and secondly, to prevent an enemy from following the line of the trace back to its origin and thus discovering the location of the weapon.

DIGLYCOL POWDER A propellant powder which is essentially a *double base* powder but in which the nitro-glycerine has been replaed by a nitric ester, such as diglycol-dinitrate, triglycol-dinitrate or methy-trimethylol methane-trinitrate. The powder gives more power than single base, but without the high flame temperature of double base. This means less erosive wear in the gun barrel and thus longer life for the weapon, as well as reduced flash at the muzzle.

Two compound bullets (for the 7.62mm Soviet Nagant revolver and the .223 Armalite rifle) showing the construction of core and jacket.

'Dignity and Impudence', or the extremes of small-arms ammunition: a .22 short rimfire round on top of a Soviet 14.5mm machine-gun round.

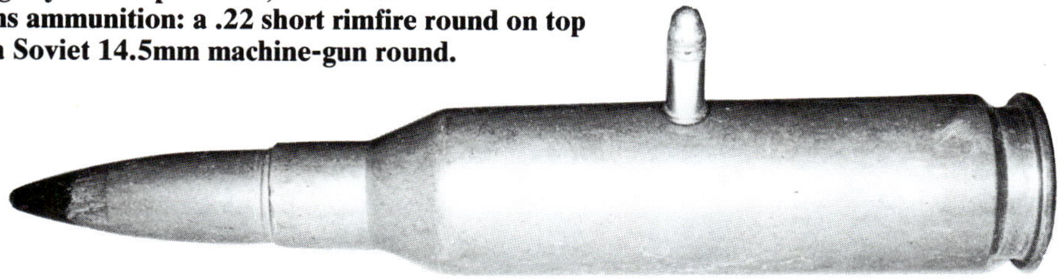

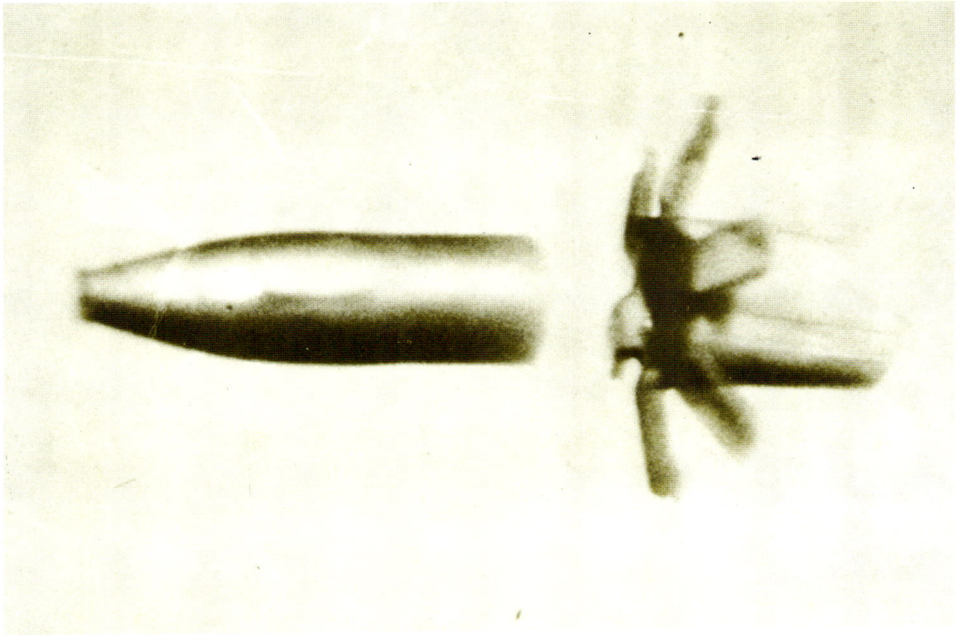

A 7.62mm discarding sabot in flight, showing how the sabot peels away from the bullet.

DISCARDING SABOT A projectile consisting of two units, the sub-projectile and the sabot. The sub-projectile is much smaller in diameter than the bore of the weapon and is supported in the sabot, which is of bore diameter. The sabot is of plastic or alloy material, and thus the complete projectile is much higher than a conventional projectile for the weapon. On firing, it therefore accelerates rapidly and attains a high muzzle velocity. On leaving the muzzle the action of centrifugal force and air pressure causes the sabot to separate from the sub-projectile and fall to the ground, leaving the sub-projectile to fly to the target.

The object is to attain the highest possible velocity, higher than can be achieved with a full-bore projectile, in order to give the bullet greater range or to ensure higher striking velocity at a given range, as in the case of armour-piercing bullets. At present commercial discarding sabot ammunition is made by Remington, under the name 'Accelerator', while the US Army have recently announced discarding sabot armour-piercing bullets in 7.62mm and .50in calibres under the acronym SLAP.

DOUBLE BASE Type of propellant in which the principal constituents are nitro-cellulose and nitro-glycerine. It is more powerful than single-base propellant, because of the incorporation of the nitro-glycerine, but for the same reason it is also much hotter and has a flame temperature which melts away the steel of the weapon's barrel more rapidly. The British military propellant 'Cordite' was among the best known of double-base propellants.

Double-base propellants are not used today to any great extent, having been replaced by triple-base types. Another drawback to double base is that the nitro-glycerine is prone to decompose and eventually exude in a pure form over long periods of storage.

DRIFT Sideways movement of a bullet as it flies through the air. The movement (the Coriolis Effect) is caused by the rotation of the bullet leading to differing air densities on either side, the bullet tending to move into the area of lesser density. Drift is compensated somewhat by the positioning of the weapon's sights, but fine adjustment must usually be done by the firer.

DRILL AMMUNITION Cartridges which are completely inert, with no explosive components, and which are used by military trainees to practise loading and manipulation of the weapon. Sometimes called 'dummy' though in strict terminology there is a difference. Drill cartridges may be made of service components which have failed inspection for some minor faults, of once-fired components which have been salvaged and refurbished, or they may be of non-standard materials (ie, a light alloy case and a wooden bullet). The cap chamber is either empty or filled with a plug of plastic or rubber to prevent damage to the firing pin of the weapon. The cartridge case is usually pierced with holes, fluted or impressed with *cannelures* so that it can be positively identified as non-functional.

Three rifle drill cartridges: left, British .303 Lee-Enfield (steel); centre, German 7.92 Mauser (plastic); and, right, US .30 Springfield (brass and wood).

DUM-DUM Type of bullet devised at Dum-Dum Arsenal in India in the 1890s by Captain Bertie Clay, the Superintendent of Small Arms. It consisted of the standard .303in British Service ball bullet with the jacket trimmed back at the nose to expose the lead core. The object was to have a bullet which deformed rapidly and dealt a powerful shock on impact so as to knock down a tribesman or other 'savage' enemy. They were first used at the Battle or Omdurman in 1898, but displayed a serious defect in that since the bullet jacket did not entirely cover the base, there was a tendency for the core to blow out and leave the jacket in the rifling to hinder the loading of the next round. The design was therefore abandoned and replaced by the 'Ball Mark III' bullet which had a full jacket with a hole bored in the ncse and filled with a short metal tube. This had the same expanding effect and was more reliable in the rifle.

In 1899 the Hague Convention outlawed the use of any expanding bullets in military service and the Dum-Dum and the Ball Mark III were withdrawn. Since that time no expanding bullets have been used by military forces, but the word 'Dum-Dum' is loosely used to describe any expanding bullet, particularly one which has been converted from its original non-expanding design by unauthorized means, (eg, by cutting a cross into the nose of the lead bullet).

The armourer's tools: three dummy cartridges with, in front, a 'chamber gauge' for measuring worn guns.

A crudely-manufactured Dum-Dum bullet, made by cutting into a normal lead revolver bullet.

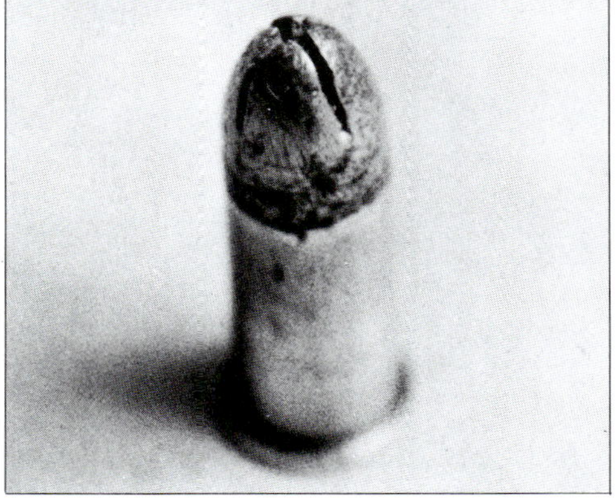

DUMMY AMMUNITION Similar to *drill* ammunition, and the words are often used interchangeably. In British terminology, however, dummy ammunition is a very carefully made replica of a cartridge, usually of tool steel and carefully dimensioned, to be used by weapons inspectors and repairmen when checking the functioning of weapons.

DUPLEX American military term for *multiple bullet* cartridges, developed in the early 1960s and used to some extent in Vietnam. They were produced in 5.56mm and 7.62mm calibres.

Two experimental US Duplex and Triplex ball cartridges, 7.62mm calibre, photographed by X-ray.

E

ENVELOPE Sometimes used interchangeably with the word 'jacket', but strictly it means the outer metal coating of the *jacket* where such coating is used. Thus a mild steel jacket coated with gilding metal could be said to have a gilding metal envelope.

EXPLOSIVE An explosive is a substance which, upon being suitably initiated, is capable of exerting a sudden and intense pressure on its surroundings. This pressure is produced by the extremely rapid decomposition of the explosive into gas, the volume of which at ordinary atmospheric pressure and temperature would be many times greater than that of the original explosive. The pressure is greatly intensified by the heat which is liberated during the process. (*Treatise on Ammunition*, 1926)

Explosives are classed as low or high. A low explosive burns extremely rapidly, the flame passing through it at speeds up to 3,000 metres per second (9,840fps). A high explosive detonates, or undergoes molecular disintegration, the shock wave passing through it at speeds as high as 10,000 metres per second (32,800fps) or more. Generally speaking, low explosives can be initiated by flame, whereas high explosives require a shock wave to detonate them.

In firearms, low explosives are used as propellants, while high explosives are confined to use as bursting charges in projectiles. In consequence, high explosive is rarely seen in the small arms ammunition field. High explosive is used as the basis for the propellant in some caseless ammunition, but it has been moderated or treated so as to convert its characteristic action from detonation to burning, and it must therefore be reclassified as a low explosive.

EXPLOSIVE BULLET A term having two meanings. In the context of the Hague Declaration of 1899, an explosive bullet is any 'bullet which expands or flattens easily in the human body' and its use for military purposes is forbidden.

In military usage an explosive bullet is a small arms bullet containing a charge of explosive which will detonate on impact and give off a visible flash and puff of smoke. Such bullets were designed for use in long-range machine-guns so that the strike of the bullet could be seen, allowing the fire to be adjusted. At present, explosive bullets are used with aiming rifles or machine-guns attached to anti-tank artillery weapons; the aiming weapon is locked to the barrel of the weapon so that the strike of the bullet coincides with the strike of the artillery projectile at some specific range. The gunner fires the aiming weapon, adjusting his aim, until he sees a strike on the target; he then fires the major weapon, confident that the projectile will strike in the same place as did the bullet. This system allows very fast determination of range or corrections for wind, when rangefinding apparatus is unavailable. It is now being adapted to the fire control of anti-tank rocket launchers.

F

FIRST CATCH That point in the flight of a bullet where it falls close enough to the ground to strike the head of a standing man. The beginning of the *beaten zone.*

FIRST GRAZE That point in the flight of a bullet where it will strike the ground, or a prone man. The end of the *beaten zone.*

FLECHETTE A dart-like solid projectile, stabilized in flight by fins. Originally used to describe large steel darts (about one foot long) dropped in large quantities from ground-attack aircraft during the 1914-18 war as a method of attacking marching troops. The term was revived in the 1950s for the US Army development of small darts (about 2.54cm or 1in long) as small arms projectiles. The applications varied: some were tried as single-projectile loadings for rifles, others as multiple projectile loadings for shotguns, grenade launchers and artillery shells. In small arms applications the single flechette had to be supported by a sabot in the rifle bore, and this gave rise to difficulties in discarding at the muzzle. The flechette was not particularly accurate, its lethality was irregular, and the project was abandoned. As a multiple loading in shotgun ammunition it was rather more useful, and at present the cartridge for the 12-gauge Close Assault Weapon is loaded with 20 steel flechettes.

FRANGIBLE BALL Type of ball bullet made from compressed particles of metal and paint. Used in the US Army and Air Force during and after the 1939-45 war as a training bullet for aerial gunners. When fired at another aircraft the bullet would disintegrate to dust on impact, causing no damage, but leaving a paint mark on the target so that the trainee's gunnery could be assessed.

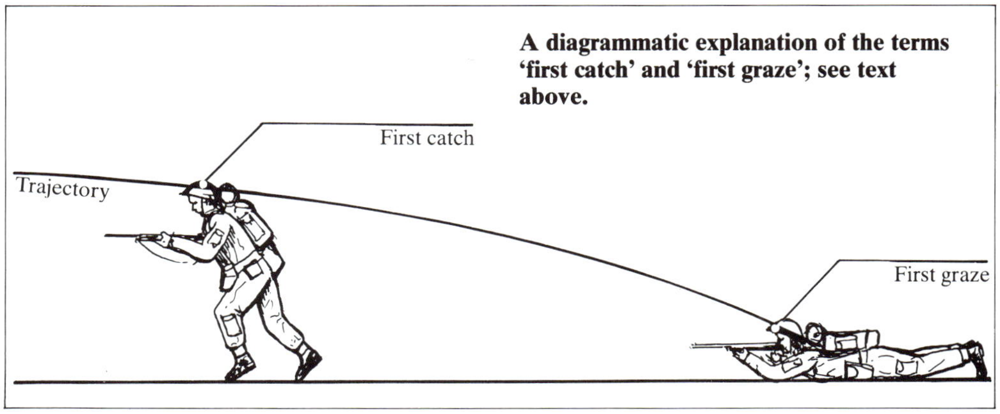

A diagrammatic explanation of the terms 'first catch' and 'first graze'; see text above.

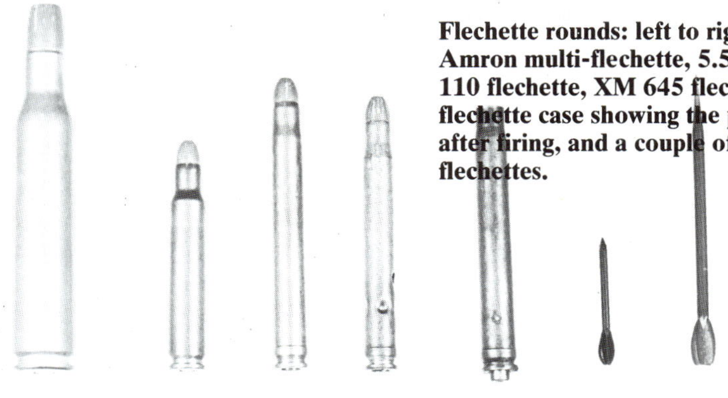

Flechette rounds: left to right, 8.35mm Amron multi-flechette, 5.56mm case, XM 110 flechette, XM 645 flechette, a fired flechette case showing the piston primer after firing, and a couple of typical flechettes.

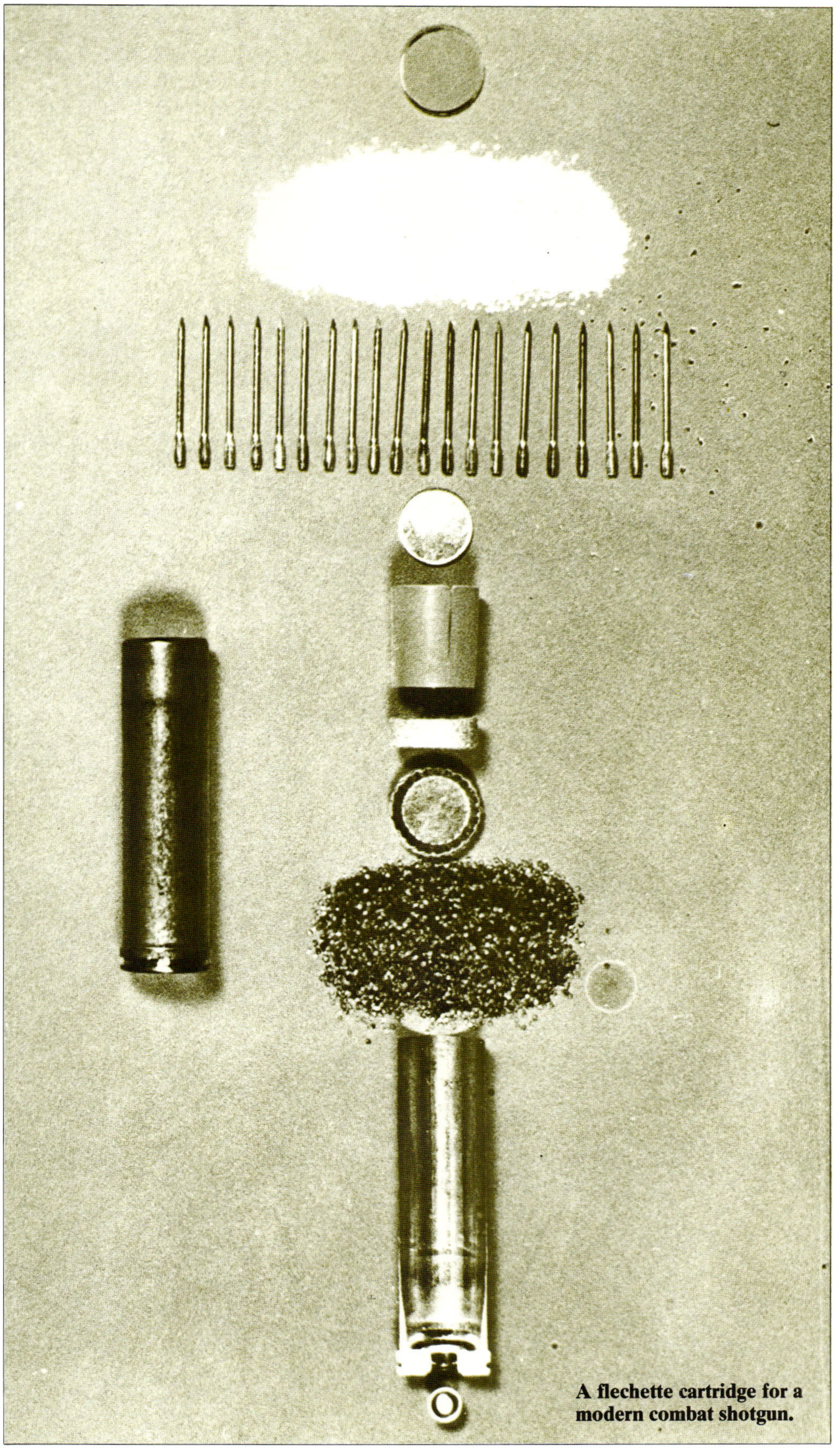

A flechette cartridge for a modern combat shotgun.

G

GILDING METAL Alloy of approximately 90 per cent copper and 10 per cent zinc, used to coat the mild steel jacket of a bullet. Used in order to protect the steel from corrosion and to develop less friction in the rifle barrel during the bullet's passage.

GRENADE CARTRIDGE Type of *blank cartridge* used to launch grenades from a rifle. The requirement is to develop a large volume of gas very quickly, but not so quickly as a black powder blank, and therefore grenade cartridges are generally loaded with fast-burning smokeless powder. The grenade can be launched in one of three ways: by attaching a rod which is inserted into the barrel of the rifle; by loading the grenade into a cup attached to the muzzle of the rifle; or by the grenade having a hollow tail unit which can be slipped over the muzzle of the rifle. The first method is no longer used, the cup discharger is confined to use with riot-control munitions, and most modern rifles have two 21mm diameter rings formed on the exterior of the muzzle to accept standard grenade's.

Current practice eliminates grenade cartridges by having a bullet trap in the grenade's tail unit. This permits the use of standard ball cartridges and does away with the need to unload the rifle, load a grenade cartridge, then re-load with ball after firing the grenade. It also removes the need for the soldier to carry grenade cartridges. The only drawback is that the propellant loading of a ball bullet is not the optimum for grenade launching and therefore bullet-trap grenades cannot usually attain the same range as conventional patterns fired with the special cartridges.

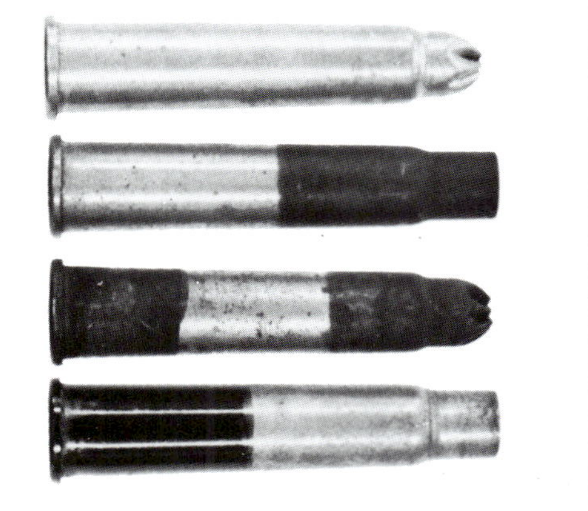

Grenade-launching cartridges.

H

H-MANTEL German term, used universally, for a rifle bullet having a solid internal partition about half-way up its length, with cores at front and rear of the partition, thus giving the bullet an 'H' appearance in section. The jacket in the front section is somewhat thinner than that in the rear. The object is to have controlled deformation on impact, the forward portion splitting open and mushrooming, while the partition prevents the rear of the bullet deforming. Used only for hunting.

HEADSTAMP Markings impressed on the base of a cartridge case in order to identify it. The information carried by headstamps usually includes the calibre and manufacturer; it can also include trade names, date of manufacture, batch or lot number, details of the case manufacture, details of the case metal composition and trademarks or emblems. Military headstamps are frequently coded to conceal the identity of the manufacturing plant, but usually have the year of manufacture and calibre plainly marked. Commercial ammunition usually carries the name or trade name of the maker and the calibre. Headstamps are a specialized study and there are a number of books on the subject.

HEEL BULLET A pistol bullet in which the rear portion is reduced in diameter to fit into the mouth of the cartridge case, so that the rest of the bullet appears flush with the mouth of the case.

A typical military headstamp, seen on a German 7.92mm Mauser cartridge.

Heel bullets, showing the reduction in diameter which fits inside the cartridge case.

Only used with rimmed ammunition because it prevents the mouth of the case being used to position the cartridge in the weapon chamber.

HOLLOW POINT Type of bullet in which the nose is drilled out to about 2 calibres depth, leaving a hole. The tip of the bullet may be jacketed to the edges of the hole, or it may be of lead, and if jacketed the hole may be lined with jacket metal. The object is to cause the head of the bullet to deform — or 'mushroom' — on impact, as a result of the compression of the column of air inside the hollow as it strikes the target. Hollow-point bullets are usually used for hunting, in order to ensure a kill with a single shot; they have much greater 'stopping power' than conventional bullets and are occasionally used by police and security forces, though most police authorities have forbidden their use. Hollow-point bullets are completely forbidden to military forces by the Hague Convention and this is observed universally.

A .303 British hollow-point bullet compared with round-nose, Spitzer and soft-point types.

I

INCENDIARY Type of military bullet used to cause fire in a target; generally confined to use by aircraft armament in order to ignite fuel tanks, though they can be used in ground roles. The most usual construction was a jacketed bullet with the front half of the core removed and the space filled with white phosphorus, which has the property of spontaneously igniting when coming into contact with air. When the bullet was crushed on impact and the phosphorus was liberated, there was immediate ignition. This method was suitable for hard targets only. Another method was to drill holes in the bullet jacket and fill them with solder which melted by friction during the bullet's passage along the weapon barrel. The bullet emerged with phosphorus leaking from the holes, and it was intended that sufficient would remain to leak out as the bullet passed through the target. Another method was to use a fusible plug in the base of the bullet, entirely filling the jacket with phosphorus, but lining it with a lead sleeve to give the necessary weight.

Phosphorus ammunition is disliked; invariably some of it leaks and many fires in ammunition dumps can be traced to such leaks. During the period 1939 to 1945, incendiary compositions of barium nitrate and powdered aluminium and magnesium were developed. Filling the forward end of a half-cored bullet, they relied on impact to split the jacket and friction to ignite the mixture. The effect of an incendiary bullet can be obtained by using tracer, and the line between tracer and incendiary fillings is fairly thin. Incendiary bullets are not common today because rifle-calibre weapons are no longer in general use in aircraft.

J

JACKET The outer metal component of a bullet, surrounding the core, used in order to prevent lead fouling of the barrel. It may be of steel, steel coated with gilding metal or steel coated with copper, or of cupro-nickel, the choice depending largely on the whim of the designer. Steel coated with gilding metal is almost universal today.

L

LEAD Metal used for bullets or for bullet cores. Lead originated because of its density, giving a good weight in a small size, and its ease of melting and casting. Lead is usually mixed with a small percentage of antimony to make it slightly harder and thus resist 'lead fouling', in which flakes of lead are deposited on the rifling of the gun during the bullet's passage, largely as a result of the lead melting from friction. Lead bullets are still used with revolvers because they do not develop high velocities and do not suffer from severe fouling, but some of the more powerful revolvers now use semi-jacketed bullets in which a copper jacket engages with the rifling. The jacketed bullet was developed in order to interpose a metal of higher melting point between the lead and the barrel, since it was still desirable to have a lead core in order to obtain the maximum density.

LETHAL BALL British term used for shotgun cartridges in which a single spherical projectile is used, and specifically applied to this type of cartridge issued to members of the Home Guard for anti-personnel use during World War Two.

LUBRICATED BULLET A lead bullet with grease or beeswax pressed into cannelures so that the lubricant is deposited on the bore of the weapon. This coating resists lead fouling from the passage of the next bullet, which in turn leaves a further coating. Modern lubricated ammunition has the grooves and lubricant concealed inside the cartridge case; older designs had them exposed in front of the cartridge case mouth.

A French lethal ball cartridge for a 12-bore shotgun.

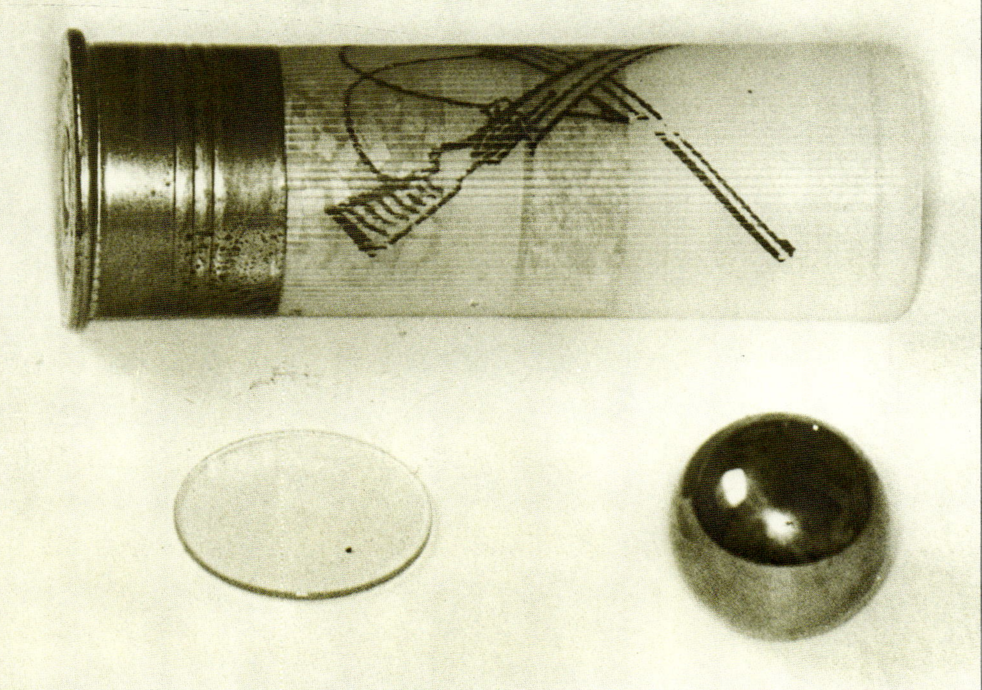

A lubricated revolver bullet, the lubricating grease being packed into the two cannelures.

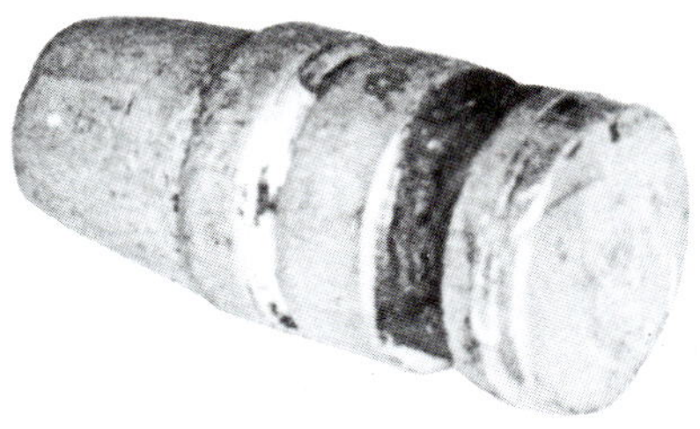

M

MANSTOPPER Trade name for a revolver bullet developed by Webley, the British pistol and gun maker. The bullet was a cylinder of lead/antimony with hemispherical recesses in nose and base; the base recess expanded under gas pressure and sealed the bullet in the bore, while the nose recess spread out on impact and deformed the bullet into a mushroom shape to give the maximum stopping power against a human target. Used with the .455 Webley it had formidable performance, but like all expanding bullets was outlawed for military use in 1899 and was forthwith abandoned.

The .455 Webley Manstopper bullet (left) compared with a conventional jacketed bullet of the same calibre.

An American 7.62mm multiple bullet cartridge.

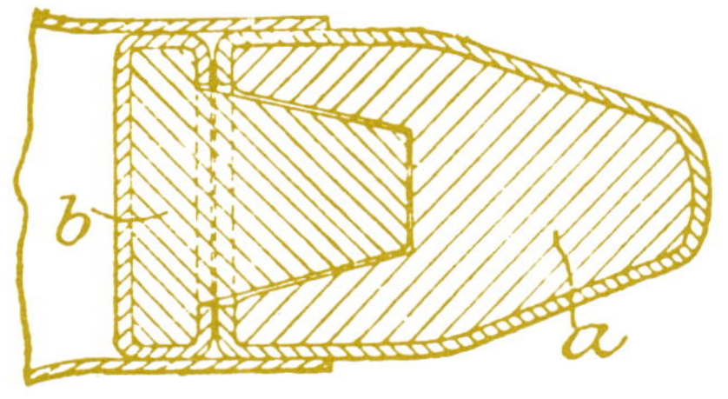

Georg Luger's 1906 patent for a multiple-bullet cartridge.

MULTIPLE BULLET A form of cartridge which contains more than one bullet, though distinct from shot cartridges. A typical multiple-bullet round will have one bullet secured in the neck of the cartridge case and a second bullet inside the case, concealed from view. The idea is not new, Georg Luger having patented such a cartridge for the Luger pistol in the early 1900s, and it reappears at intervals. Its most recent manifestation was in the US Army 'Salvo' small arms development programme of the 1950s when two-bullet cartridges in 5.56mm and 7.62mm were developed and issued for service in Vietnam. The object is to produce a better chance of hitting a target in a hurried engagement. The American multiple-bullet design achieved this by having the base of the first bullet conventionally square to the sides, but that of the second bullet slightly oblique to the sides. The first bullet followed the normal trajectory and went where the rifleman pointed his weapon; as the second bullet emerged from the muzzle, gas was able to leak out to one side and cant the bullet slightly away from the trajectory of the first. This was carefully engineered so that the second bullet struck within a 25cm (10in) circle around the first bullet at 100 metres range and proportionately greater radii at longer ranges. It was not particularly successful and was eventually withdrawn from service.

MUZZLE ENERGY Measure of the striking power of a bullet as it leaves the gun's muzzle. Calculated by squaring the muzzle velocity and multiplying by the weight of the bullet, and then applying a factor to convert the result to the desired units of measure. In the USA it is measured in foot-pounds and is obtained by expressing the muzzle velocity in feet per second, the bullet weight in grains, and dividing their product by a factor 450,420 which compensates for gravity and the different units of measurement. In Europe the Joule is the unit, obtained by expressing the muzzle velocity in metres per second and the bullet weight in grammes and applying the factor 2,000. Conversion from one to the other is simple; Joules x 0.74 = foot-pounds.

Muzzle energy is indicated by ME or by Eo, the 'o' indicating zero distance from the muzzle; the energy at any other point on the trajectory can be calculated in the same way, provided the velocity at that point is known; it is then indicated by 'E_{25}', the figure indicating the distance from the muzzle in feet (USA) or metres (Europe).

MUZZLE VELOCITY The speed at which a projectile is travelling as it leaves the muzzle of the gun. Measured in feet per second or metres per second; feet ÷ 3.28 = metres. Muzzle velocity is measured by various methods, but for small arms it is usual to use a pair of photo-electric cells wired to a high-speed counter. As the bullet passes over the first cell it starts the counter, and as it passes over the second it stops it. The time and the distance between the cells being known, a simple arithmetical calculation gives the speed, and this can be done by a computer incorporated in the 'chronograph'. Since the measurement occupies a finite distance it cannot be the true muzzle velocity, but the figure is accepted as being the velocity at a point half-way between the two cells. It is therefore necessary to make a calculation, based on air resistance and gravity losses, to determine what the velocity would have been at the muzzle. Because of this limitation many manufacturers quote the actual chronograph value and specify the distance at which it was taken; eg, instead of V_0 they quote V_{25} for a value taken at 25 metres from the muzzle.

A quoted muzzle velocity should specify the type of weapon and length of barrel which produced it; each cartridge has an optimum length of barrel in which the propelling charge develops its maximum power. Lengthening the barrel does nothing to improve matters and shortening it will reduce the muzzle velocity in a loosely-defined ratio. On the other hand the same cartridge fired from a revolver and an automatic pistol of equal barrel lengths will produce a higher velocity for the automatic since there is no leak of gas as there is between the cylinder of a revolver and the barrel. Muzzle velocity is a useful guide to performance, but published figures should be approached with caution unless there is some explanation of how they were achieved.

N

NON-STREAMLINED A bullet in which the base is the same diameter as the body; ie, it is not tapered at the rear. Usually found with bullets intended for high-velocity weapons, since the base drag which is inherent with this type of bullet is of less importance than the nose drag which occurs at supersonic velocities. May also be found on specialized military bullets (eg, tracer, incendiary or armour-piercing types), in order either to facilitate manufacture or permit the emission of the tracer or incendiary compound.

O

OGIVE Technical description of the compound curve forming the nose of most rifle and automatic pistol bullets. 'Ogive' is defined as 'a pointed arch' by most dictionaries, and this is an approximation of the curved point of a bullet. The object is to blend the parallel walls into a point as smoothly as possible and with a shape which offers the best pressure distribution; long experimental studies indicate that the ogive is the optimum shape for general use. The curvature of the ogive can be described by the distance from the commencement of the curve at which the centre of the arc lies, and this measurement is given in calibres. The curve can be 'simple' or 'compound', depending upon whether the locus of the curve is opposite the shoulder of the bullet or behind it.

P

PINFIRE Type of cartridge in which the ignition cap is concealed inside the cartridge case and has a pin resting upon it. This pin protrudes from the side of the case, and the gun chamber has a notch which allows the pin to stand proud of the weapon when it is loaded, so that it can be struck by a falling hammer. This drives the pin into the cap and initiates the explosion of the charge. It originated in France in the 1830s in shotgun ammunition, and later spread to pistol ammunition; it was unsuited to high-pressure ammunition which was liable to blow the pin out. Although generally superseded by *rimfire* and *centre-fire* cartridges by the 1880s, pinfire ammunition was still manufactured, in small quantities, until the 1930s and can still be encountered.

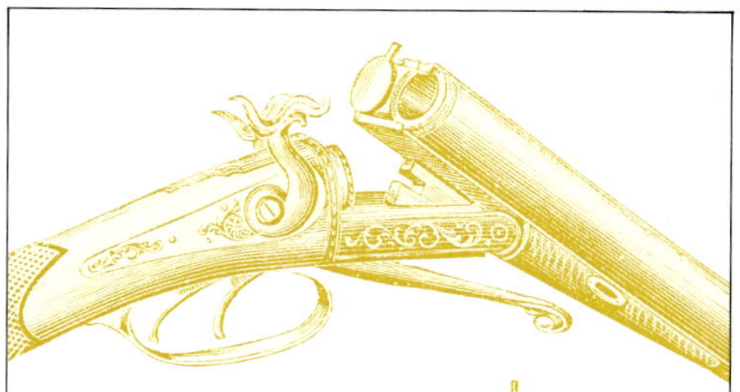

Above: The Lefaucheaux pinfire shotgun. Below: The interior of a pinfire shotgun cartridge.

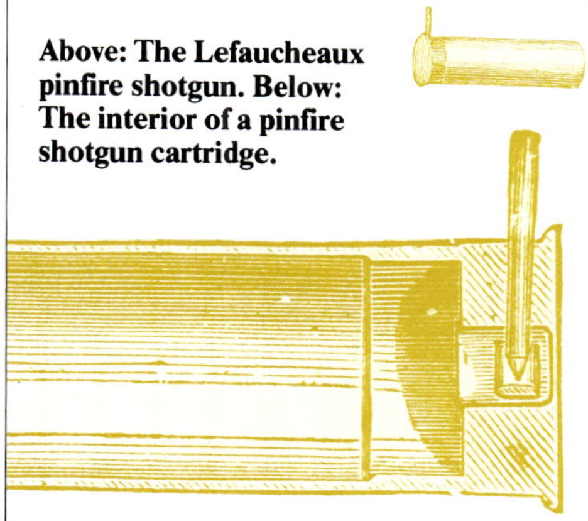

A collection of non-streamlined bullets. The largest is for the .450 Martini-Henry rifle of the 1880s.

PLASTIC AMMUNITION General term covering any small arms ammunition using plastic components. The most usual type is practice or training ammunition using a metal case and plastic bullet, though some brands use a cartridge case of plastic which has the base and extraction groove made of metal and moulded into the case. The propelling charge is small and the plastic bullet generally conforms to the trajectory of the standard metal bullet to about 25-30 metres range, after which it loses velocity and accuracy very quickly and falls to the ground after about 100-150 metres flight. The object is to provide cheap ammunition for use in indoor ranges which will not require the use of expensive and strong back-stops to stop the bullet. One type of plastic ammunition for revolvers uses a plastic case and bullet with no propelling charge other than the cap; this is sufficient to send the bullet for about 15 metres, and the case can have a fresh cap loaded in so that by re-inserting the bullet the round can be used several times.

Plastic ammunition of the training type can also be used as an anti-personnel weapon, as is currently the practice by certain special security forces who require a bullet capable of injuring an adversary at short range, but which is not lethal at longer ranges (eg, against innocent bystanders), is unlikely to ricochet, and will not cause other damage such as piercing the pressure cabin of an aircraft.

PLASTIC BULLET General term for the *baton round* used for riot control. It is a cylinder of hard plastic and is fired from a shotgun-type cartridge.

PRACTICE Term used in military ammunition to describe items used solely for training. Only found where the service item is particularly expensive or difficult to manufacture (eg, in some types of armour-piercing ammunition for major-calibre weapons), in which case a replica, of cheaper material and without any piercing ability, is used for shooting against targets. Rarely found in small arms ammunition because it is usually more expensive to make practice rounds than service rounds. If a factory is tooled up to make ball, for example, it can produce a ball round far cheaper, as part of a mass-production run of several millions, than a practice bullet for which special tooling and assembly facilities would need to be set up.

Plastic practice ammunition is now coming into use, however, because bullets can be made more cheaply, but their use is confined to short indoor shooting ranges, their accuracy is not sufficient for ranges in excess of about 25 metres and the light bullet is easily affected by wind if fired outdoors.

PRIMER Alternative term to describe the percussion cap of a cartridge.

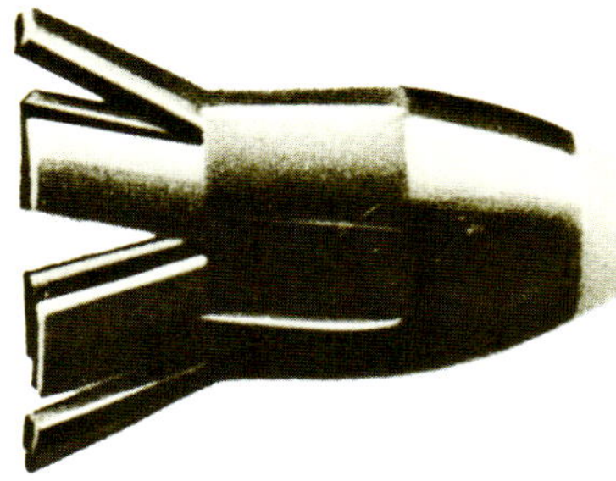

The West German 'Aero-flap' 9mm bullet is used for training. The split skirt is thrown out by spin.

PROPELLANT The low explosive which makes up the charge to fire the projectile from a weapon.

The principal requirement of a propellant is that it should explode rapidly, generating a mass of gas to propel the bullet, but it should not detonate since this would damage the weapon. Black powder (or gunpowder) was the first propellant and remained the only one until the discovery of smokeless powders in the latter half of the 19th century. The first such powders were produced by treating cotton or wood fibres with nitric acid, and although they worked, they were mostly chemically unstable, decomposed in storage and often exploded spontaneously, which led many countries to ban their manufacture. Further experimentation developed nitrocellulose powders which, as a result of their chemical composition and method of manufacture, were of a hard and impervious texture so that flame did not enter the individual grains and break them up, giving rise to spontaneous high-pressure waves, but ignited the powder grain and burned it in a regular and predictable manner. This allowed the grain size to be tailored to suit the specific requirement of the weapon/ammunition combination; thus a pistol with short barrel and light bullet could use a small-grained, fast-burning powder to ensure that the entire charge was burned to useful gas before the bullet left the barrel. Longer-barrelled weapons, such as rifles, could use larger grains which burned for a longer time and thus were able to

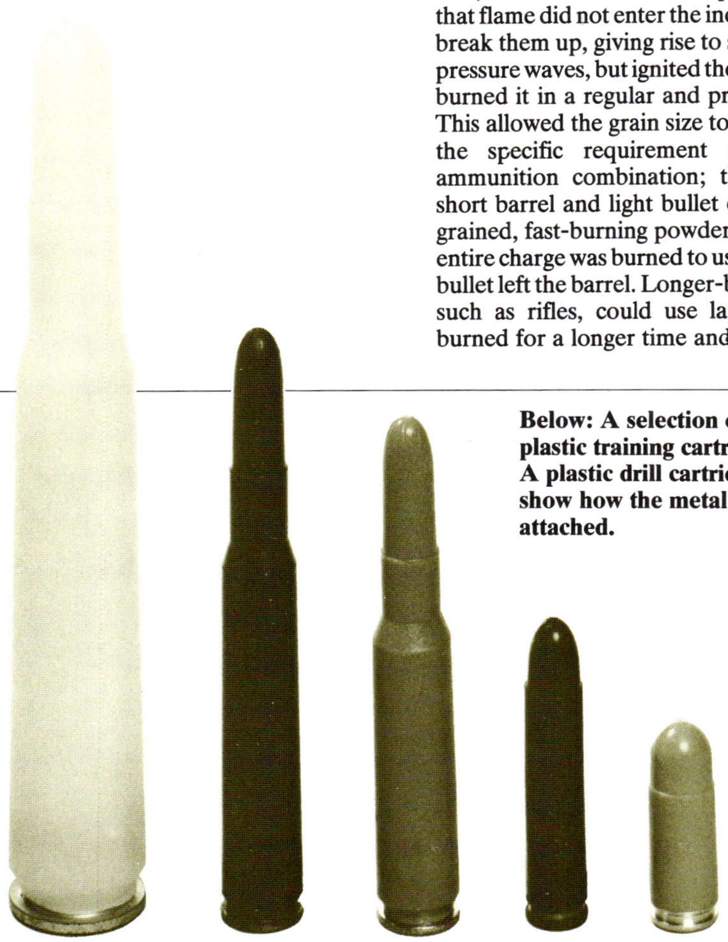

Below: A selection of Austrian plastic training cartridges. Bottom: A plastic drill cartridge cut open to show how the metal base is attached.

generate gas steadily during the bullet's travel up the bore, but still be completely consumed before it left the muzzle.

The original nitro-cellulose *single base* powders were soon accompanied by more powerful *double base* powders, but the latter tended to erode the weapon's barrel because of the extreme heat of their combustion and *triple base* powders were then developed. There are now a variety of formulations, intended to give better resistance to damp, less flash, no smoke, cooler burning or whichever particular attribute the designer feels is desirable.

Smokeless propellants come in a variety of shapes and sizes; most pistol propellant is in the form of thin flakes, since these burn quickly; rifle powders may be in short cylindrical or tubular grains or may be composed of small spheroids, known as 'ball powder'. The shape has a bearing on the rate of burning: simple shapes — flakes, cylindrical grains, balls — burn with a gradually decreasing surface so that the amount of gas, and hence the pressure in the weapon, decreases in proportion. A tubular grain, burning on the inside and outside, has a burning surface which remains almost the same, since the decreasing external surface is balanced by the increasing inner surface, and thus can develop a near-constant pressure. By combining a certain shape with a certain size of grain it is possible to tailor the burning speed and thus the generation of gas to produce whatever ballistic performance is required.

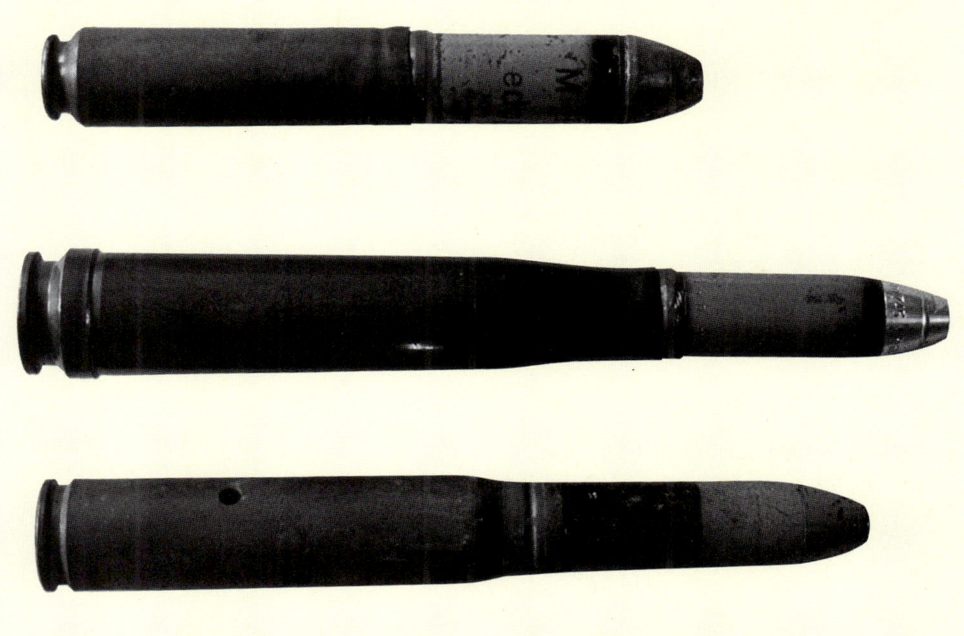

Three cannon cartridges: top to bottom, rebated rimless, belted and rimless cases.

R

REBATED RIMLESS CASE A cartridge case, generally of the *rimless* pattern, but which has a base of smaller diameter than that of the body of the case. An uncommon form, it is used in weapons where it is necessary that the case be entirely enclosed in the gun chamber, but still held by the bolt and extractor at the moment of firing, or where it is convenient to use a stock size of bolt with an oversized cartrdge case and chamber. The former is seen in some Oerlikon 20mm cannon cartridges developed during the period 1925 to 1945; the latter, in some custom-built sporting rifles of unusual calibres.

REMAINING VELOCITY The velocity of the bullet at any particular point on the trajectory. Given that the bullet leaves the muzzle at some specific velocity, thereafter it slows down, as a consequence of the pull of gravity and the drag of the air through which it passes. With knowledge of certain of the bullet's characteristics, notably its weight and shape, it is possible to calculate the loss of velocity and arrive at a figure for any point on the trajectory. More reliably, it is possible to fire bullets across chronographs at different ranges so as to have a basis of positive measurements against which the theoretical figures can be checked.

Knowledge of remaining velocity is of value in calculating the likely performance of, for example, an armour-piercing bullet. Knowledge of the fall in velocity throughout the flight, and the effect of gravity and drag on the path of the bullet, are necessary to the designers of sights so that they can calculate the difference between the line of sight and the actual path of the bullet and thus determine the amount of elevation to be applied to the sight for a given range. It also allows ballistic tables to be calculated.

RIFLED SLUG Lead or steel-and-lead projectile for shotguns, with wing-like helical ribs on its outer surface which, due to the passage of air during flight, give it a rotational movement and so produce a spinning projectile from a smooth-bored barrel. It requires a cylinder-bored barrel (ie, not choked at the muzzle) and good sights, which shotguns rarely have as standard, but given these two conditions it is possible to put a rifled slug into a 15cm (6 in) square at 100 metres' range with regularity. The rifled slug is intended for large game, such as deer; it is also used by police and security forces for such tasks as stopping fleeing automobiles, taking locks off doors and demolishing explosive devices at high speed.

A 12-bore rifled slug.

Rimfire cartridges: foreground, 9mm saloon rifle; left, .22 shot; rear, 10.4mm Swiss military cartridge of the 1880s; right .22 long; and, right foreground, .22 short.

A selection of cartridges with rimmed cases. The largest is for the .450 Express big game rifle.

RIMFIRE Type of cartridge in which the ignition is performed by a layer of *cap composition* packed into the hollow rim of the case. When the cartridge is loaded, the rim butts against the rear face of the gun's chamber and the firing pin is aligned so as to strike the rim and crush it against the chamber edge. This causes the composition to ignite and the flame passes immediately to the propelling charge.

Since the metal of the case has to be rather soft, to enable it to be crushed between pin and chamber face, it follows that the case metal is insufficiently strong to withstand high pressures, and the rimfire principle has always been confined to low-velocity ammunition. It was particularly popular in the United States from 1860 to 1880, when calibres up to .44in were loaded in rimfire. Today it is entirely confined to the .22 calibre and the 9mm shotgun, though a few years ago an entrepreneur in Sweden tried, without much success, to interest the world in a high-velocity rimfire cartridge for military rifles.

RIMMED CASE A cartridge case which has a well-defined upstanding rim around the base. This serves to butt against the chamber and so positively locate the cartridge in the correct relationship to the rifling and the firing pin, and also affords a purchase for the weapon's extractor to grip the case and remove it from the chamber after firing. Rimmed cases have the disadvantage of being liable to have their rims jam together when feeding from a box magazine; this was the reason for the characteristic curved magazine on the Bren light machine-gun, for instance. When the Bren was modified to 7.62mm NATO calibre, which uses a rimless case, it was able to accept a straight magazine.

RIMLESS CASE A cartridge case in which there is no upstanding rim, but which has an 'extraction groove' cut around the base which, in effect, leaves a rim for the extractor to grasp. This rim, though, is of the same nominal size as the base of the case. The advantage of a rimless case is that the absence of a rim allows cartridges to slide across one another when feeding into the gun from a magazine. They can also be thrust straight through a cloth ammunition belt and be loaded into a machine-gun chamber, whereas rimmed cases need to be pulled out of the belt backwards before being loaded into the chamber.

The drawback, such as it is, to rimless cases is that the location of the case in the chamber relies entirely upon the interfacing of some part of the case and some specific part of the chamber. In a rimless straight case, the mouth of the case comes up against a step in the chamber and this then places the bullet in the correct place to enter the rifling, and places the base of the case correctly at the chamber mouth, ready to be struck by the firing pin. With a bottle-necked case, the interface is between the neck of the case and the corresponding necked portion of the chamber. In both instances the cartridge will tend to creep forward, as wear takes place in the chamber, until it is too far forward for the firing pin to strike the cap. For this reason it is usual for weapons which use rimless ammunition to make some provision for adjusting the 'headspace' (ie, the space between the face of the bolt and the rear of the loaded cartridge).

RIOT-CONTROL AMMUNITION Term describing those types of ammunition used by police and security forces to combat riots and other types of civil disorder. The term is,

Rimless cased cartridges, from the 9mm pistol to the 14.5mm machine-gun.

obviously, subjective; there are some parts of the world where a main battle tank is considered an acceptable riot-control weapon. But in Britain, the USA and western Europe generally, it is taken to mean such things as baton rounds, tear-gas cartridges and grenades, cartridges which discharge rubber balls, and similar non-lethal but punitive loadings. This type of ammunition can be discharged from shotguns, but it is more usual to have special large-bore weapons of 25mm or 37mm calibre which are able to discharge large projectiles at low velocity, but which are entirely incapable of firing more lethal types of ammunition.

RUBBER BULLET Another, and perhaps the most common, name for the *baton round.* Rubber is still used by some police forces, but because of the liability of rubber projectiles to bounce uncontrollably in any direction, they have largely been replaced by plastic. The term is also used to describe some riot loadings for shotgun and 26mm cartridges in which the projectiles are a number of small rubber pellets which, on impact, sting severely without causing injury.

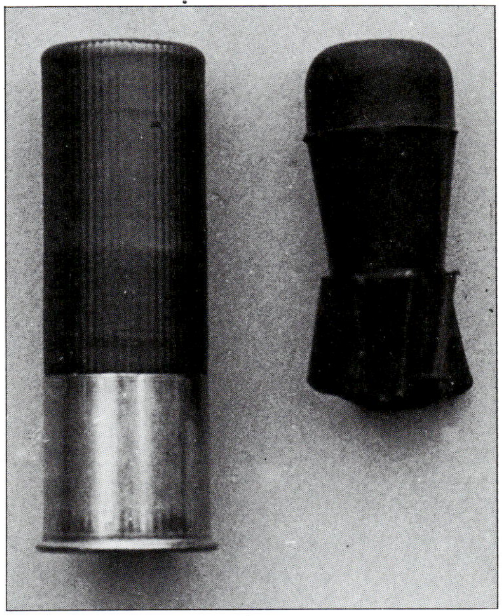

An American 'rubber rocket' riot control projectile, fired from a 12-bore shotgun.

S

SEMI-RIMMED CASE A cartridge case, basically of the rimless pattern, but which, on close examination, will be seen to have the rim of slightly larger diameter than the body of the case. Invented by John M. Browning for an automatic pistol in 1903, the purpose is to have sufficient rim to allow the case to be located correctly in the chamber by the meeting of rim and chamber face, but not so large as to cause jams in feeding from the pistol magazine. Commonly seen in 6.35mm, 7.65mm and 9mm Browning Long automatic pistol cartridges; it

The 'Round, Anti-Riot, 1.5in, Baton', showing the nose of the rubber bullet.

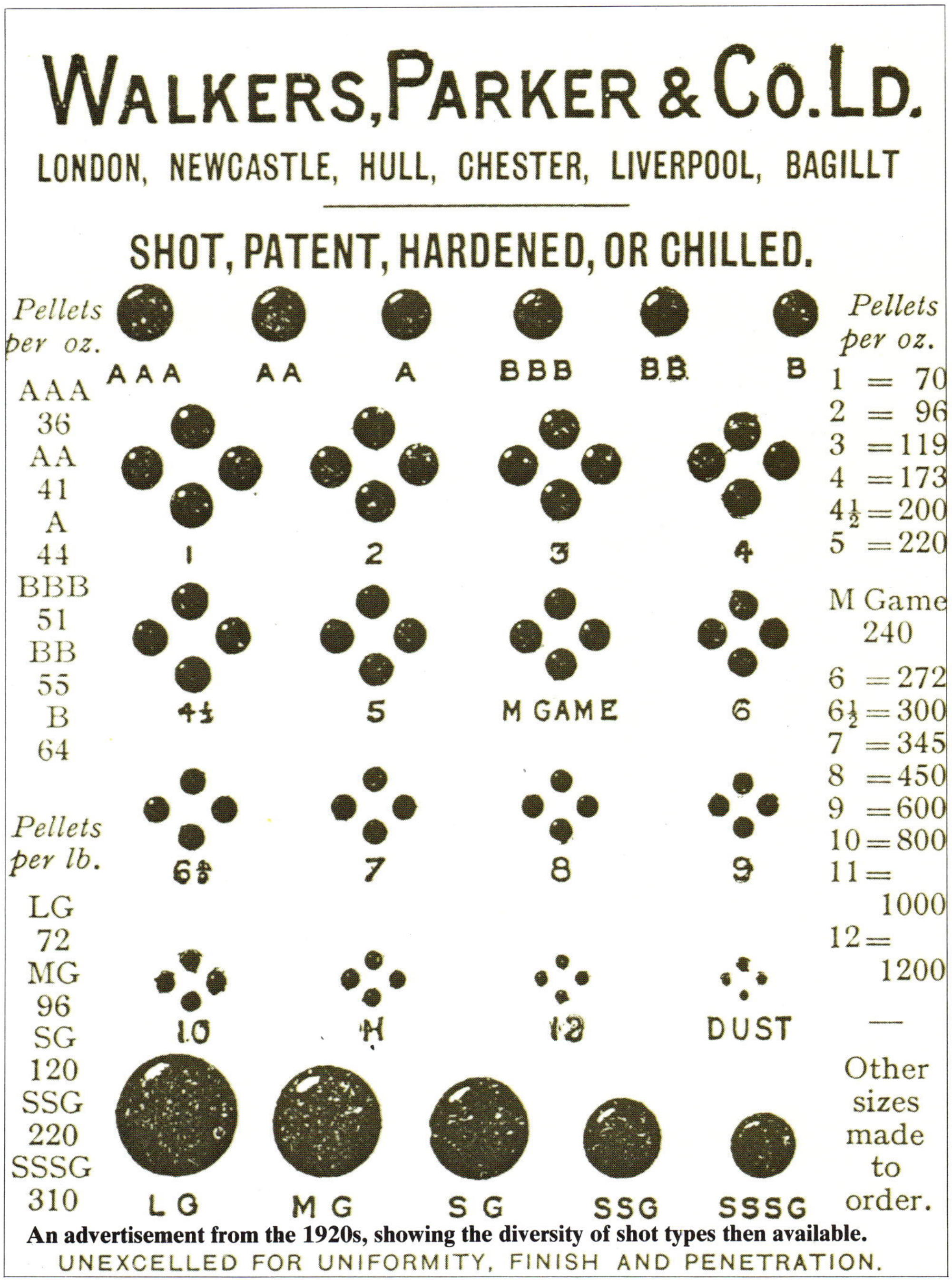

An advertisement from the 1920s, showing the diversity of shot types then available.

has also been used on some less well-known pistol cartridges and on a Japanese rifle cartridge.

SHOT In small arms, this refers to any loading containing a number of smaller spherical projectiles; eg, in the common shotgun cartridge. Shot is generally of lead, though in recent years a certain amount of ecological agitation has lead to the development of steel shot, particularly for wildfowl hunting where the shot inevitably falls into water. It varies in size from BBB (0.19in diameter, 50 shot per ounce) to 'Dust Shot' (0.04in diameter, 4,565 shot per ounce). Probably the most common for general purpose small game shooting — pigeon, rabbit, small birds generally — is No 6 (0.11in, 223 shot per ounce) or No 4 (0.13in, 136 shot per ounce).

SHOTGUN AMMUNITION Cartridges for shotguns take the form of cylinders containing the cap, the charge, and the load of shot. The case has a brass head and rim, with a cardboard or plastic body securely attached. The cap chamber is in the brass portion, together with the charge, and the remainder of the body is taken up by a number of wads, the shot, a closing wad, and the closure of the case. The contents are held in place by rolling the mouth of the case inwards to form a rim which holds the closing wad in place.

The loading of the cartridge varies enormously from maker to maker and offers the widest possible choice for the game shooter. One man may demand 1¼ ounces (35gr) of No 6 shot; another, for the same type of target, may feel happier with 1½ ounces (45gr) of No 4 shot, and the manufacturers are prepared to humour them. A great deal depends upon the individual shooter and what gun he is using, its weight, length of barrel and so forth. Generally speaking, the heavier the charge the greater the recoil, and the shooter needs to compromise between power and controllability; it is of little use to have a loading which will kill an elephant if the recoil is so severe that the shooter cannot shoot straight with it.

The bore or gauge of a shotgun is defined by the number of lead balls that will exactly fit the bore and that make one pound in weight. Thus a 4-bore has a ball of ¼lb weight which will exactly fit the bore. The exceptions to this rule are the .410 and the 9mm guns. The standard bore dimensions are:

4-bore	0.935 in
8-bore	0.835 in
10-bore	0.775 in
12-bore	0.729 in
14-bore	0.693 in
16-bore	0.662 in
20-bore	0.615 in
24-bore	0.580 in
28-bore	0.550 in
32-bore	0.501 in

The most common bores in use today are the 12 and 20; the 10, 16 and 28 still have some adherents, the former being more popular in the USA, while the 14, 24 and 32 are virtually obsolete. The .410 is universally popular, but the 9mm, a Continental size, is rarely seen today.

Shotguns can also fire solid projectiles, calibre-size balls of lead or steel, or *rifled slugs.* These are used principally for larger game such as deer and boar; the ball is more popular in Europe than in the USA, while the rifled slug is used in all countries.

SINGLE BASE Term used to describe propellant powders which are made from nitrocellulose with the addition of very small quantities of chemicals to promote chemical stability, flashlessness, or to assist in the manufacturing process. It is probably the most common type of commercial powder since it combines simplicity, adequate power, good keeping properties and a low flame temperature which does not cause excessive erosion in the gun barrel. Single base

Specimens of shotgun ammunition, including rifled slug and buckshot loadings.

power is more susceptible to damp than other kinds, and needs to be adequately protected during storage. For this reason, British military ammunition loaded with single base powder (in the days when the standard loading was cordite, a *double base* powder) was always marked on the cartridge base with the letter 'Z'.

SLAP American acronym for 'Saboted Light Armor Penetrator', meaning a *discarding sabot* armour-piercing cartridge. Developed by the Remington company from 1978 to 1982, it is now being developed for use in 7.62mm and .50in machine-guns in two versions: one with a conventional AP bullet as the sub-projectile, the other with a fin-stabilized, dart-like 'penetrator' as the sub-projectile. At the time of writing, both types are undergoing evaluation trials and it is anticipated that service ammunition will appear in about 1986.

SOFT-POINT Any type of bullet in which the jacket is made short so that the front of the core is exposed. Being of lead, this bullet deforms readily on impact and inflicts a greater wound than would a normal full-jacketed bullet. Used solely for hunting.

SPIN The rotational motion of a bullet in flight resulting from its passage up a rifled barrel. Spin is necessary in order to stabilize the bullet in flight — to keep it travelling point-foremost; an elongated bullet fired from a smooth barrel would simply tumble over and over and would be incapable of following an accurate trajectory.

The amount of spin required by any given bullet depends almost entirely upon its length in relation to its calibre; there are several formulae for calculating the optimum spin for a given bullet, but perhaps the simplest (if not the most accurate) method is to take the diameter of the bullet, square it, divide it by its length, and multiply the result by 180. This will give the length in which the bullet should make one complete revolution; if the dimensions are in millimetres, the length will similarly be in millimetres; if in inches, then in inches. Having

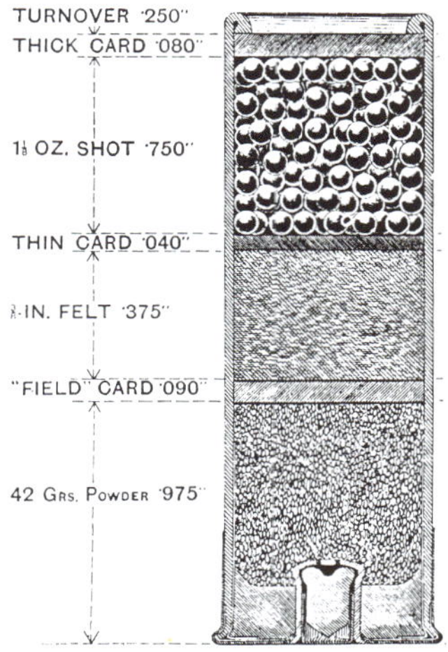

An engraving from the 1900s showing the contemporary shotgun cartridge.

Examples of 7.62mm and .50in (12.7mm) SLAP rounds alongside a conventional 5.56mm ball cartridge.

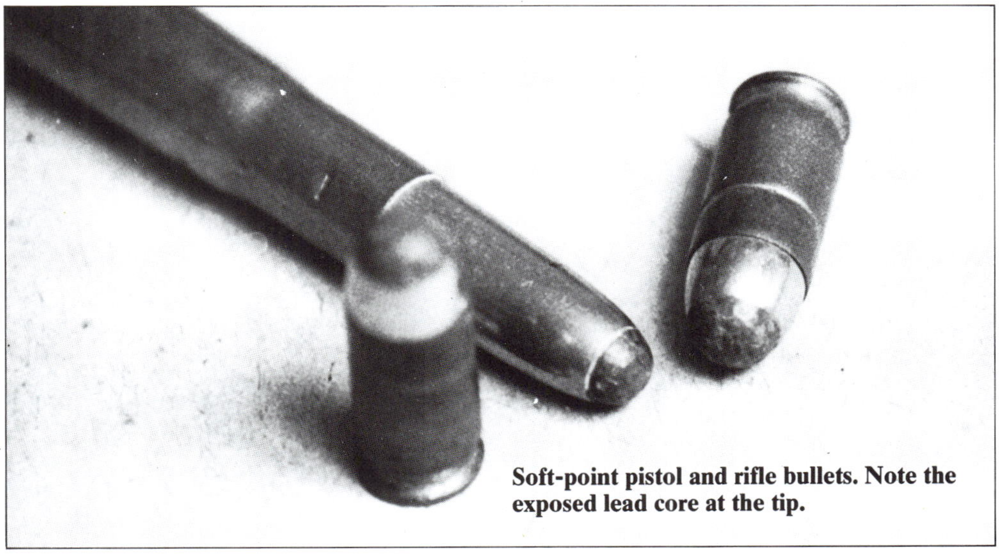

Soft-point pistol and rifle bullets. Note the exposed lead core at the tip.

thus determined the optimum rifling pitch, it must be converted to angular measurement, the angle between the line of the rifling and the axis of the bore; this is done by first dividing the length in millimetres by the calibre of the bullet, then dividing the answer by 3.14159, the result being the cotangent of the rifling angle. Having thus found the angle, divide the muzzle velocity (metres per second) by the bullet calibre (in metres), multiply by a constant 19.09, and multiply the result by the tangent of the angle. The final answer is the spin of the bullet in revolutions per second. This all sounds very complicated, but with a modern pocket calculator it is very easily done.

Once this optimum spin rate has been discovered, it is possible to examine different weapons and calculate what the actual spin is, by using the last calculation and basing it on the actual rifling of the weapon in question. It will often be found that the optimum and actual spin rates do not agree by a considerable amount, and the reasons for this vary. The most common reason is simply that weapon manufacturers base their design on one particular bullet, and fashions in bullets may have changed in the intervening years. Another reason, found in military rifles, and particularly in the 5.56mm calibre, is that the optimum spin is best for flight but not best for allowing the bullet to topple over and yield up its energy on impact, and so it is common to under-spin a military bullet so as to cause more severe wounding.

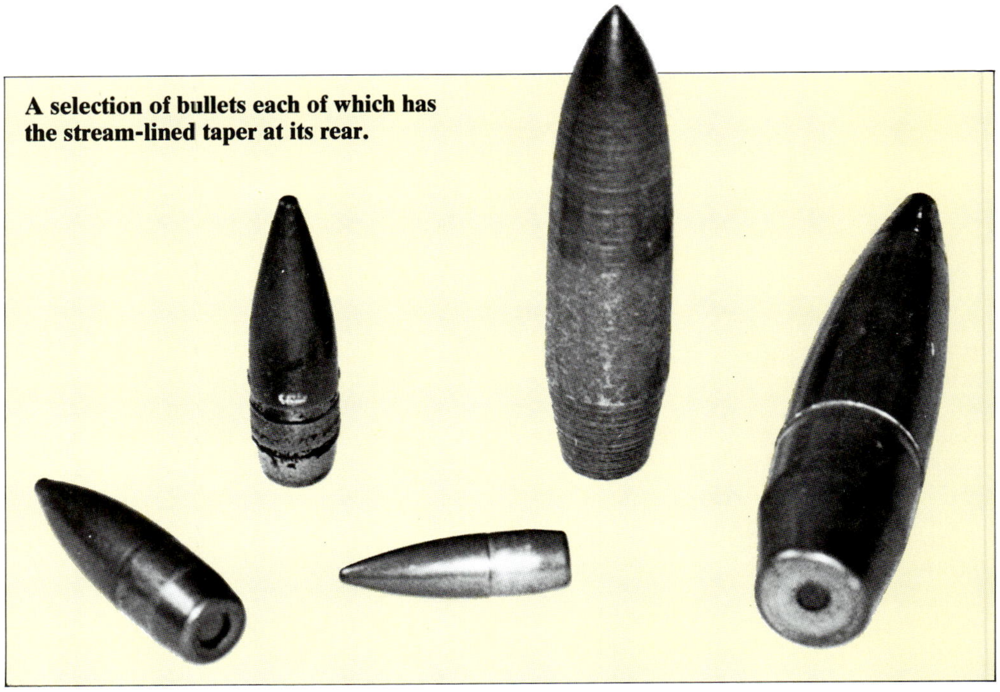

A selection of bullets each of which has the stream-lined taper at its rear.

SPITZER BULLET 'Spitz' means 'pointed' in German, and the term 'Spitzer' began to be used after the German Army adopted a pointed military bullet in the early 1900s. Prior to this almost all military bullets were blunt-nosed and straight-sided, but the invention of spark photography enabled ballistic experts to see for the first time a bullet in flight and the formation of air waves and eddies around it. Clearly the blunt bullet made a poor job of cleaving the air, and the pointed bullet became the norm. For convenience, since the Germans called it the 'Spitzgeschoss', American and British sportsmen began calling pointed bullets 'Spitzer bullets' and the name has stuck.

STOPPING POWER Stopping power is the ability of a bullet to stop an assailant in his tracks and make him fall down; wounding or killing him have no bearing on the matter, only stopping his advance. The question was seriously examined by the US Army in the 1890s when it was found that Moro tribesmen in the Philippines were not stopped by .38 revolver bullets; they would take several bullets in their bodies and keep coming. The British had encountered similar problems when fighting Sudanese warriors and other well-motivated and warlike tribes in various places. Broadly speaking, a heavy but slow bullet stands a better chance of stopping a man than a light bullet of high velocity, but, of course, a great deal depends upon the victim's physical condition and his degree of motivation.

However, it is possible to arrive at an empiric figure which is a useful measure of the comparative ability of bullets to stop. The most commonly used value is known as 'Hatcher's Relative Stopping Power' after its first calculation and publication by Colonel Julian Hatcher of the US Army in the late 1920s. In its simplest form Hatcher's value is arrived at by multiplying the energy of the bullet (see *Muzzle Energy*) by the cross-sectional area of the bullet and by a factor according to its shape. The 'shape factors' are: jacketed bullet 0.9; flat-nosed bullet 1.1; wadcutter or man-stopper 1.25; a plain lead round-nose bullet takes a factor of zero, because this was the basic standard on which Hatcher worked.

This formula gives a result which is simply a number, and the only thing which can be done with it is to compare it with the number calculated for some other bullet. Thus the Relative Stopping Power (RSP) value for the bullet from the US Army .45 automatic is 46.8; that for a 9 mm Parabellum is 38.9. Since the higher the figure the greater the stopping power, consideration of these two values suggests that the .45 will stop a man more efficiently than a 9 mm, and this is borne out by actual experience.

STREAMLINED BULLET A bullet which has the front section tapered to a point and the rear section tapered to a smaller diameter than the main portion of the bullet. The shape is derived by consideration of the action of the air as it passes over and around the bullet in flight. As mentioned under *Base Drag*, the air sweeps around the bullet and swirls into a low-pressure area behind the base. This area, and the consequent swirl, is reduced by tapering the rear section of the bullet so as to guide the air more smoothly into the base area. The actual degree of taper is a compromise; experiments suggest that the best taper is about 10°, but the actual taper used is based as much on ease of manufacturing as on ballistic theory, so that 8° is about the usual value. Careful choice of length of the tapered section and degree of taper could improve a bullet's performance by as much as 25 per cent, but this has to be weighed against manufacturing problems and against the possibility of accuracy being diminished; too long a taper at the base of the bullet might result in too short a mid-section and thus a bullet which was unstable inside the bore because of insufficient bearing surface.

SUBSONIC AMMUNITION Cartridges intended for use in silenced weapons. There is more to silencing a weapon than muffling the noise which comes from the barrel; a bullet moving at speed above the speed of sound makes a sharp crack (the sonic boom) as it passes, and there is little point in silencing the discharge if the bullet continues to advertise its presence. For absolute silence, therefore, it is necessary to ensure that the bullet is travelling at a velocity less than that of sound 332mps (1,089fps). Since few pistol bullets normally travel at subsonic velocities, it is necessary to provide special subsonic ammunition for most weapons; a combination of heavier bullet and lighter charge so as to bring the velocity down to the desired level. Of course, the bullet will not range so far, but in silenced weapons, which are invariably used at short range, this is of little consequence. It is necessary to mark such ammunition distinctively, so that it is not confused with standard ammunition.

T

TERMINAL VELOCITY The velocity of a bullet when it reaches the end of its trajectory and strikes the ground or when it strikes a target at any point in its flight; much the same thing as *remaining velocity*, though RV is usually taken to mean a value during flight rather than at the termination of flight.

TOMBAC Continental term for *gilding metal*. Used for bullet envelopes and occasionally for caps.

TRACER Type of bullet which leaves a visible mark or 'trace' while in flight so that the gunner can observe the strike of the shot or make adjustments in the event of a miss. It was originally developed for used by aerial machine-gunners who had no way of determining where their shots had gone in relation to the target if they failed to hit. Its use was later extended to ground firing with machine-guns for similar reasons.

Tracer bullets resemble ball, but have the rear portion of the core removed and the space filled with a chemical compound, which may be pressed directly into the bullet or loaded into a copper tube which is inserted into the bullet. It is commonly believed that phosphorus is used for tracers, but in fact this is never so; tracer mixtures are based on barium nitrate and powdered magnesium, with strontium nitrate added to give a red colouring; other metallic salts can be used to give other colours, but tracers other than red are exceptionally rare. The mixture is ignited by the flash of the propellant and burns during the bullet's flight, shedding red sparks which, because of persistence of vision and the speed of the bullet, appear as a continuous line of red. Another method used a mixture giving off coloured smoke, but this is obsolete because it is useless in darkness.

The brightness of the trace can be varied to suit the conditions in which it is employed. Air service ammunition was usually made in two degrees of brightness: the brightest for use in daylight, the less bright for use at night so as not to blind the aircrew. The advent of sophisticated fire-control systems and the fact that small-calibre machine-guns are not used in aerial combat means that the need for such refinement has lapsed. Today the only problem with brightness is the possible dazzling of the gunner by his own tracer, and this is countered by the adoption of *dark ignition*.

A ballistic problem with tracer is that as the bullet flies and the tracer composition is burned away, the weight and balance of the bullet changes. So that while it may leave the gun travelling on the same trajectory as a ball bullet, this soon changes and the tracer describes a trajectory of its own. Since it was customary to mix tracer and ball bullets in a machine-gun ammunition belt — eg, one tracer to four ball — this meant that the guide to the trajectory, upon which the gunner was relying, was not telling him the truth. In order to establish accurate relationship, tracer bullets are now designed so that they do not exactly follow the ball trajectory at the start, but will cross it at some specified distance. This distance is usually a figure chosen as being the likely 'fighting range' of the weapon; in practice it is usually 300 or 500 metres depending upon the calibre and tactical use of the weapon.

TRACER COMPOSITION As mentioned above, this is the chemical compound used to fill tracer bullets. It is a mixture of magnesium powder and barium nitrate, with metallic salts to give the desired colour of flame or smoke.

TRIGLYCOL POWDER A triple base powder in which the nitro-glycerine is replaced by one of the nitric esters — diglycol-dinitrate, triglycol-dinitrate, or trimethylol-methane trinitrate. As with diglycol powders, the nitric esters lower the temperature, but provide a more powerful powder than would be obtained by simply omitting the nitro-glycerine. The powders have good mechanical properties (they can be shaped easily and keep their shape), but so far they have been little used in small arms and are something of a novelty.

TRIPLE BASE A type of propellant powder which uses three principal ingredients: nitrocellulose, nitro-glycerine and nitro-guanidine. It was devised in an attempt to compromise between the low power of single base powders and the high power but excessive heat (and thus excessive bore wear) of double base powders. The percentage of nitro-glycerine is small, but sufficient to give added power; the nitro-guanidine lowers the flame temperature while still adding an active explosive constituent. One of the virtues of triple base powder is that it is entirely flashless, though it does generate rather more smoke than the other types.

TWEEDIE BULLET A jacketed rifle bullet in which the jacket is cut and slotted from just behind the nose to the shoulder. The bullet retains its shape in flight, but on impact the weakened jacket allows the lead core to distort very rapidly, and so cause severe wounding and heavy shock effect. It was originally developed as a potential military bullet in the late 1890s, but the Hague Convention's ban on expanding bullets saw the end of that idea, and it was later adopted as a hunting bullet. It is not common today because other designs, easier to manufacture, give equally good target effect.

W

WADCUTTER Flat-nosed pistol bullet so designed as to punch a sharply-defined hole in a paper target so that arguments about whether or not this shot cut that line are avoided.

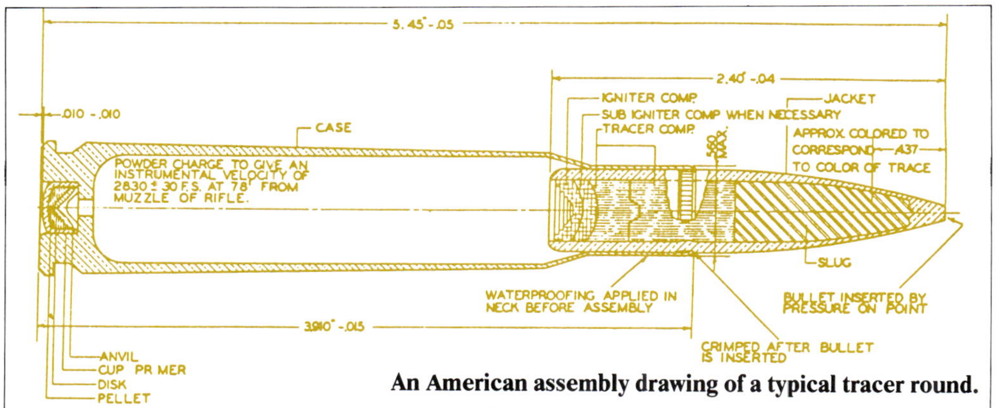

An American assembly drawing of a typical tracer round.

'Wadcutter' pistol cartridges with a lubricated bullet.

Loading the Austrian 81mm. One man holds the mortar steady while the other drops the bomb down the barrel.

MORTARS & GRENADES

American marines firing a mortar. The picture catches the bomb as it emerges, followed by the propellant gas.

A

ALLWAYS FUZE Type of impact fuze for grenades which will function irrespective of the attitude the grenade is in when it hits the ground. Usually consists of a heavy ball or pellet held between two conical surfaces, the lower of which carries a firing pin mounted above a loose detonator. If the grenade lands on its head, the loose detonator is thrown forward, strikes the firing pin, and sets off the grenade. If the grenade lands at any other angle, the ball will be thrown sideways or downwards, and will thus force the lower conical plate down, driving the firing pin into the detonator. Prior to throwing, movement of the ball is prevented by a safety pin which locks the movable cone unit and keeps the loose detonator secured away from the firing pin. This safety pin is usually attached to a strip of tape with a weight on the end. After throwing, the weight flies free and pulls out the safety pin.

AUGMENTING CHARGE Major part of the propelling charge for a mortar. The mortar charge consists of two components: the primary charge and the augmenting charge. The augmenting charge is in the form of packages of propellant clipped or otherwise fixed around the tail boom or fins so that they can be removed. When ignited by the primary charge the augmenting charge explodes, generating gas to thrust the bomb from the mortar barrel. The number of augmenting charges varies, though six is a common and convenient number, and the mortar achieves various ranges by using different numbers of augmenting charge units at various elevations. Because a conventional mortar operates only at elevations between 45° and about 80°, this limited vertical arc would restrict the ranges available unless the charge were made variable.

B

BASE-EJECTION SMOKE Type of smoke bomb in which the smoke chemical is contained in one or more canisters which are ejected through the base of the bomb during flight. It demands a cylindrical bomb, with the tail unit attached by shear pins or a light screw thread.

An Allways fuze used with Stokes mortar bombs. The shatter unit releases on firing and the ejector spring throws the safety bar out.

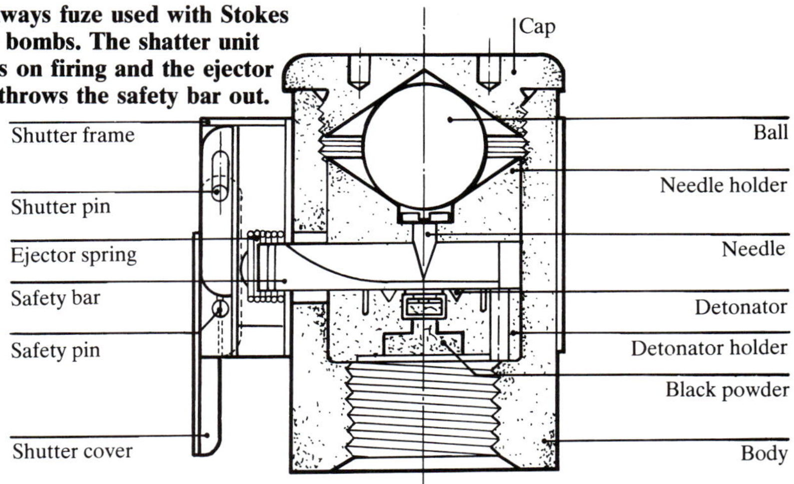

An involved augmenting charge assembly on a Brandt 120mm rifled bomb. The tube holds propelling charge and drops off outside muzzle.

The base ejection smoke bomb for the wartime British 4.2in mortar. Note the sabot around its waist — narrower existing smoke canisters were used.

The Blacker Bombard, one of the first spigot mortars to be adopted. A steel spigot was driven into the cartridge. The bomb blew off and the spigot was recocked automatically.

Inside the bomb is a small compartment containing a charge of gunpowder, and beneath this is the canister(s). When ignited by a time fuze the powder explodes; the flash ignites the smoke canister and the explosive force drives it back in the bomb body, against the tail boom. This breaks the joint and throws the tail boom clear, allowing the canister to fall free. Not commonly used in mortars because it is frequently inconvenient to have canisters of near-mortar calibre (necessitated by the need for a tubular body) and because the type of smoke chemical used is slow to build up an impenetrable screen; infantry, the principal users of mortars, want quick smoke, and prefer to use white phosphorus bombs.

BLACKER BOMBARD Type of *spigot mortar* developed in Britain in 1942 and issued to Home Guard units. It consisted of a short, wide barrel on three legs which lay flat on the ground, though the barrel unit could also be spindle-mounted on some permanent and solid object such as a concrete road-block. The bomb was a 13kg (29lb) high-explosive warhead on the end of a long, finned tail boom which contained a propulsion cartridge. The bomb was inserted into the barrel, and when loaded the warhead was visible in the muzzle. On releasing the trigger a steel 'spigot' was driven into the tail boom, striking the cartridge and exploding it. This blew the bomb off the spigot, and re-cocked the spigot at the same time. The Bombard was surprisingly accurate and had a maximum range of about 500 metres, though its operational range was about 200-250 metres.

BOMB General term for the projectile fired from a mortar, though some countries occasionally use the word 'shell'. The usage probably stems from history; the mortars of the 17th century fired hollow projectiles containing powder which were usually termed 'bombs', and when the British and French armies began using mortars in 1914 some of these antique weapons were put to use until more modern types could be developed.

Mortar bombs are of two basic shapes: streamlined (ie, shaped like a teardrop, rounded at the front, tapering at the rear to a tail boom carrying a set of stabilizing fins); or cylindrical, in which there is a distinct parallel-walled section between the rounded nose and the tapering rear. Cylindrical bombs can carry more payload, but they are generally heavier than streamlined bombs designed for the same mortar, and they will not range as far because of their extra weight and less efficient aerodynamic shape.

The principal problem with bombs fired from smoothbore mortars is to try and trap as much gas behind the bomb as possible, so that the explosion of the propelling charge is used efficiently. The basic requirement is opposed by the need to make the bomb smaller in diameter than the bore of the mortar so that it can be drop-loaded from the muzzle; if it were a tight fit, the air trapped beneath would cushion its fall, and could seriously slow up the rate of fire. The difference between the interior diameter of the bore and the exterior diameter of the bomb is known as *windage* and is kept to a minimum by the designer; nevertheless, some gas is bound to

Loading a British 81mm mortar. The design problem is to trap as much propellant gas as possible behind the bomb.

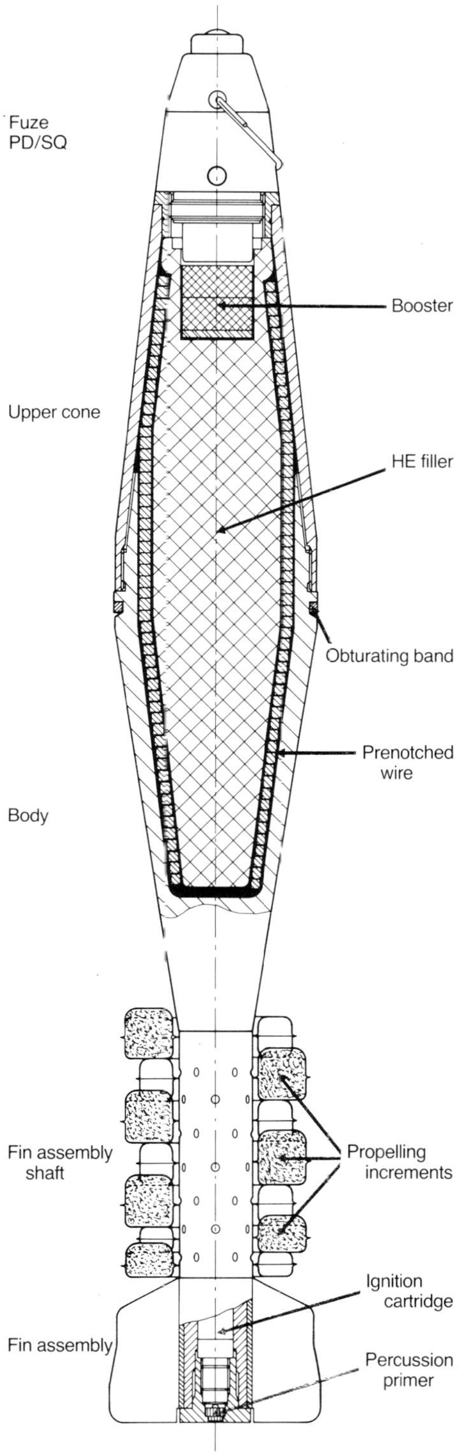

A modern bomb, using an obturating ring and with a notched wire liner to give controlled fragmentation.

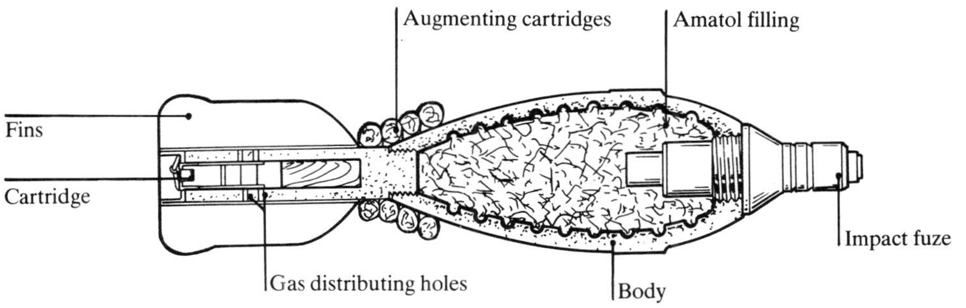

Above: The first streamlined bomb for the World War I Stokes mortar. Below: A World War II British 3in mortar bomb.

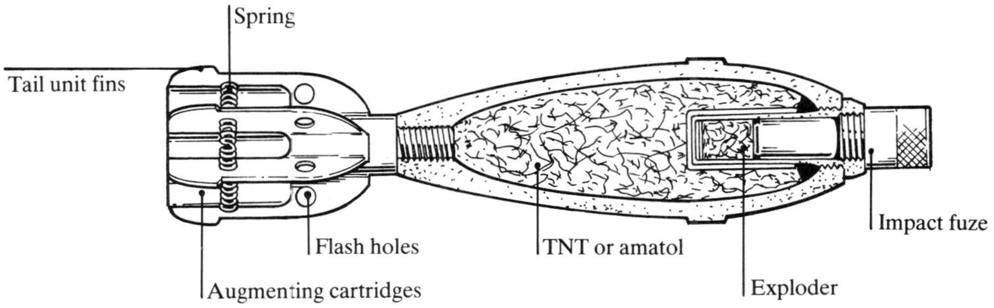

escape through the windage when the bomb is fired. One method of countering this is to cut a number of circumferential grooves in the maximum diameter of the bomb; the rushing gas gets into these grooves and sets up turbulence which tends to block the windage and prevent further gas escaping. A far more efficient, though more expensive, method is to cut a single groove and install an expanding plastic obturating ring. This is so shaped that the first rush of gas lifts it outward to press against the bore of the mortar, forming a seal.

It is possible to have rifled mortars, in which case the problem becomes one of allowing the bomb to fall freely when loaded but having it engage the rifling when fired. Three systems have been used in practice. The American 4.2 inch, developed in 1924 and still in service as the 107mm, uses a system invented by an Australian, Captain R.H.S. Abbot, in 1919. The projectile is shaped like a shell, with a short tubular cartridge container sticking out of its base. Around this, and attached to the base of the shell, is a saucer-shaped copper plate with a thick rim. In the normal state this rim is of the same diameter as the body of the shell and thus permits drop-loading. On firing, the propelling gas flattens the saucer and forces the rim out to engage in the rifling, so that it spins the shell and acts as an efficient gas seal.

The second method, used by the Japanese Army in a 5cm mortar developed in the 1930s, also had a shell-shaped projectile in which the rear end was a hollow compartment containing the propelling charge and having a percussion cap in its centre. Around the periphery of the shell which surrounded this compartment, was a copper driving band recessed into a groove so that it was of shell diameter and permitted drop-loading. Behind the driving band were small holes through which, when the propelling charge was ignited by the percussion cap, gas pressure forced the driving band outwards to engage in the rifling. More holes in the base of the shell allowed the gas to escape rearward into the mortar barrel in order to propel the bomb.

The third method, employed today by the French Army's 120mm Hotchkiss-Brandt mortar, fits a normal type of driving band to the bomb, pre-engraved to fit the rifling. This means that, on loading, the driving band indentations have to be fitted into the rifling of the mortar, which tends to slightly slow the rate of fire.

Various unconventional types of bomb have appeared from time to time. One unusual design was a German bomb which, unusually, had to be loaded nose-first into the mortar, and carried its cartridge container and propelling charges on its nose. The reason for this unusual system was that the bomb contained an illuminating flare and parachute, and by loading nose-first it became possible to ignite a delay fuze from the propellant flash, thus doing away with the need to fit and set a time fuze. On firing, the bomb left the barrel tail first, but because of the air drag on the fins and the effect of the bomb's centre of gravity it soon turned around and flew nose-first until the delay element burned through and blew off the nose, allowing the flare and parachute to deploy.

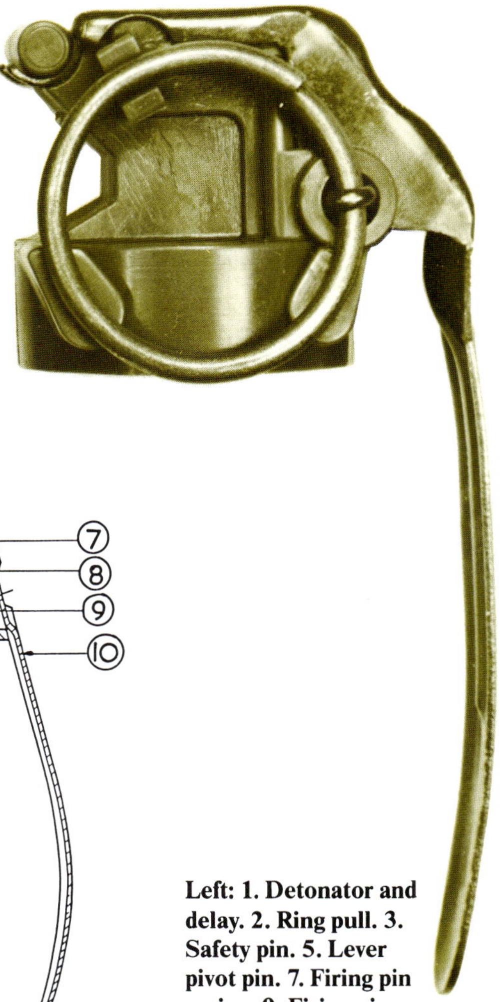

Above: Typical Bouchon igniter units ready to be fitted to grenades. Correctly fitted they are safe and reliable, but the temptation to hang the handle from webbing can have fatal results.

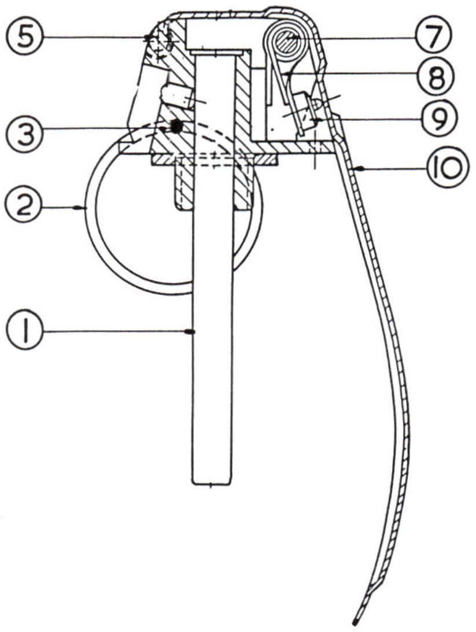

Left: 1. Detonator and delay. 2. Ring pull. 3. Safety pin. 5. Lever pivot pin. 7. Firing pin spring. 9. Firing pin. 10. Fly-off lever.

BOUCHON IGNITER Type of hand-grenade ignition system, commonly called the 'mousetrap' method; the name 'Bouchon' comes from the World War I French inventor. The fuze unit, which screws into the grenade body, consists of a tube with a detonator and, at the top, a casting which holds a percussion cap in its centre. Hinged to one edge of the casting is a flap with a firing pin, so designed that when folded over the pin aligns with the cap. This flap is propelled by a powerful coil spring around its pivot. On the opposite side of the casting is a lip around which the end of the safety lever clips. The lever is clipped on and the firing pin flap forced back against its spring, and the safety lever is then closed down so that it holds the firing pin, whereupon the lever is locked in place by a safety pin which passes through two ears on the fuze head casting. The fuze is then screwed into the grenade. To operate, the grenade is gripped so that the safety lever is held flush with the body of the grenade and the safety pin is removed. When the grenade is thrown, the coil spring forces the flap across so that it flings the safety lever off, and the pin is driven into the cap to ignite the grenade fuze.

When correctly fitted the Bouchon igniter is a perfectly safe and reliable device; unfortunately, the lever tends to lie away from the body of the grenade, thus encouraging the practice of hooking it into web equipment, pockets and so on, a practice made popular by US troops during Word War II. If the handle is flimsy the weight of the grenade will eventually distort it to the point where it will unhook from the lip and allow the firing pin to slam across with fatal results, even though the safety pin is still in place. Although this risk has been known and publicized to soldiers for the past thirty years or so, the practice persists.

BOUNCING BOMB German 81mm mortar bombs, the Wurfgranate 38 and 39, developed during World War II. They resembled conventional streamlined bombs, but the rounded nose section was a separate unit, pinned to the body. Inside the nose, which carried an impact fuze, was a container with a small charge of smokeless powder; beneath this was the flat head of the bomb body, filled with high explosive. In this flat head was a hole filled with a delay composition, leading to a detonator unit.

The bomb was fired in the normal way, and when it struck the ground at the target the impact fuze exploded the charge of smokeless powder. This explosion sheared the pins holding nose and body together and blew the body back up into the air. At the same time, the flash from the charge ignited the delay filling which burned through and then fired the detonator which in

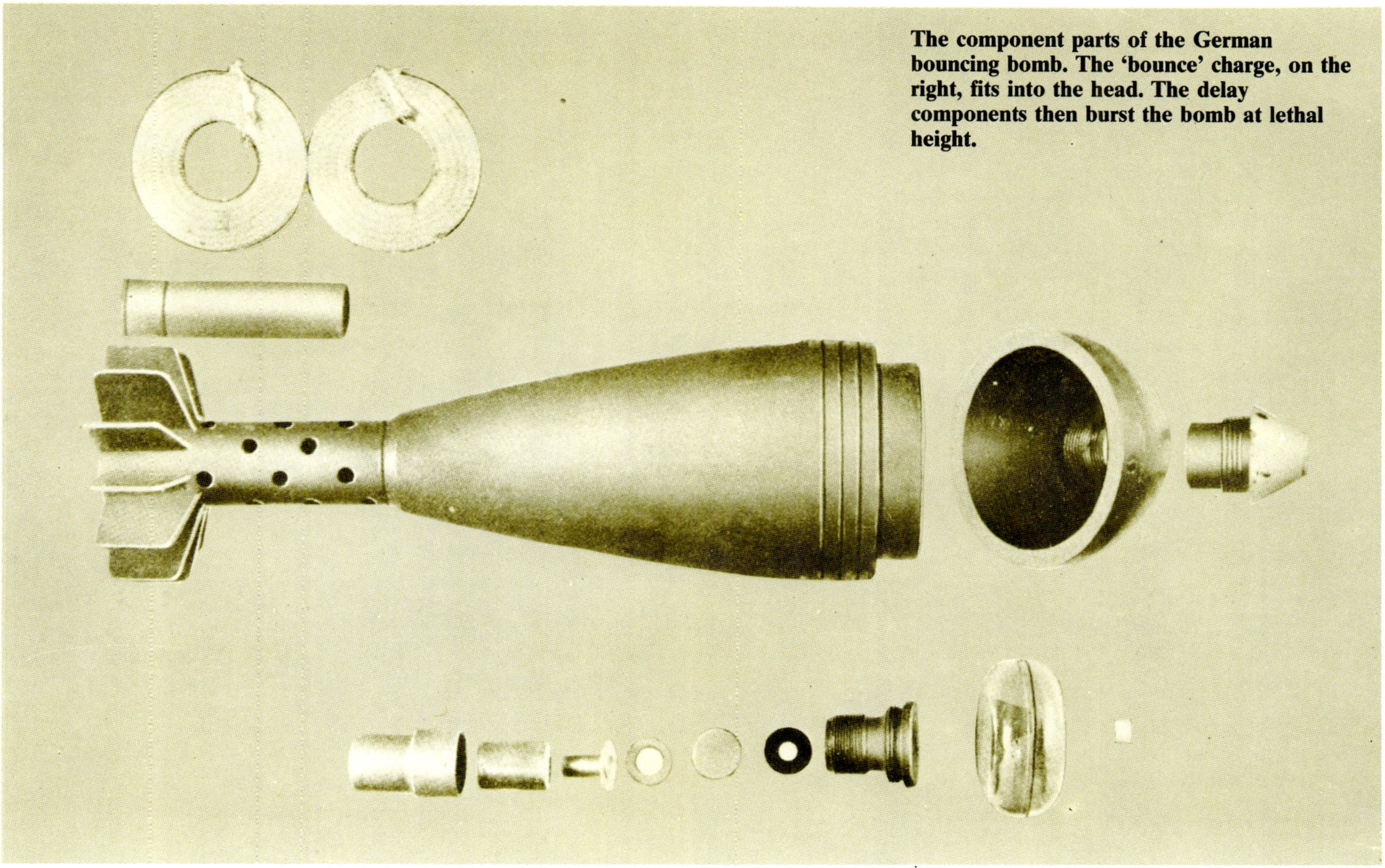

The component parts of the German bouncing bomb. The 'bounce' charge, on the right, fits into the head. The delay components then burst the bomb at lethal height.

turn detonated the main filling of explosive and shattered the bomb body. The delay was about 0.7 second and the bomb would be at a height of about 4–6 metres (15–20 feet) when it detonated. This gave a very effective airburst, far more lethal than a ground burst because the fragments would strike downwards into trenches, without the complication of a time fuze. The height to which the tail unit 'bounced' varied with the nature of the ground; on hard ground it could go as high as 18.5 metres (60 feet), on soft ground it might burst at only 1.5 metres (5 feet).

The bouncing bomb was quite effective on firm ground, less so on soft ground. The design was copied by the British and Americans, but both considered that the variable performance resulting from the ground factor, was unacceptable and neither army placed their design in service. So far as is known no bouncing bomb is used at present.

BOURRELET The machined surface at the maximum diameter of a bomb. This is a carefully controlled dimension, because of the need for windage. In a cylindrical bomb there will be two bourrelets, one at each end of the parallel-walled section of the body.

A selection of mortar bombs, illustrating the form taken by the Bourrelet.

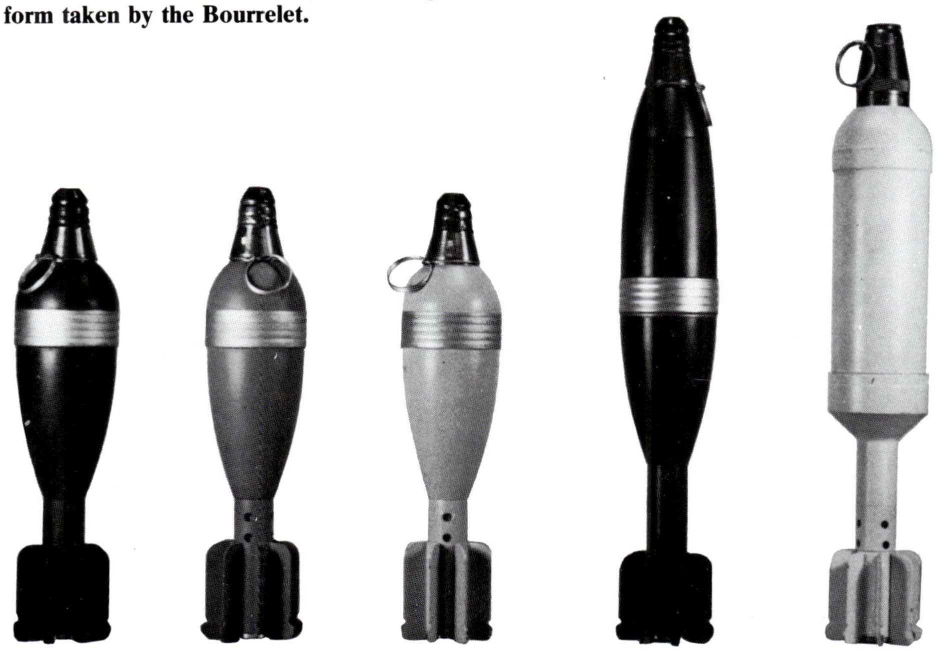

BRANDT Edgar Brandt was a French ordnance engineer who adopted the basic Stokes design of mortar and made various detail improvements during the 1920s. By the 1930s he had a well-perfected design which he sold to various countries including France, Italy and the USA. After the war the company was absorbed by Hotchkiss to become Hotchkiss-Brandt, and later became part of the Thomson-CSF organization and is currently known as Thomson-Brandt. It still manufactures mortars in all calibres, which are in use throughout the world.

The company has also pioneered various types of ammunition, notably rocket-assisted bombs and APFSDS projectiles for gun-mortars.

The Brandt 120mm rifled mortar as perfected by Edgar Brandt and in use all over the world.

BULLET TRAP GRENADE Rifle-grenade designed to be launched by firing a ball cartridge instead of a blank cartridge. Externally, it is a conventional modern rifle-grenade, with warhead and long tail tube carrying fins. But inside the tail tube is the bullet trap, a series of steel baffles which are designed to yield under certain forces. The rifleman slides the grenade on to the launcher and fires the ball cartridge; the bullet goes up the barrel and enters the tail of the grenade where it strikes the bullet trap. The baffles break under the impact; each baffle reduces the amount of energy left in the bullet and usually, after three baffles, the bullet is stopped. The energy of the bullet is transferred to the grenade, and the propelling gases which were pushing the bullet now lift the grenade from the launcher and propel it to the target.

The only drawback is that the propellant charge which is optimum for a ball bullet is not optimum for launching a grenade, and therefore bullet trap grenades do not have quite so much range as non-bullet trap types. However, it is generally considered that this is a small price to pay for the convenience of not having to carry and load special cartridges (which is particularly irksome with modern automatic rifles) and for the removal of the danger which previously existed when the soldier forgot to change from ball to blank when launching a grenade.

BURSTING SMOKE Type of smoke bomb which uses a small central charge of high explosive to break the body open and liberate a smoke-producing compound. Using this system obviates the need of a time fuze, as is required with base-ejection smoke bombs. The filling is usually white phosphorus (WP) which does not require to be ignited; it is self-igniting on coming into contact with the air and develops a dense white smoke very quickly. The only drawback to WP is that it burns with considerable heat; this warms up the surrounding air which rises and carries the smoke upwards, so causing a 'pillaring' effect, rather than permitting the smoke to hug the ground as is preferable for screening purposes. There is also a grave risk of igniting grass and vegetation around the point of burst; this is often immaterial in war, but unwelcome in training, and for this reason there are other fillings which are sometimes used in preference to WP. The most common alternative is titanium tetrachloride (FM) which produces smoke by chemical reaction with the water vapour in the air. It is perhaps slightly slower to produce smoke than is WP, but burns cooler, gives a dense and low-lying screen, and carries no fire risk.

An American filling used in World War II was a mixture of sulphur trioxide and chlorsulphonic acid, known by the code 'FS', which gave good smoke but was difficult to fill into bombs since it was highly corrosive. The German Army used 'Oleum', a solution of sulphur trioxide in fuming sulphuric acid, and simplified the filling problem by soaking the liquid up in pumice powder. When distributed by the bursting of the bomb, both these compositions reacted with the water vapour in the air to produce smoke. It might be thought that this type of composition might be of marginal utility in, say, a desert, but even the dryest atmosphere contains sufficient water vapour to react with these fillings.

The bullet trap assembly inside the tail unit of a rifle grenade. The baffles collapse on the bullet's impact.

A white phosphorus bursting smoke bomb for the US 4.2in (107mm) rifled mortar. A small explosive charge to break open the body — the phosphorus ignites on contact with air.

The 60mm canister shot used with the Brandt gun- mortar.

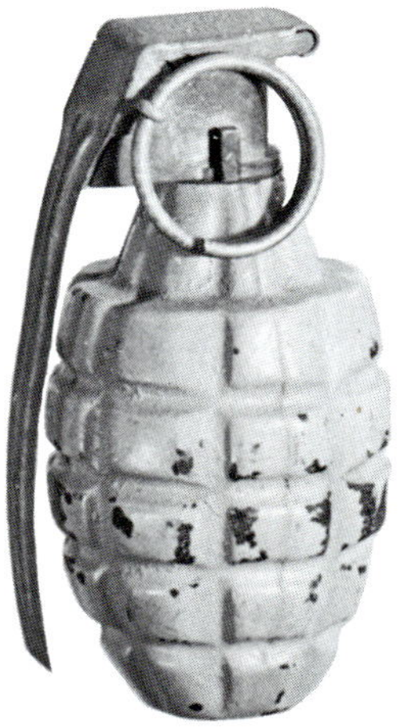

The typical World War II defensive grenades. Above: The American Mark IIA1. Below: The Russian F1.

C

CANISTER SHOT A special anti-personnel projectile developed for use with the Hotchkiss-Brandt gun-mortar in armoured vehicles. The canister shot resembles a shotgun cartridge in its effect. It consists of a metal casing with a cap and small explosive charge at the rear, the remainder of the body being filled with steel balls. This round is loaded into the breech of the weapon, from inside the vehicle, and is fired against personnel in the open at close range. When fired, the charge blows the balls out of the casing and out of the gun, spreading them in a cone to form a screen of missiles which have lethal effect to a range of about 60–70 metres. A few seconds after firing, a second expelling charge fires and the empty canister is ejected from the muzzle of the gun-mortar, so allowing a very rapid rate of fire.

The balls are harmless to armour, and thus the canister shot can be used safely by an armoured car against enemy personnel who may be clambering over a companion vehicle in an attempt to attack it.

D

DEFENSIVE GRENADE A hand-grenade which delivers large fragments for a considerable distance around the point of burst. The term comes from the assumption that such a grenade is best suited to use by a defender who is in a protected position, so that he can throw the grenade and take over. Thus the projected fragments may well come back as far as his own position, but because of his cover this does not matter. Defensive grenades are usually heavy, with heavy metal bodies or fragmenting sleeves; prime examples are the British Mills Bomb or Grenade No. 36, and the American Mark 2 pattern.

DELAY FUZE Method of initiating a munition by a system which interposes a delay between the delivery of the firing impulse and the detonation of the device. Specifically, a fuze which detonates the bomb a short time after striking the target, the delay allowing the bomb to penetrate the target so as to cause more damage. For example, a mortar bomb which strikes a house will do relatively little damage to the occupants if it detonates outside the wall, on impact. By using a delay, the fuze starts operating when it strikes the wall, but the momentum carries the bomb through the wall and into the house, at which time the delay ends and the explosive filling is detonated.

The delay is usually performed by a short column of compressed black powder inside the fuze. When the bomb strikes the target a needle is driven into a detonator as the fuze is crushed; the detonator produces a flash which ignites the black powder. This burns at a rate which can be regulated to some degree of accuracy by the amount of consolidation applied to the powder when it is pressed into place. When the column of powder has burned through, it ignites another detonator which initiates the main explosive filling. The amount of delay is very small — generally in the order of 0.15 to 0.35 second, but at the speed the bomb is travelling when it hits, this is quite sufficient to allow it to penetrate several feet into the target.

A typical defensive grenade — an American Mk IIA1 (left) shown with a M26 controlled fragmentation type (centre) and a drill grenade.

DISCARDING SABOT Means of firing a projectile of smaller calibre than the bore of the mortar by fitting it with a belt or 'sabot' of bore diameter, this sabot being so designed that it breaks away from the bomb and falls clear after leaving the muzzle. The system has been put to practical use only once, in a base-ejection smoke bomb issued for the British 4.2in mortar in 1943 and used until the middle 1950s. The construction of the base-ejection type of projectile makes it necessary to have a slender, tubular bomb, and because of the economic requirement to use smoke canisters already in production for smaller-calibre weapons, the 4.2in bomb was slightly smaller in diameter than the mortar's barrel. It therefore had wide fins, of 4.2in diameter, and a two-piece sabot around the centre of the bomb; the combination of sabot and fins kept the bomb central in the bore as it was fired. The gas pressure behind the sabot caused two retaining screws to break, so that after leaving the muzzle the two pieces fell away and the bomb continued in flight.

A similar design was used in a 160mm Tampella (Finnish) mortar projectile in the 1950s, though this was a very slender, dart-like bomb with large fins and a large sabot, the bomb being filled with high explosive. The intention was to fire a bomb weighing considerably less than a full-calibre conventional bomb, so as to achieve a higher velocity and greater range. The 160mm Dart Bomb is believed to have reached ranges of 11–12 kilometres (about 7 miles) on trials, but the accuracy was poor and the long, slender shape gave problems with initiating the long, interior column of high explosive. The design was never placed in service.

Currently the discarding sabot principle is used in fin-stabilized armour-piercing projectiles designed to be fired from the Thomson-Brandt 60mm and 81mm gun-mortars as direct-fire anti-tank weapons. These projectiles are breech-loaded (the mortar being used in the turret of armoured vehicles) and work in the same way as APFSDS projectiles for conventional guns. The sub-projectile reaches a high velocity and has good penetrative ability against armour.

A typical drill grenade.They are drilled to show that they contain no explosive and are only used to train recruits.

DRILL GRENADE A completely inert grenade used for the purpose of teaching recruits. Contains the standard firing mechanism though without any cap, fuze or detonator, and the body is usually drilled so that it can be seen to be empty. Drill grenades can be provided for any hand- or rifle-grenade; they should not be confused with practice grenades.

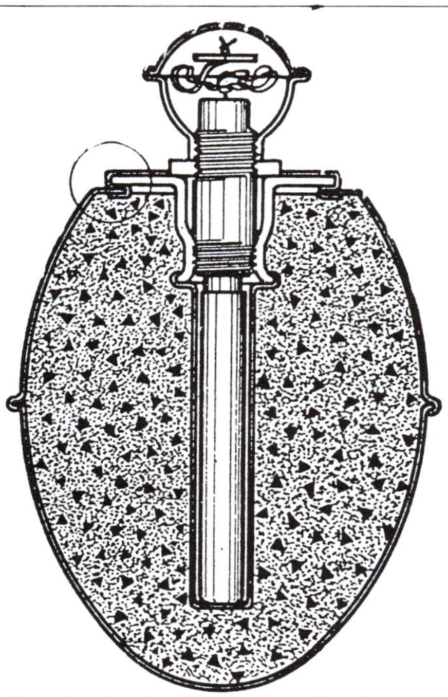

The World War II German 'egg' grenade, showing the thin-wall construction and the pull-type friction igniter.

E

EGG GRENADE Any type of grenade which is smooth and roughly ovoid in shape is likely to be called an 'egg grenade', but the name is most properly applied to the German Eihandgranat 39, as used during World War II. It had an egg-shaped metal body about 76mm (3in) long and weighed about 250gm (9oz). An igniter set was screwed into one end, and the grenade was operated by unscrewing the cap from this igniter; the cap was attached to a short cord, the other end of which was attached to a friction igniter unit. The cord was pulled sharply, to light the delay fuze, and the grenade was then thrown. The delay was about five seconds; on the Russian Front it is reported that special booby-trap igniters were often inserted into these grenades. These had zero delay, so that if a Russian soldier found an unused grenade left behind by the Germans and attempted to use it, as soon as he pulled the cord it would detonate in his hand.

ESPERANZA Spanish company which manufactures a wide range of mortars and ammunition from 50mm to 120mm calibre. The mortars are used by the Spanish Army and are widely exported. The British Army adopted an Esperanza design of 50mm mortar as their '2 inch Mark 1' in 1937. The full name of the company, Esperanza y Cia, is often abbreviated to 'Ecia'.

F

FIN ASSEMBLY The tail unit of a mortar bomb, intended to give stability in flight. Originally, cheap and rudimentary, experience during World War II showed that tail fins have a vast influence on the accuracy and range of a bomb and that money spent on developing an efficient fin assembly gives results out of proportion to the expenditure. Today's tail fins are carefully contoured and machined and are a major factor in the performance of modern mortars.

The air flow around the conventional streamlined bomb is such that placing the tail fins too close behind the body puts them into a low-pressure area where they will have little effect, and for this reason they are usually mounted at the end of a tubular tail boom so that they are in a location where the air flow across them gives stability to the bomb. Some designs, eg, the US 81mm mortar, have the fins slightly canted at an angle to the axis of the bomb, so that air flow gives the bomb a slow roll, adding to stability and accuracy.

Ideally the fins should be greater in diameter than the body of the bomb, in order to get the operating surface of the fins out into undisturbed air where they will have the greatest effect, though this is only really necessary at supersonic velocities. However, the improvement in accuracy and stability achieved by this is not generally thought to be worth the additional mechanical complication. A good example of this approach was a design adopted with the Brandt 81mm mortar in the 1930s by both Italy and the USA; the fins were hinged in the middle and folded inwards, against spring pressure, and were held by a shear wire. Thus when loaded they formed a tight bundle at the end of the tail boom and were well below barrel diameter. On firing, the shear wire broke, the springs forced the fins out to ride against the bore, and after leaving the muzzle they sprang out to a full diameter which was about twice the diameter of the bomb. It was a good theoretical idea, but the mechanical design frequently failed to withstand the conditions in the bore, and the explosion of the propelling cartridges often mangled the fins so that they opened erratically and upset the flight of the bomb.

A more reliable system is to hinge the fins at their broad edge, fold them sideways, and allow them to open outwards under spring pressure. This has been used in some rocket-assisted bombs when it is desirable to discard the normal fins (used to stabilize the bomb during the initial

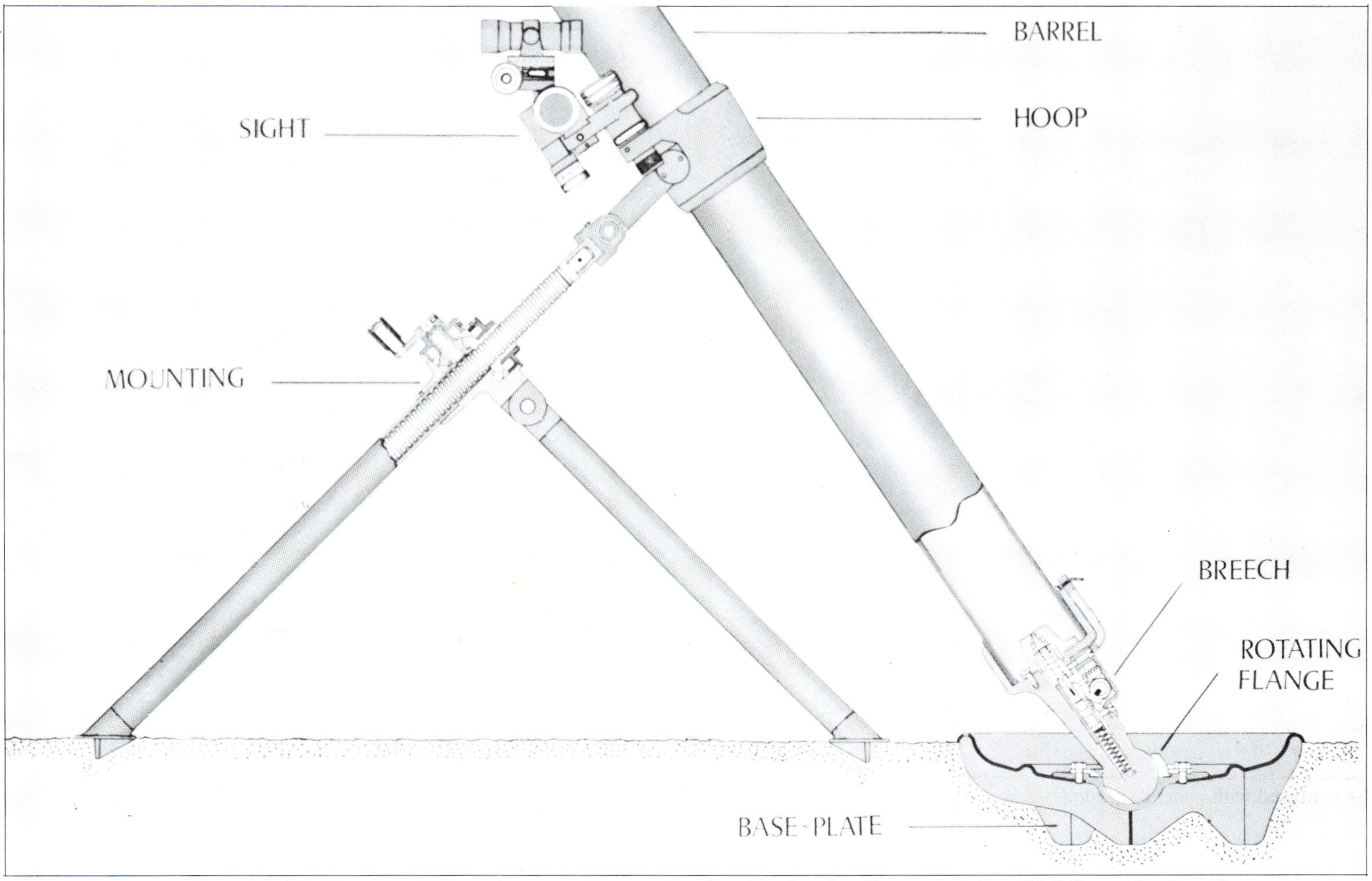

Above: Esperanza 60mm Model 'L' mortar. Below: A group of mortar bombs, showing the divergent styles of tail fins which have been adopted by different designers (left to right): Italian, American, German and British.

part of its flight) so as to allow the rocket motor to come into play. As the conventional tail boom and fins fall away, they release a set of folding fins which spring out to continue stabilizing the bomb during the remainder of its flight.

The principal drawback with fin stabilization of a bomb is that the fins have a distinct 'sail effect' in windy weather. Thus a bomb fired when the wind is blowing from the right will tend to turn into the wind, because the wind blows against the flat fins and pushes them to the left, turning the nose of the bomb to the right.

FLY-K MORTAR Trade name for a spigot mortar developed by PRB of Belgium and previously known as the 'Jet-Shot'. The particular feature of the FLY-K is that the propelling cartridge, inside the tail boom, is located in front of a captive piston. The bomb tail is placed over the spigot in the usual way, and dropped; the impact fires the propelling cartridge which explodes and drives the captive piston downwards with great speed and force. This impacts on the spigot and the bomb is thrown off into the air; but the piston is trapped before it can leave the tail tube, so that the gases of the charge, and almost all the noise of the explosion, are trapped inside the tube. As a result the FLY-K is almost silent and can scarcely be heard more than one or two hundred metres away; it is also flashless and smokeless, and is thus an ideal weapon for clandestine forces. The weapon's nominal calibre is 52mm and its maximum range about 700 metres.

FM Code designation for titanium tetrachloride, a chemical used in bursting smoke bombs. It is a colourless liquid which, when released from the bomb by a small explosive charge, combines chemically with the water vapour in the air to form a dense white smoke; when the combination is in the order of five parts water to one part FM the smoke is extremely dense, but less water vapour gives less satisfactory screening. The smoke is rather acrid, but in the concentrations used in the field does not give rise to coughing or respiratory problems. As little as 5gm (0.18oz) of FM can produce 30 cubic metres (39 cubic yards) of smoke.

FRAGMENTATION SLEEVE A sleeve or casing of iron or steel which can be placed around an offensive grenade in order to convert it to a defensive grenade with greater fragmenting power. Sometimes in the form of two halves which fit over the top and bottom of the grenade and lock together by a bayonet joint. Some modern designs use a thick plastic sleeve which has several hundred pre-formed fragments or steel balls embedded in the plastic.

FRANKFURTER BOMB British code name for the discarding sabot base-ejection smoke bomb for the 4.2in mortar. Invented by Colonel B.J. Murphy, RA, the name came from its sausage-like appearance.

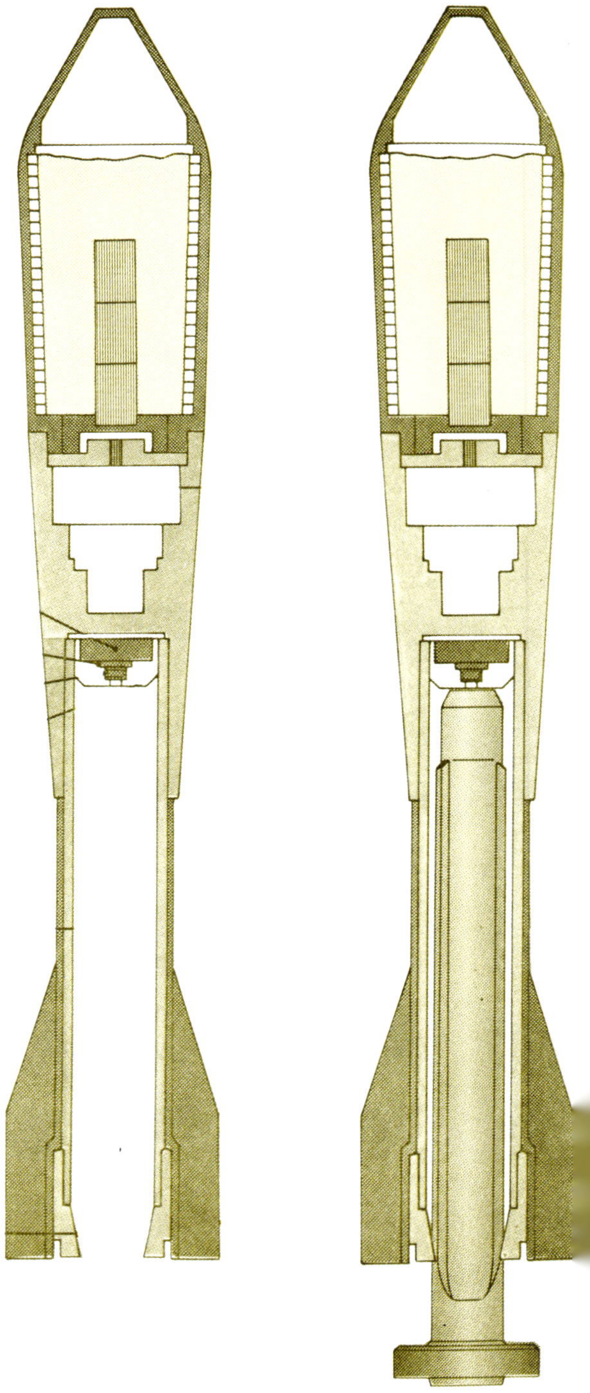

The operation of the Fly-K spigot mortar. The bomb is dropped onto a spigot. The impact fires the propelling cartridge which explodes, driving the captive piston down. This pushes on the spigot and forces the mortar off into the air. As all the propellant gas is trapped in the cavity, the weapon is flashless, smokeless and virtually noiseless.

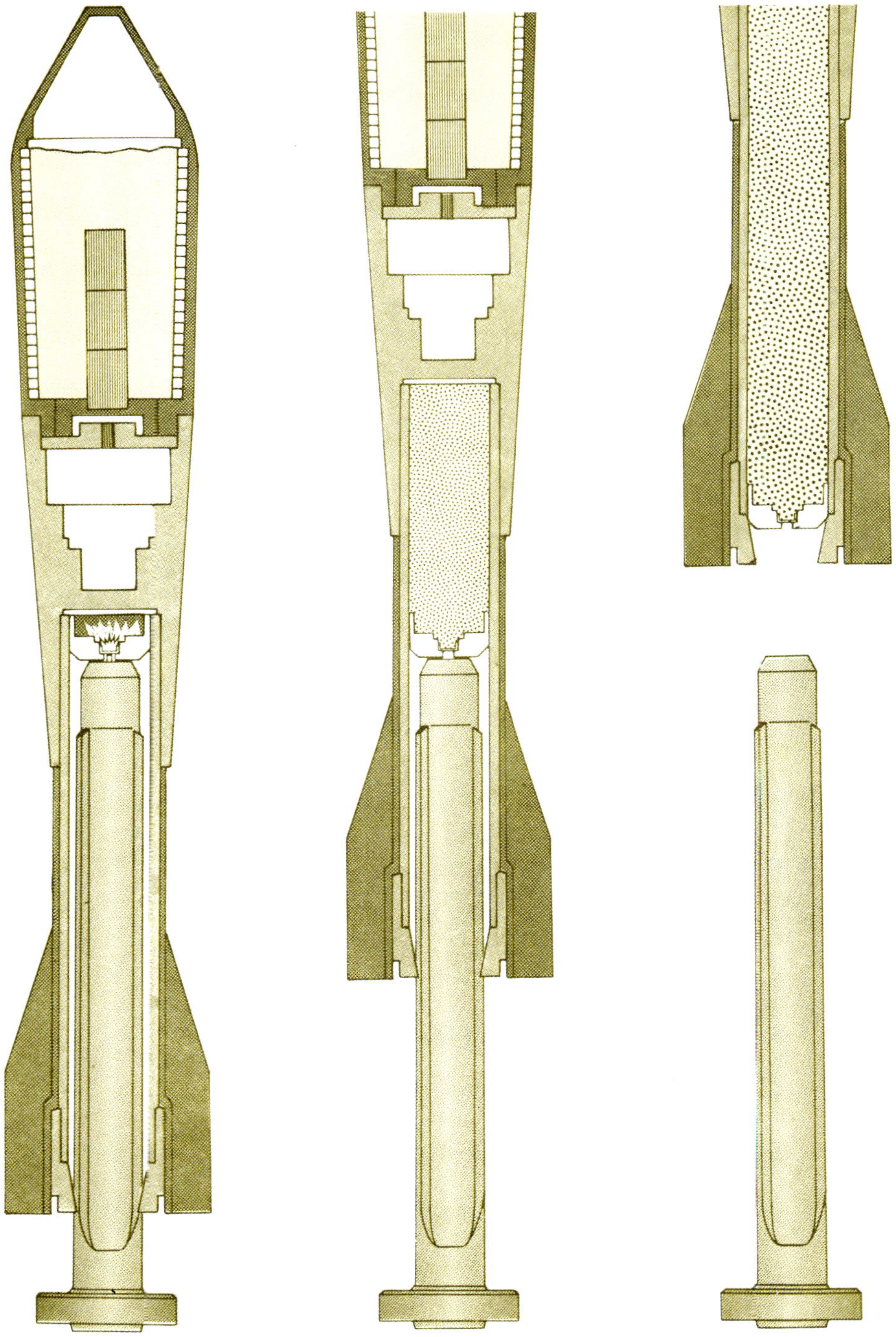

Top: A plastic-bodied offensive grenade. Above centre: A fragmentation sleeve of plastic containing about 6,000 small steel balls. Above: The two combined to make a defensive grenade.

FUZE Device fitted to a projectile in order to cause it to function in accordance with its tactical purpose, ie, to detonate on impact with the target or burst at some predetermined time or place.

Mortar fuzes have, traditionally, been fairly simple devices, because the essence of a mortar system was cheapness and speed, and an expensive fuze, or one which required much manipulation by the firer before loading, was not welcomed. Today, however, there is no cheap ammunition and the enhanced tactical effect gained by using a well-designed fuze is considered to be worth the expense. Moreover, modern technology permits the use of complex fuzing systems which do not require time-consuming setting or adjustment before firing.

Impact fuzes are the most usual type found in mortar ammunition and are used with high-explosive bombs and with bursting smoke bombs, both of which are required to function on striking the ground at the target. Most modern impact fuzes can be set to give a slight delay so that the bomb will penetrate light cover before detonating. Time fuzes are used when it is desired to make the bomb function while still in flight, above or short of the target; they are principally used with illuminating bombs so as to eject the star unit over the area it is desired to illuminate. Time fuzes could also be used with high-explosive bombs in order to burst them in the air just above the target, in order to drive fragments downwards to get behind cover or into trenches. However, the necessary setting and adjustment of a time fuze is unwelcome in the infantry mortar; it is desirable to produce a bombardment very rapidly, before an enemy can take cover, and the slow business of firing a sequence of single bombs to arrive at the correct fuze setting advertises the bombardment long before it can take effect. Where airburst fire is required, proximity fuzes are the optimum choice; these operate on radio principles, carrying a small transmitter and receiver unit. As the bomb descends towards the target the fuze emits a radio signal which strikes the target and reflects back to be picked up by the receiver circuit. When the strength of the returned signal indicates that the bomb is at the optimum height for lethal effect, the circuitry fires an electric detonator and detonates the bomb.

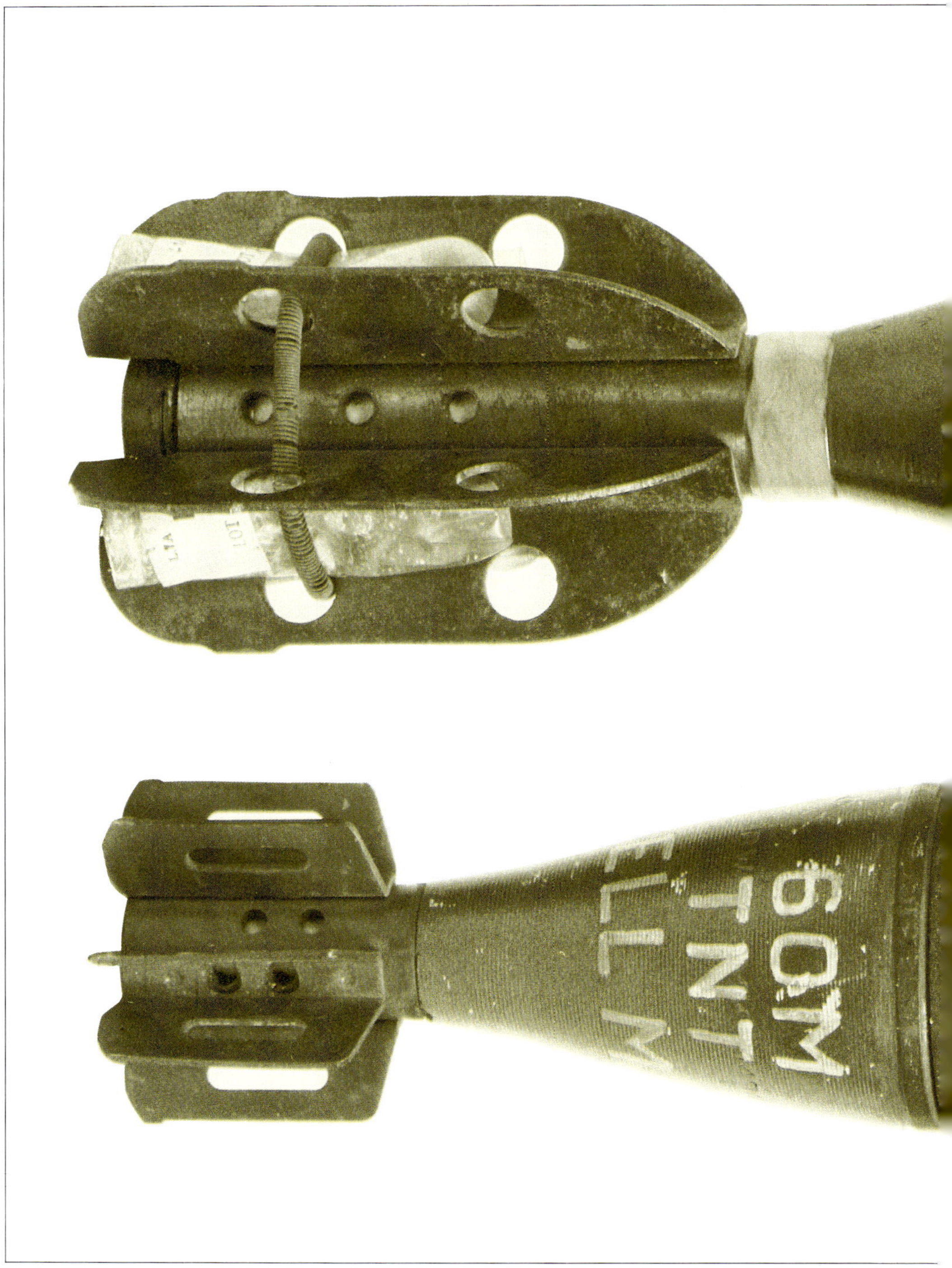

Fuzes are a contradiction; they must be sensitive enough to function rapidly on impact or at the end of their set time, but must be sufficiently robust to withstand handling during transport, loading into the weapon, and the shock of firing, without detonating. In order to reconcile these conflicting requirements, mechanisms are employed which take advantage of the peculiar environment in which the fuze works. In general terms, the fuze is so arranged that the internal mechanism is securely locked until the instant of firing. When the bomb is fired, the sudden acceleration of the bomb tends to leave behind anything inside the fuze which is loose, and the safety device is allowed to move backwards so as to begin unlocking the components. Its movement is restricted by an escapement — a toothed wheel operating against a restricting mechanism — which slows the removal of the safety device so that the bomb is several metres outside the muzzle of the mortar before the fuze is placed in an operating condition. Once this primary lock is removed, other locks may operate sequentially, so that completion of the 'arming' takes several seconds of flght, and if, at the end of that time, the fuze should malfunction it is so far away from the mortar and so high in the air that the consequences will be negligible.

Proximity fuzes, being electronic, are much easier to control: simple resistor-condenser delay circuits can prevent the charging of the firing circuit and can delay the emission of radio signals without requiring any mechanical devices. A mechanical lock may be used to keep

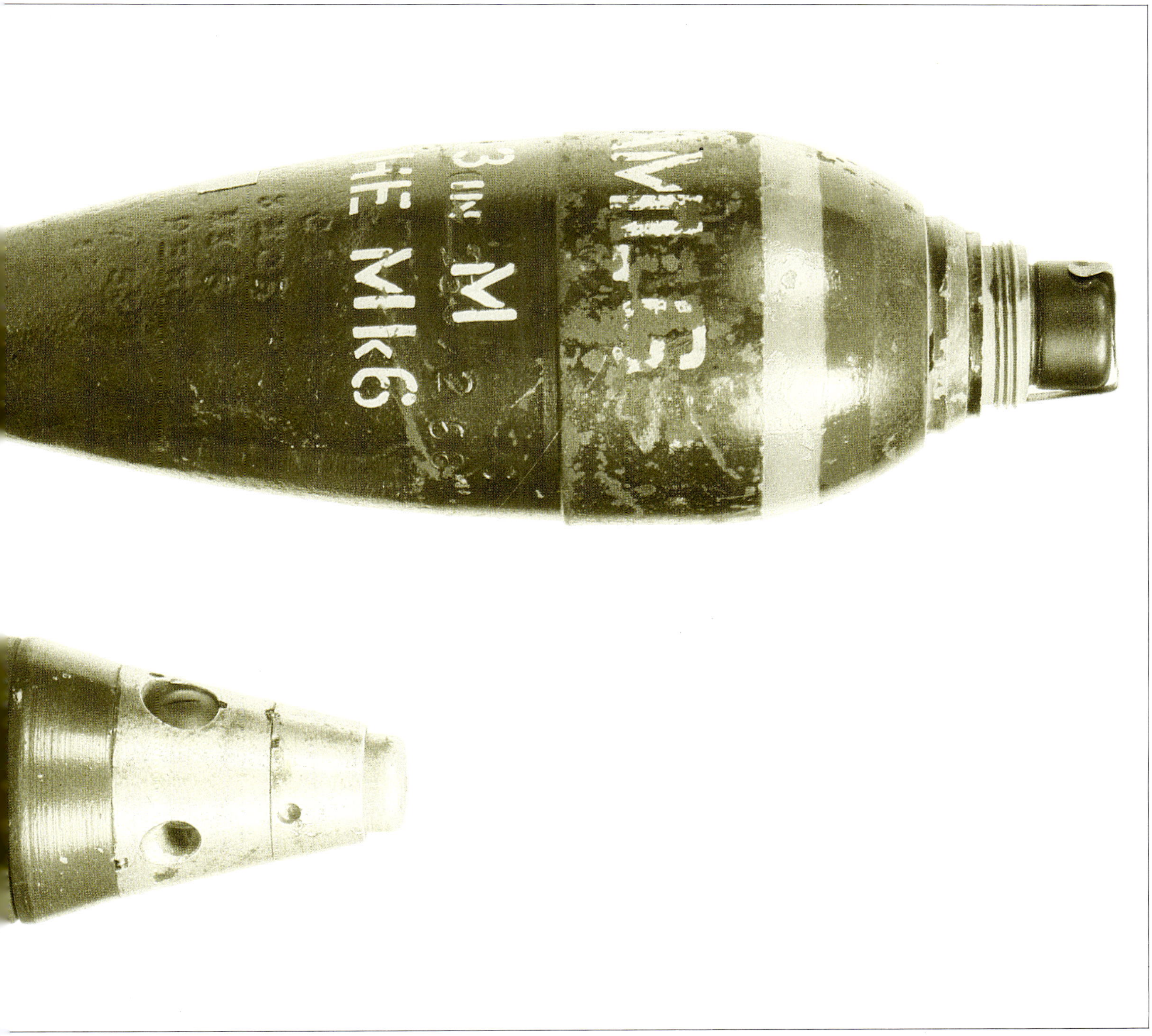

a detonator out of alignment with the explosive charge for some distance, but this is mandatory with any fuze intended to be used with high-explosive bombs. Theoretically it should be possible to detect the signal of a proximity fuze and jam it, but there are immense difficulties to be overcome in practice, and modern proximity fuzes use new 'frequency hopping' techniques to vary the frequency of the signal emitted several times a second and so rendering any attempt at jamming futile.

FUZE PROBE A rod extending from the tip of an impact fuze and coupled to the firing pin. The object is that the rod should strike the ground and fire the fuze while the bomb is some distance above the ground (perhaps 25-30cm 10-12in) which would permit the best distribution of fragments for anti-personnel effect. Allowing the bomb to hit the ground before detonating tends to permit a proportion of the fragments to be buried. Although theoretically sound, fuze probes are awkward things to handle and adversely affect the flight of the bomb, and they are rarely used.

Top: A British 3in mortar bomb. Above: An American 60mm bomb.
Both are fitted with impact fuzes. The British fuze has its safety pin in place.

The Gammon grenade, filled and empty states, as carried by airborne troops in World War II.

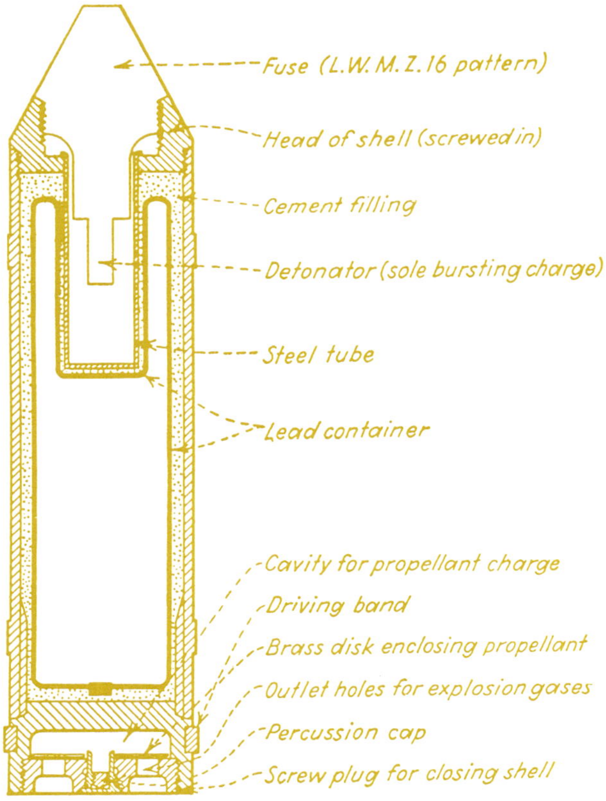

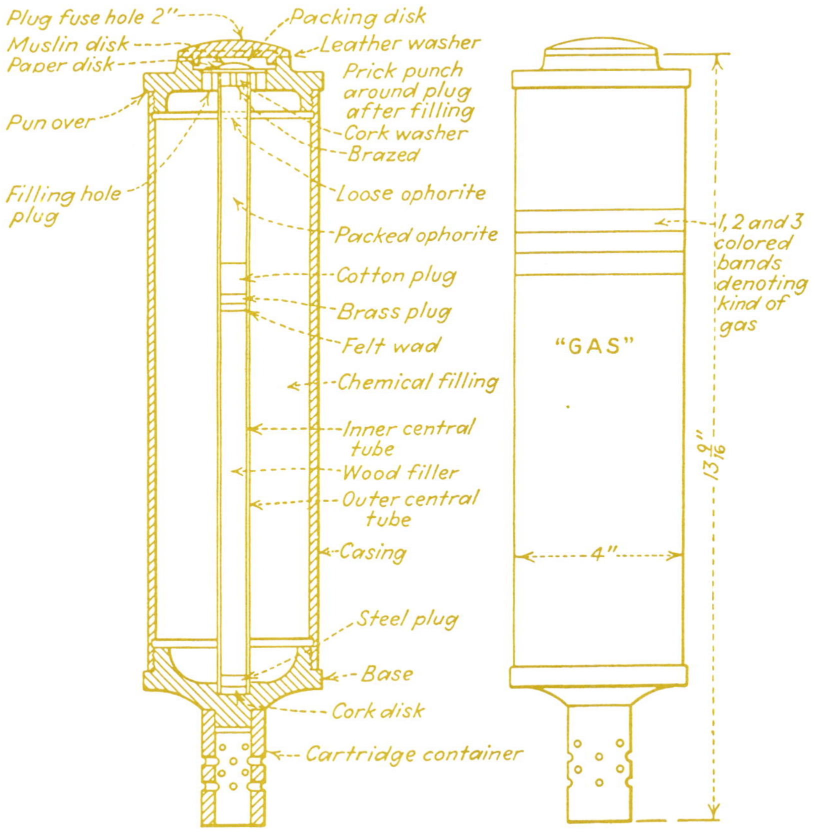

Above left: A German 77mm mortar gas bomb. This was a rifled mortar and carried its augmenting charge in a separate compartment in the rear of the shell. Above right: The American 4in Stokes mortar gas bomb, used in 1918.

G

GAMMON GRENADE Grenade invented by a Captain Gammon of the British Army during World War II and extensively used by airborne troops. The grenade was simply an Allways fuze attached to a stockinette bag; the plastic explosive filling was put in by the user to suit his requirements. Plastic explosive was issued for demolition purposes and was carried by airborne troops; it therefore made sense to give them these 'Gammon Grenades' which could be carried rolled up in a pocket or ammunition pouch. They would assess the amount of explosive required, according to the task, insert that amount of plastic into the bag, remove the safety cap and throw the grenade. For casual anti-personnel use, half a stick of plastic would suffice; for something more difficult, such as breaching a wall or attacking an armoured car, perhaps two or three sticks would be wadded together and put in the bag. It was officially introduced as the 'Grenade, Hand, No. 82' in 1943 and was declared obsolete in 1954.

GAS BOMB General name for a mortar bomb filled with a chemical warfare agent. The name is incorrect, since the agents are normally liquids, though most of them vaporize into gas when released from the bomb.

The usual design is a cylindrical bomb (for greater capacity) with a central tube of high explosive running down the bomb's axis. There is an impact fuze at the nose which, on striking the ground, detonates the column of explosive and breaks open the bomb to release the contents. A violent detonation is not required, it would probably destroy the chemical agent in the explosion. All that is required is that the bomb be opened, which also has the virtue of making the arrival of the bomb less obvious —the slight explosion often being lost in battle-field noise — so that the gas can take effect before the recipients have been alerted.

Some chemical agents are in solid form, notably the various types of arsenical smoke, and these can be packed in canisters and liberated from base-ejection bombs or packed around the central tube of a cylindrical bomb and distributed by the detonation. Of the two methods, the former is more efficient because it utilises the burning of the composition to develop a smoke which carries the minute particles of agent; blasting the substance into the air by exploding a bomb is less efficient in covering an area.

Mortar bombs have not been used for chemical warfare since World War I, though research has been carried out in order to keep up to date. It is probable that designs exist for modern bombs containing nerve gas, but nothing has ever been publicly said about them.

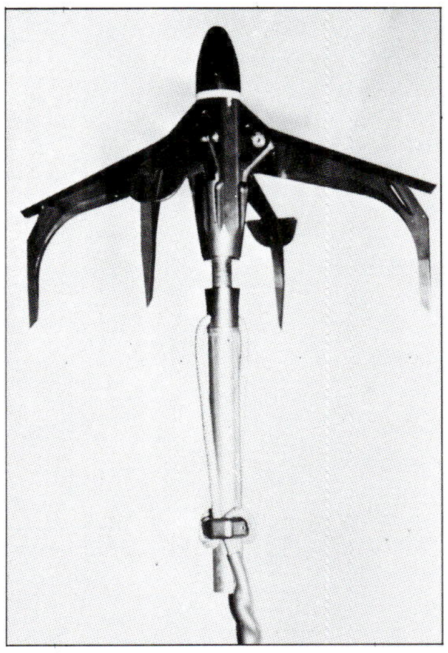

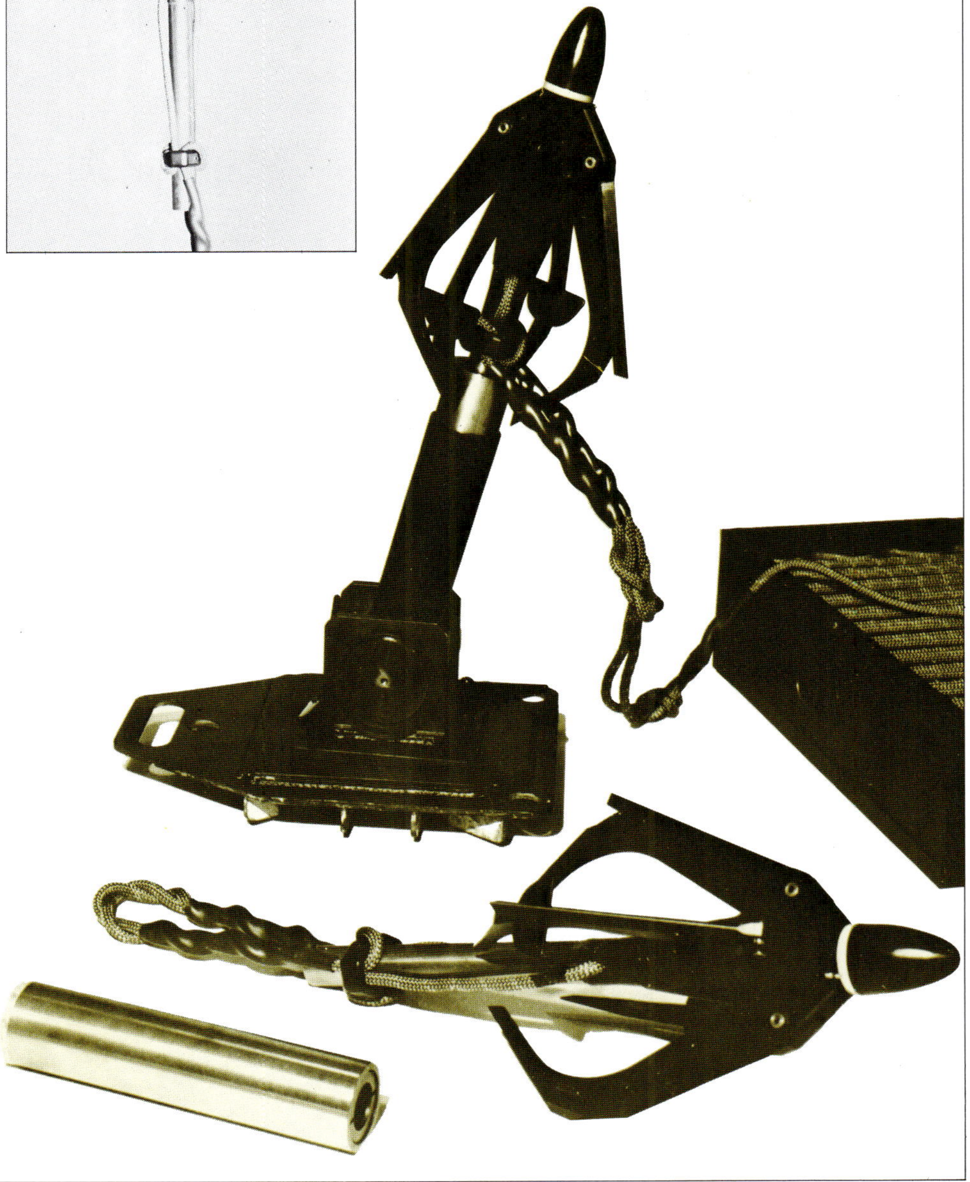

A grapnel ready to be fired from a special short-barrelled mortar. The inset shows the grapnel fully open. At the end of its flight it can dig-in to almost any surface.

GRAPNEL A form of hook or folding anchor device with a rope attached, which can be fired from a mortar so as to fall across a wire entanglement. The rope is then hooked to a vehicle which pulls the wire and tears a gap in the barrier to allow troops to pass through. A grapnel can also be used for cliff-climbing; it is fired vertically so as to land at the top of the cliff and dig into the ground, after which an assault party can clamber up the rope.

GRENADE A small, short-range, explosive munition to be thrown by hand or launched from a rifle or special-purpose weapon. Hand- and rifle-grenades are detailed separately elsewhere. It is appropriate to use this entry to describe the third class of grenade, those designed to be fired from special weapons, because these are becoming more common.

The first specially designed grenade system was developed in the USA in the 1950s and was introduced as the '40mm grenade Launcher M79', together with its special grenades, in the 1960s. The weapon is a simple single-shot device with a short rifled barrel, and is broken open like a shotgun to load the grenades one at a time. The grenade is a short, blunt-nosed cylinder of sheet metal concealing a spherical fragmentation grenade and a fuze. Attached to the rear of the grenade is a short aluminium cartridge case containing a percussion cap and the propelling charge. The propulsion system is the 'High-Low Pressure System', necessary in order to permit a grenade to be fired from a shoulder weapon. The cartridge case contains a strong central chamber in which the propellant is packed. On being exploded, the gas, which is at high pressure inside this chamber, is permitted to 'bleed' out through carefully dimensioned holes, into the main cartridge case where it provides a low-pressure impulse which fires the grenade from the weapon.

The fuze fitted to the grenade is an impact type with safety features which prevent it arming until it has travelled some distance from the weapon. There are several types of grenade: the high-explosive pattern detonate on impact, though one model uses a small charge underneath the fuze to throw the fragmenting grenade unit up into the air where it detonates to give an 'airburst' effect. Smoke, parachute illuminating, tear gas and skytrail grenades are other available types.

Shortly after the M79 launcher was introduced, the US Navy developed an automatic launcher which had the grenades loaded into a belt and could fire them at a high rate. The idea was not followed up, but in the 1970s the Soviet Army revealed a similar weapon, the 30mm AGS-17. This caused a resurgence of interest in the USA and a new automatic model, the Mark 19, is now being issued. Various armies have also developed their own versions of the M79 and a recent introduction is a six-shot revolver weapon developed in South Africa.

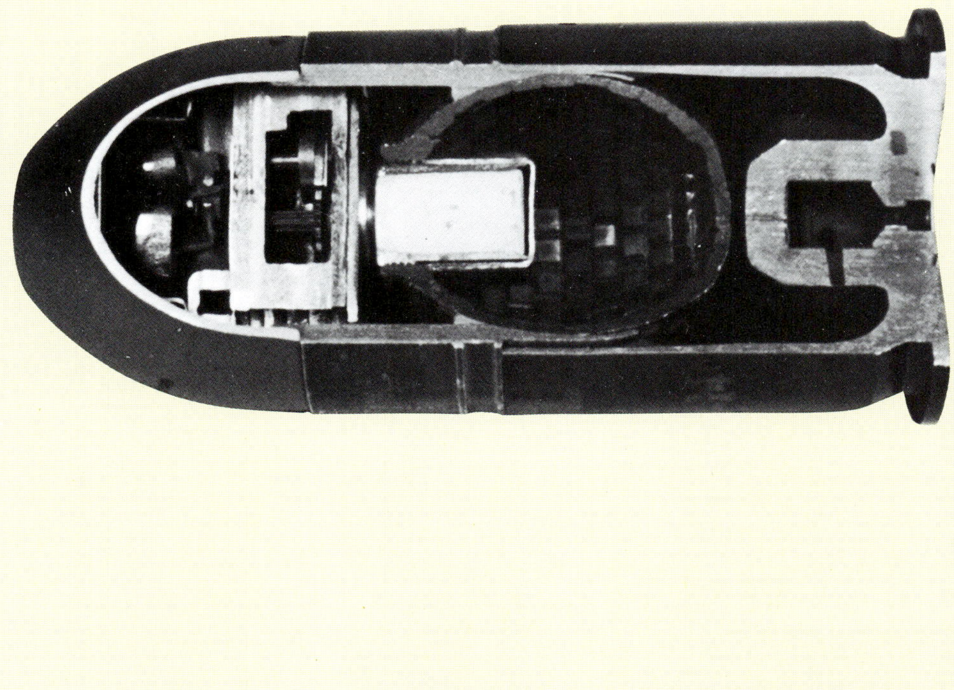

American HE 40mm grenade for the M79 launcher, showing the fuze — which is not armed until the grenade has travelled some distance from the launcher — and (centre) the pre-fragmented explosive unit. The bleed chamber in the cartridge can also be seen.

GRENADE FUZING The requirements for fuzing a hand-grenade might appear simple, but in fact are formidable to a designer. The grenade must function irrespective of what angle it hits the target; it must function reliably; and should the thrower be shot in the act of throwing and drop the grenade, it should not detonate. It must also be weatherpoof, easy and cheap to make and, preferably, silent so that it does not advertise its arrival.

There are two approaches to grenade fuzing; time or impact functioning; of the two, time is generally considered the best for hand-grenades, impact for rifle- and other types of projected grenade.

Time functioning is almost always performed by burning a short length of fuze or a column of tightly pressed gunpowder. This delay element is ignited by a simple percussion cap, initiated by a spring-actuated firing pin (see *Bouchon Igniter* for an example of the mechanism used), and thereafter burns at a predetermined rate until it ignites a detonator which then detonates the main filling of high explosive or, in the case of a pyrotechnic grenade, ignites the smoke or other filling. The delay element is manufactured from a form of high-grade fine gunpowder which is compressed to a very precise density to give the desired burning rate; even so, it cannot be guaranteed to give a repeatably exact time, and most delay systems have a timing tolerance of about 15 per cent. The customary delay is five seconds and this is usually quoted by manufacturers as being plus or minus one second. The time of delay has become more or less standardized from experience; five seconds gives sufficient delay for the grenade to be thrown to its maximum distance, but not sufficient time for a recipient to be able to pick it up and throw it back, which frequently happened when longer delays were used.

In recent years electronic timing has been suggested, and one or two designs have been put forward; they have been tested by various armies but none has yet seen large-scale adoption. In these the grenade mechanism contains a source of electric power — either a small battery or a simple generator — which charges a condenser through a resistance. This resistor-condenser circuit is a well-known electronic method of obtaining delay and by careful selection of components can be designed to give extremely precise and consistent timing. At the end of the delay period the condenser is fully charged and is then discharged into an electric detonator to fire the grenade filling. An advantage of this type of system is that it is possible to build-in various electrical safety circuits which, for example, ensure that the grenade is not armed until it is some distance from the thrower.

Impact fuzes for hand-grenades have to be of the *Allways* type, since the angle at which the grenade lands cannot be predetermined. This type of mechanism is described elsewhere, but they tend to be less safe than time mechanisms and are not liked by many armies.

Rifle-grenades are invariably fuzed for impact because they are stabilized and thus always land nose-first; this means that the Allways system is no longer demanded and more conventional mechanical techniques can be used. The most usual method is simply to have a firing pin poised above a detonator so that on impact the pin is driven back, fires the detonator, and thus initiates the grenade. A refinement of this is to have the detonator contained in a heavy 'inertia pellet', held back by a spring, so that should the grenade land at a low angle of impact, or against a sloping surface, such that the nose is not directly struck, the deceleration of the grenade will allow this inertia pellet to run forward and impale the detonator on the firing pin. In both cases it is necessary to incorporate safety devices which securely lock both the firing pin and inertia pellet until the grenade is actually fired; the sudden acceleration of the grenade as it leaves the rifle can be used to free an axial locking bolt, this actuation being the reverse of the inertia pellet's operation.

As with hand-grenades, electronic systems have recently appeared in rifle-grenade fuzes. In this case the object is to develop a proximity function so that the grenade can be fired over the heads of an enemy force and burst so as to shower fragments downwards, thus defeating any frontal cover behind which they may be sheltering. Grenades being small, these proximity fuzes use very simple electro-optical circuitry to detect the optimum height above the ground and then operate by discharging a condenser into an electric detonator.

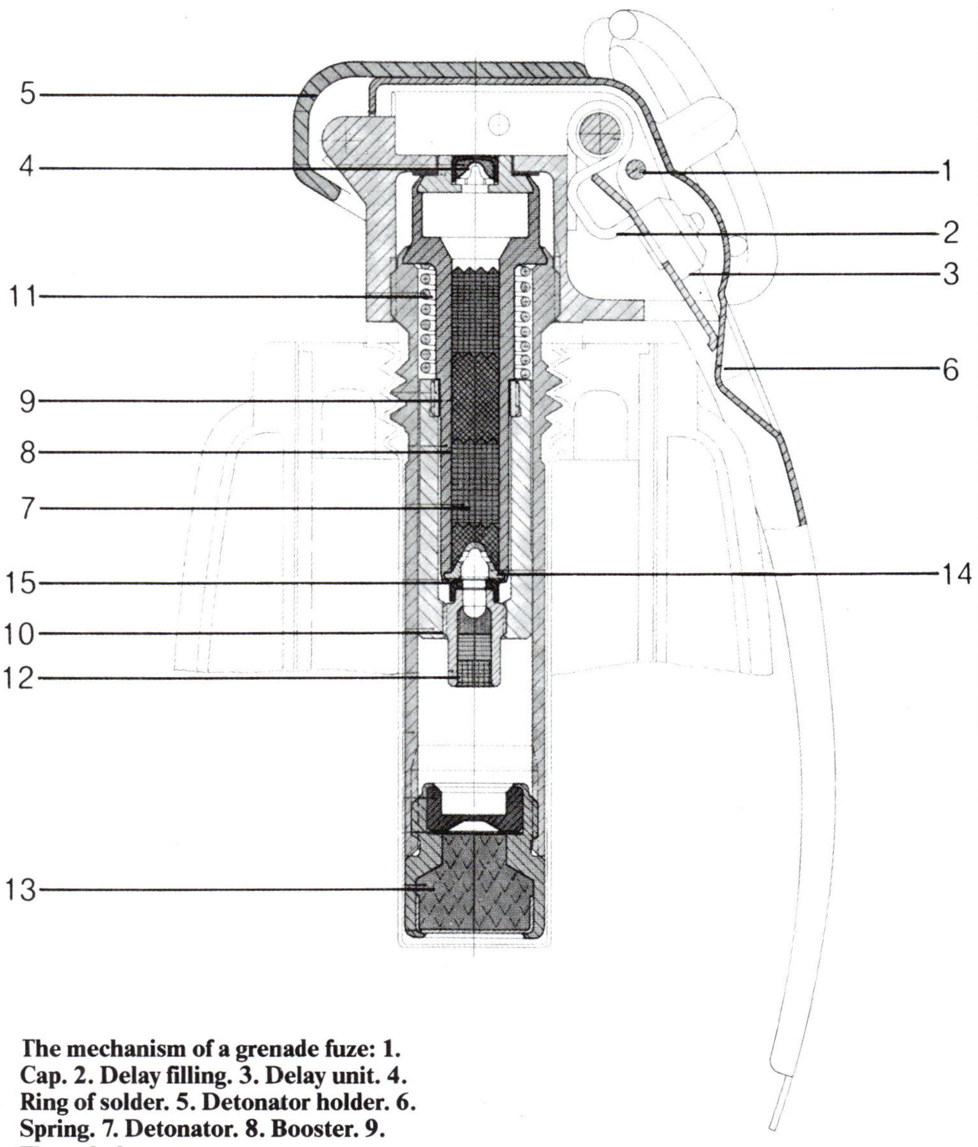

The mechanism of a grenade fuze: 1. Cap. 2. Delay filling. 3. Delay unit. 4. Ring of solder. 5. Detonator holder. 6. Spring. 7. Detonator. 8. Booster. 9. Flame holes.

H

HAIRBRUSH GRENADE A pattern of grenade developed by almost all combatant armies during 1915-16 when grenade designs were in their infancy. The object was to have a powerful grenade which was capable of being thrown easily, and the universal method was to make a wooden paddle, resembling a hairbursh or hand-mirror in shape (ie, having a thin handle which widened out to a large back-board), and then securing a cast-iron or sheet metal box to this backing. The box was filled with explosive; in the case of sheet metal boxes a quantity of scrap metal was also inserted in order to provide lethal fragments. A simple igniter, usually a friction type or even a length of safety fuze which had to be physically lighted before throwing, was fitted and that was that. The thrower grasped the handle, fired the igniter by whatever method was in use, and then pitched the grenade in an overarm throw. They were usually quite heavy, but the handle allowed them to be thrown to about 30 or 40 metres distance, and on bursting they had a substantial effective radius. By the middle of 1916, they had been entirely superseded by more efficient and neater designs.

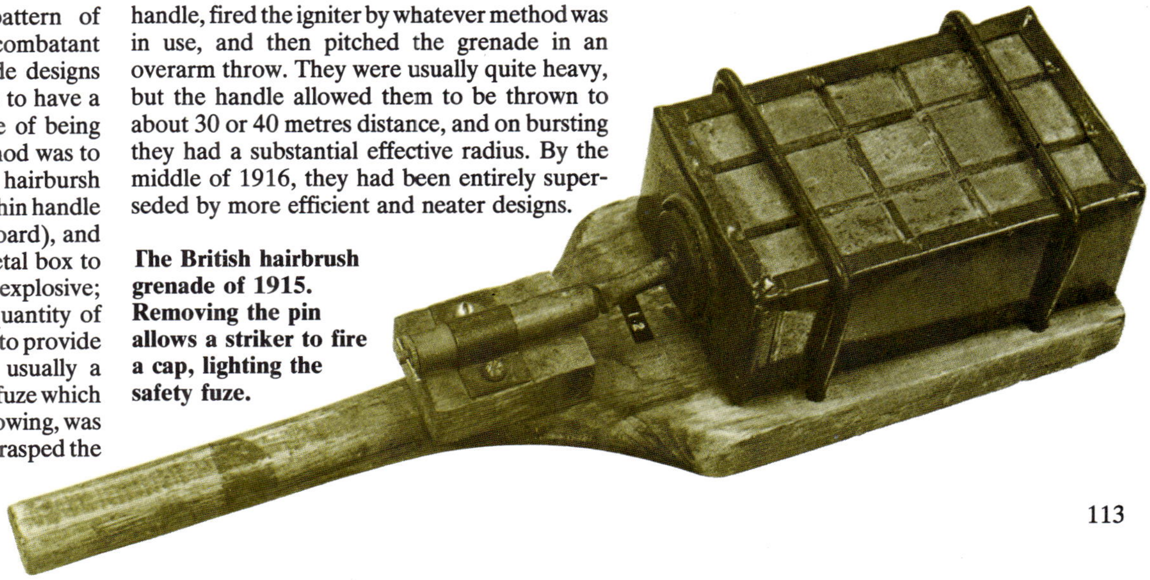

The British hairbrush grenade of 1915. Removing the pin allows a striker to fire a cap, lighting the safety fuze.

HALE'S GRENADES Prior to World War I various grenades were designed and patented by an English inventor, Martin Hale. Hale was connected with the explosives industry and interested the Cotton Powder Company in his designs in about 1907. Various models of hand- and rifle-grenades were demonstrated to military attachés in 1908. All were basically of the same design, differing only in the method of getting them to the target. They were simply brass tubes carrying a 113.4grm (4oz) charge of 'Tonite', a patented explosive developed by the Cotton company and consisting of guncotton and barium nitrate. Inside the charge was a tube holding a weighted firing pin, restrained from touching the detonator by a spring. The brass outer casing was surrounded by a segmented cast-iron ring designed to be fragmented by the explosion.

The construction of this fuze made it necessary that the grenade land nose-first, and to ensure this the hand versions were fitted with a rope 'tail', by which they were thrown, and which acted as a drag brake to ensure that the grenade arrived nose-first. The rifle-grenades were fitted with a thin steel rod at their base, and this was inserted into the barrel of a rifle. A special blank cartridge was loaded into the breech, the butt was rested on the ground, and the rifle was fired. The gas generated by the blank cartridge caused the rod and grenade to be blown from the rifle, and the rod acted as a stabilizing device, again ensuring that the grenade arrived nose-first.

The original Hale's rifle grenade, from a contemporary magazine.

Below: The British Grenade, Hand, No. 1, showing the cane handle and ribbons which ensured that it arrived head-first. Right: Typical defensive hand grenades. They are heavier than offensive grenades — the assumption is that the thrower of a defensive grenade can take cover behind defences to protect himself from the blast and fragments.

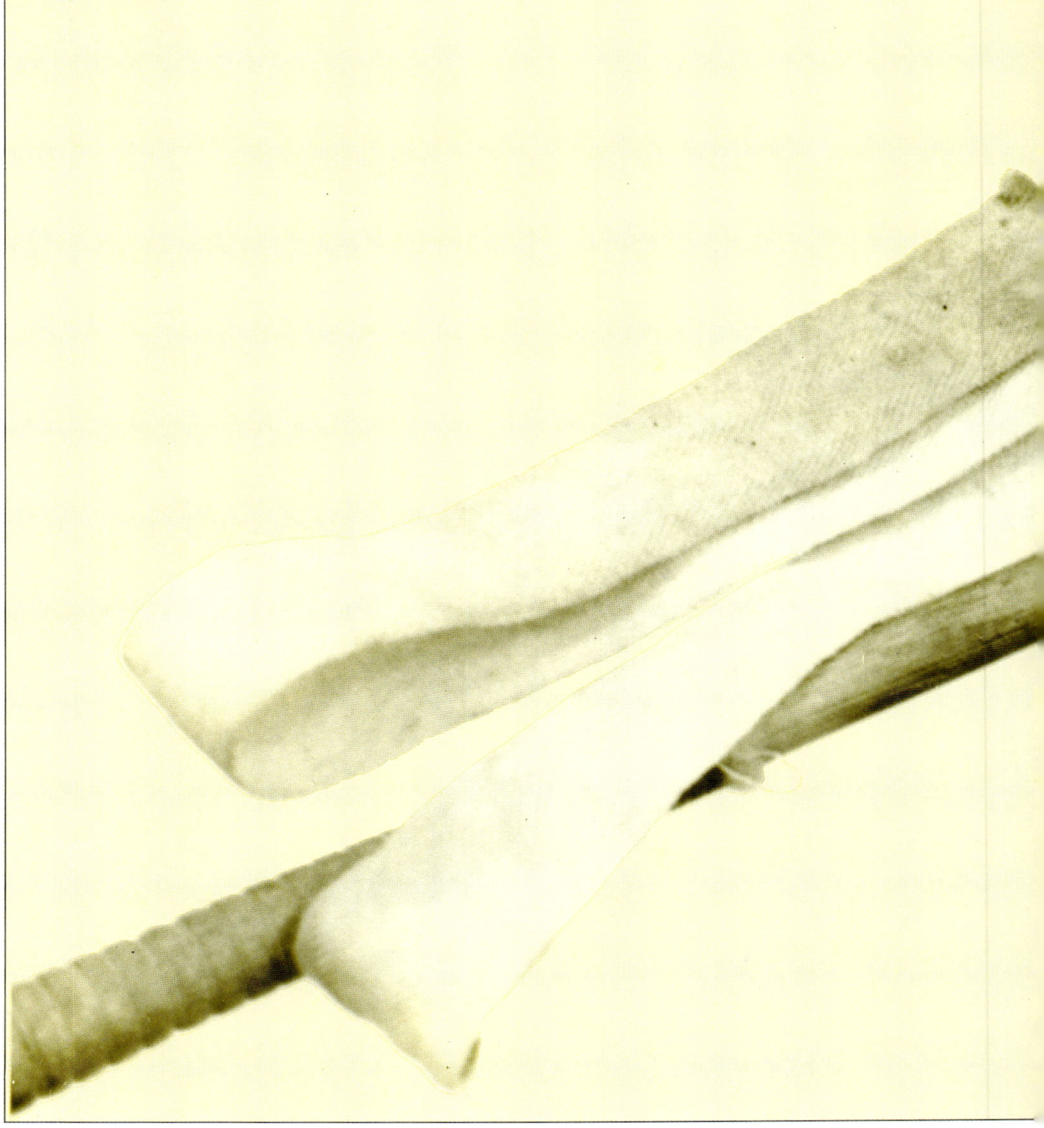

The demonstration was successful and the Spanish Government purchased a quantity of both types which were used on operations in Morocco in 1909-10. The German Army also ordered a quantity for trials and eventually introduced a 'Model 1913', which was essentially Hale's design modified by removing the cast-iron ring and using a body tube of serrated cast iron. The British Army ignored Hale's ideas, preferring to attempt their own design, but when war broke out in 1914 they rapidly adopted the Hale's pattern, issuing it as the 'J Pattern' grenade, later known as the 'Grenade Short Rifle No 3', in 1915.

Hale's rifle-grenades remained in use by both Britain and Germany until late in 1917 when cup-discharged designs began to appear. The principal drawback to the Hale design was that the sudden rise in pressure inside the rifle barrel when the expanding blank cartridge's gases struck the rod tended to bulge the barrel and render the rifle useless for shooting bullets. Special rifles were set aside for grenade launching, but once better systems appeared the use of rod grenades ceased and has never been revived.

HAND-GRENADE A hand-grenade is the general name for any small explosive munition

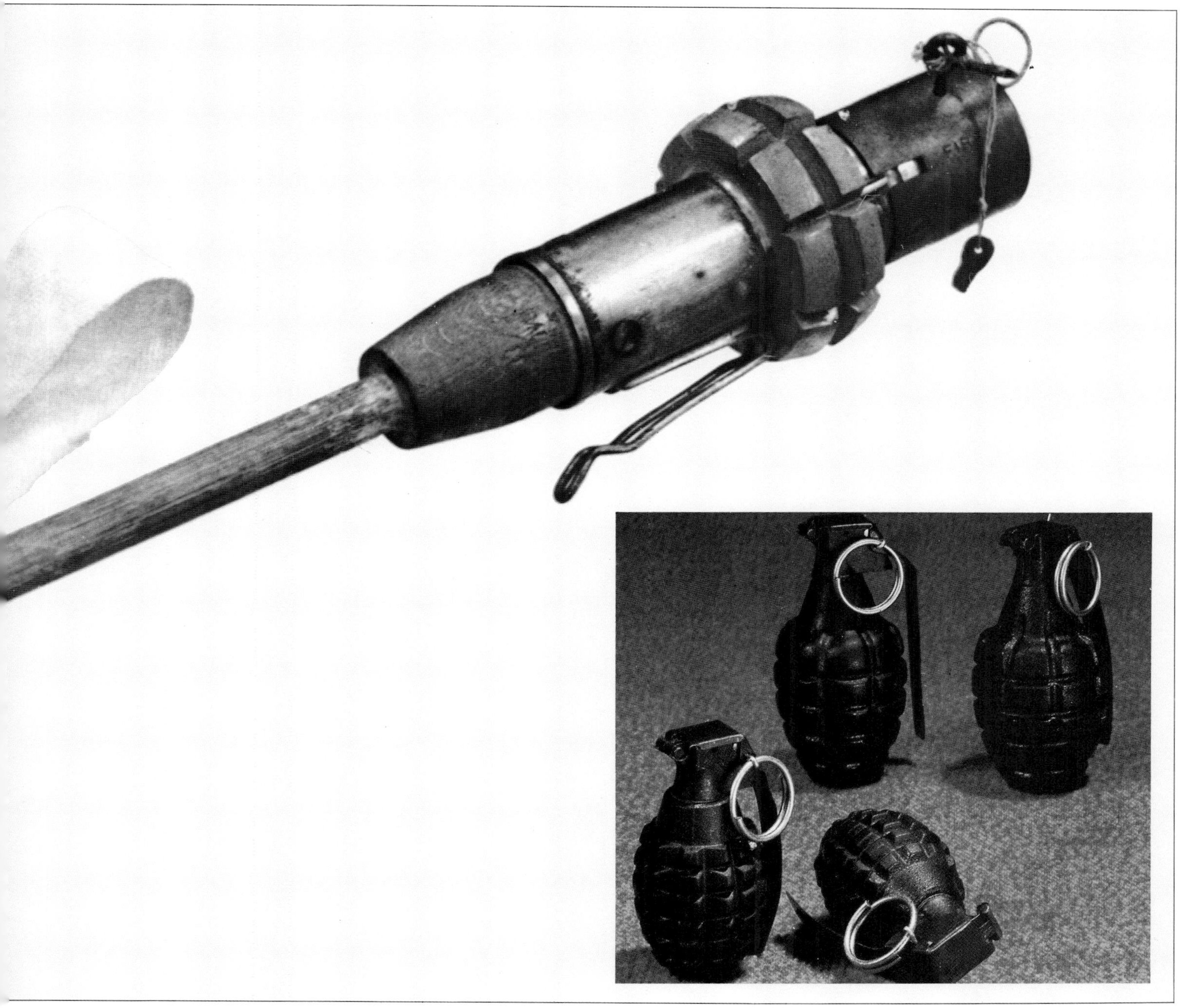

designed to be thrown by hand, whether or not it is designed for anti-personnel effect.

The hand-grenade is one of the older weapons of warfare, being used as early as the first half of the 16th century; one of the first records is of their effective use at the Siege of Arles in 1536. At that time they were generally earthenware pots filled with gunpowder and stones and with a piece of slowmatch (a primitive form of fuze) thrust into the neck and lighted before throwing. Their tactical application was principally for the defence of fortifications, by throwing or dropping them into the surrounding ditch to deter assault parties.

In later years the casing was made of brass or roughly cast iron, and by the 17th-century they were sufficiently common to warrant each infantry regiment having a special company of 'Grenadiers' to carry and throw them. The Grenadier Company soon become the élite unit of the regiment, usually regarded as the leading company and taking pride of place on parades and reviews. Eventually these companies were formed into battalions, particularly in France and Prussia, and from them sprang the Grenadier Regiments, some of which are still in existence. The British Grenadiers are among the most famous and their cap-badge is a stylized representation of the 17th-century flaming grenade.

By the middle of the 19th century the grenade was losing its popularity, and for convenience the smaller types of howitzer shell — 3-pounder and 6-pounder — were used as grenades. The British *Treatise on Ammunition*, 1887 dismissed the grenade fairly summarily, saying that 'They are used chiefly for the defence of places against assault, being thrown among the storming parties in the ditch. They are useful in the defence of houses. They can be thrown by hand for 20 or 30 metres. They are now rarely demanded. When men are using them they

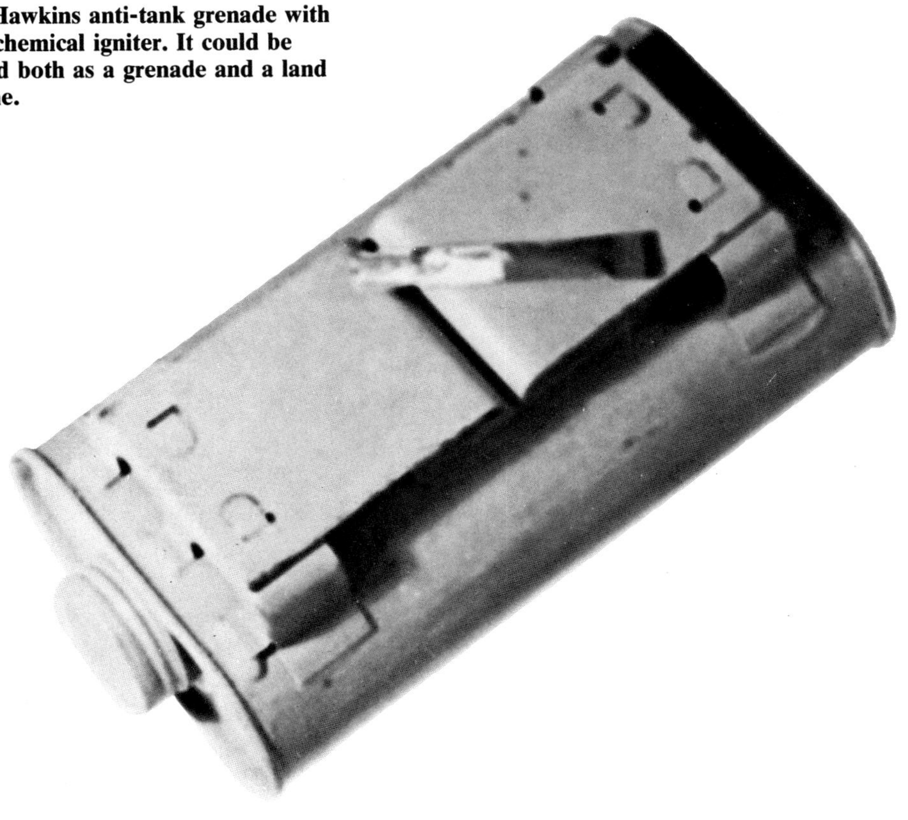

A Hawkins anti-tank grenade with its chemical igniter. It could be used both as a grenade and a land mine.

should be cautioned not to retain the grenade too long in their hands.'

The grenade made a sudden revival in popularity in the course of the Russo-Japanese War, particularly during the siege of Port Arthur. The precise details of the first revival are unknown, but it seems likely that some Japanese soldier in the trenches around the fortress, with time on his hands, passed his spare time one day by filling a stone bottle with gunpowder, wedging a piece of mining fuze into the neck, lighting it and throwing it at the opposite trench. The amusing pastime caught on and very soon both sides were busy manufacturing and throwing home-made grenades, until the Japanese regularized the position by issuing a manufactured grenade, a short length of cast-iron tube filled with blasting powder and fitted with a length of safety fuze. It was little more sophisticated than the home-made variety, but was more satisfactory as a producer of lethal fragments.

World War I, with its trench conditions, led to a massive adoption and use of grenades, every combatant army producing several designs in vast quantities; by the middle of the war British factories were producing more than one million grenades per week. The variations in design stemmed principally from problems of supply; it was not a question of functioning, more a question of what materials were easily available in the vast quantity needed. The war also brought about the distinction between offensive and defensive anti-personnel grenades, saw the first anti-tank grenades, and also saw the first smoke, gas and illuminating designs.

Little development work was done on grenades during the 1920s and 1930s, and World War II began with the combatants using much the same types of grenade with which they had ended the 1914-18 War. The changed nature of operations, with its accent on open rather than trench warfare, led to the development of new types, particularly of hollow-charge anti-tank grenades and blast grenades designed for house clearing.

Since 1945 the principal area of development has been in the control of fragmentation so that the lethal area of the grenade is fairly tightly defined and is thoroughly covered by a large number of small but effective fragments. Among the methods of achieving this are the use of pre-notched wire coils inside the grenade casing and the incorporation of steel balls or pre-cut fragments held in a plastic matrix inside the grenade casing. The casing itself is of thin sheet steel, and the grenade is filled with powerful high explosive such as TNT or TNT/RDX mixtures to give the small fragments a very high velocity and thus considerable wounding power. However, the fragments are so light that they soon lose their velocity and at distances greater than 10 metres are relatively harmless.

HAWKINS GRENADE An unusual anti-tank grenade developed in Britain during World War II; it was unusual in being as much a land-mine as a grenade, and could be used in either role. Indeed, it was better as a mine because it was difficult to throw with any accuracy.

It consisted of a flat, one-pint canister with a screwed stopper on one end, through which it was filled with ¾kg (1¾lb) of Nobel's 704 high explosive. One side of the canister had a tin plate fitment which formed two pockets, and above this was attached a flat pressure plate with a sharp ridge in the middle. Two chemical igniters were inserted into the two pockets, and tabs turned over to keep them in place. The grenade was then thrown or placed so that a tank would run over it. This forced the pressure plate down and the sharp ridge crushed the chemical igniters. These contained a glass ampoule of acid which, when crushed, mixed with a chlorate salt to generate flame which fired a detonator forming part of the igniter assembly. The detonator then fired the main explosive charge, and the blast was sufficient to cut the track of most wartime tanks. It was also useful as a demolition device, being just the right size to fit into the side of a standard railway track; in this role it could be set off by a length of safety fuze and a detonator inserted into one of the pockets.

Introduced as the 'Grenade No.75 Anti-tank' in 1942, it remained in service until 1955.

HEXACHLOROETHANE (HC) Chemical composition (C_2 Cl_6) used as a constituent of smoke-producing mixtures. Usually mixed with zinc dust, zinc oxide and other oxygen-bearing salts to aid combustion, it is pressed into canisters for use in base-ejection smoke projectiles. The smoke generated is cooler than that produced by white phosphorus and therefore there is no heating of the surrounding air and no upward convection to carry the smoke from the ground; HC smokes therefore cling to the ground and give better concealment, though their actual 'obscuring power' or density is rather less than that of WP. Another slight drawback is that HC is difficult to ignite and therefore the canister must be 'primed' with some easily ignited material which will burn with sufficient intensity to ensure that the HC is set off. Three grams (0.11 oz) of HC mixture is sufficient to produce 30 cubic metres (39 cu yds) of smoke.

HIGH EXPLOSIVE An explosive is a substance which, upon being suitably initiated, is capable of exerting a sudden and intense pressure on its surroundings. This pressure is produced by the extremely rapid decomposition of the explosive into gas, the volume of which at ordinary atmospheric pressure and temperature would be many times greater than that of the original explosive. The pressure is greatly intensified by the heat which is liberated during the process. (*Treatise on Ammunition*, 1926).

Explosives are classed as low or high. A low explosive burns extremely rapidly, the flame passing through it at speeds up to 3,000 metres per second (9,840 feet per second). A high explosive detonates, or undergoes molecular disintegration, the shock wave passing through it at speeds as high as 10,000 metres per second (32,800 feet per second) or more. Generally speaking, low explosives can be initiated by flame, whereas high explosives require a shock wave to detonate them.

High explosives are used in grenades and mortar projectiles to provide the bursting charge which shatters the casing to release anti-personnel fragments, and to provide blast which has an anti-*materiel* effect. The first HEs used in these roles, during World War I, were often commercial and mining explosives, selected for their cheapness and availability and so as not to compromise the supply of more refined HEs which were needed for artillery shells and other munitions. These cheap explosives, however, were not well-suited to their task because their rate of detonation was often slow, leading to poor fragmentation of the bomb or grenade casing, and they were highly susceptible to damp, which soon made the munitions ineffectual in the conditions which frequently obtained in the trenches. As the supply situation was critical throughout most of the war, rather than abandon the attractions of cheap explosives the problem was solved by improving the sealing of the grenades and bombs.

During World War II TNT became the most usual filling for hand-grenades and mortar bombs, though British bombs were usually filled with a mixture of TNT and ammonium nitrate known as 'Amatol'. This was selected because it was less powerful than TNT, which tended to blast the cast-iron mortar bomb bodies into fine powder rather than lethal fragments. German grenades often used an explosive known as PETN (penta erythrytol tetranitrate); by itself this was too sensitive for use in munitions, but desensitized with wax it was an efficient and cheap explosive.

Today a mixture of RDX and TNT (known in the USA as 'Composition B') is the near-universal filling for both grenades and mortar bombs. Ample production facilities exist (it is the standard artillery shell filling) and it has regular and consistent performance characteristics to which controlled fragmentation techniques can be carefully matched so as to produce the designed lethal effects.

HOLLOW-CHARGE GRENADE Type of grenade used for the attack of armour or other hard targets; often known simply as an 'anti-tank grenade'. Hollow-charge grenades can be thrown or projected, the latter type being the most common.

The hollow-charge phenomenon (also known variously as the *shaped-charge*, the *Monroe Effect* or the *Neumann Effect*) requires the shaping of a charge of explosive so that its front face has a conical recess. This recess is lined with a cone of copper, steel or other suitable material. The explosive is detonated at its rear end, and as the detonation wave sweeps through the explosive it collapses the liner and melts it. The shape of the cone ensures that the molten material, together with the gaseous products of the detonation, meet to form a jet on the axis of the cavity and are propelled forward at a very high speed — about 10,000 metres per second (32,800 feet per second) — to strike the target. The mass of the jet, together with its speed, defeats the target by sheer force — it does not 'burn through' as is often supposed — and punches a hole through the target. The size of the

The British No. 68 hollow charge rifle grenade, probably the first hollow-charge weapon to enter military service.

The Energa hollow-charge rifle grenade. The clip on the tail holds a discharger cartridge.

hole is roughly that of the jet and is rarely more than about 20cm (8in) in diameter, but the energy is such that a large charge can drive its way through one metre (39in) of homogeneous armour plate. As the jet passes through the target it loses some of its heat to the surrounding material and thus the molten metal in the jet solidifies into a solid slug; this may pass entirely through and become a missile behind the target (eg, inside a tank) or it may lodge in the hole. As the jet completes penetration, it blows off several hundred fine fragments from the inside of the armour and the hot jet passes into the target where it can cause severe damage if it should strike any inflammable material; thus, in a tank, it could strike ammunition or fuel and set the tank on fire. It is also, of course, extremely dangerous to the crew.

In order to allow the penetrative jet to gain velocity and become coherent, the hollow-charge munition should ideally be detonated some distance away from the target; this distance, which is called the 'Stand-off Distance', is roughly two to three times the diameter of the charge.

The first hollow-charge grenade to see service was the British 'Grenade, Rifle, No 68', introduced in the summer of 1940. This had a bell-shaped body with cylindrical walls and four stabilizing fins and was fired from a cup discharger. A simple fuze in the tail carried a firing pin on to a detonator when the grenade struck, and the 156gr (5$^{1}/_{2}$oz) charge was capable of penetrating 50mm (2in) of armour. No hand-thrown hollow-charge grenade was ever developed by Britain, but the German Army had a most effective hand-grenade which carried three magnets around the outside of the front end. This enabled the grenade to be stuck to a tank and allowed time for the thrower to take cover before the time fuze fired the grenade. The Soviet Army developed a hand-thrown grenade with an impact fuze; in order to ensure its arriving head-first it had a number of cloth streamers at the tail end. The US Army developed an efficient rifle-grenade.

After the war a Belgian design, the 'Energa' rifle-grenade, became standard issue in many armies. This was simply a much refined advance on the British 68, longer with a more efficient tail fin unit and a much more efficient explosive warhead. It was capable of penetrating more than 200mm (8in) of armour.

The hollow-charge principle is well-suited to use in grenades for two fundamental reasons: the effect on the target relies entirely upon the explosive carried inside the grenade, so that it is entirely independent of range or velocity —

The sectioned workend of an Energa hollow-charge grenade. The hollow-charge principle is well-suited to grenades as they do not spin and force the jet apart.

provided the grenade hits the target it will do its work. Secondly, it was discovered quite early in the development of hollow-charge munitions that spinning the charge degraded its effect, centrifugal force tending to pull the jet apart. Because grenades are not spun they use the hollow-charge effect in the most efficient way. The only real drawback is their small size, necessary because of their method of projection, which restricts their performance in terms of penetration. This means that they are no longer feasible against main battle tanks, but they still have a very useful place in the infantryman's armoury, giving him the possibility of dealing with armoured personnel carriers, infantry fighting vehicles and similar forms of lightly armoured vehicle which abound on today's battlefield.

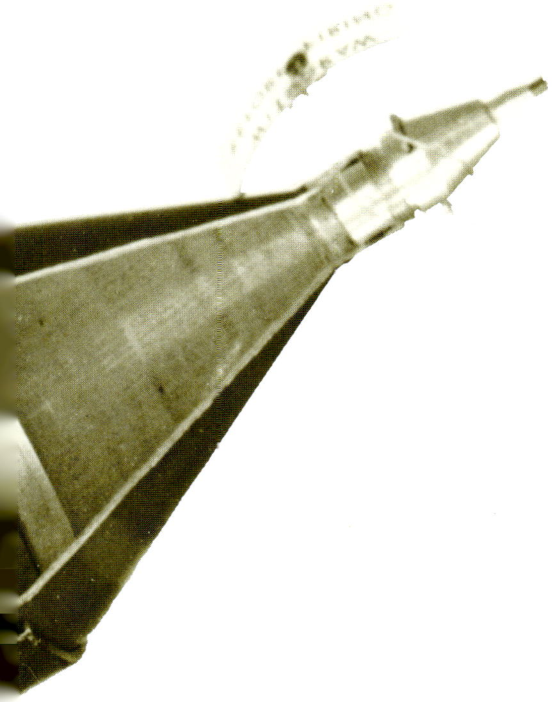

I

ILLUMINATING BOMB Mortar bomb carrying a powerful pyrotechnic flare which is ejected in the air over the target area and which burns to provide illumination for other weapons or so that observers can determine what is happening.

The design of illuminating bombs is almost universal; the bomb contains a canister packed with magnesium flare compound, and this is attached to a parachute. The only major difference between systems is in the method by which the contents of the bomb are released. The most common method is to fit the bomb with a time fuze which, at the end of the selected time, ignites an expelling charge inside the bomb. This, in turn, ignites the flare composition in the canister and then, by the gases evolved during the explosion of the charge, forces the flare canister outwards so that it pushes off either the nose or base of the bomb. This allows the canister to fall free, the parachute unfolds and fills, and the canister floats to the ground, giving off light as it does so.

The alternative, is to have an expelling charge in the base of the bomb, and to ignite a delay fuze by the explosion of the propelling charge. When the fuze burns through it ignites the expelling

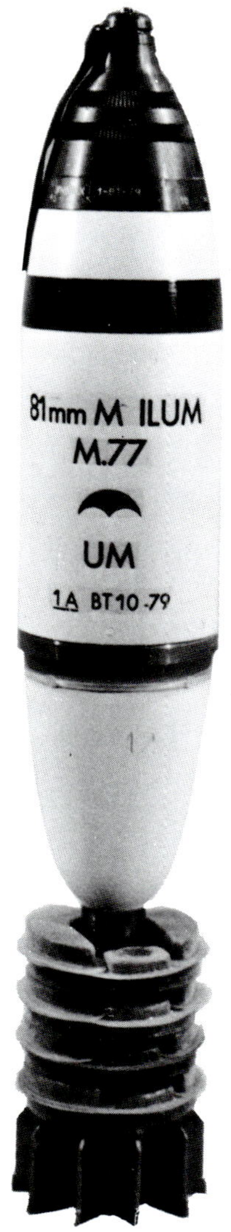

A typical illuminating bomb with time fuze — the 81mm M77 made by Brandt Armements.

A Brandt illuminating bomb sectioned to show the parachute, star unit and primary cartridge.

charge and the action is then as before, blowing the front end of the bomb away and ejecting the flare and parachute. This system has the advantage that there is no need to calculate a fuze length and set the fuze; the delay time is constant, and the bomb can be arranged to burst at different ranges by simply altering the elevation of the mortar. However, altering the elevation while keeping the time constant means altering the height of burst, and the bomb may not always burst at the best height to obtain maximum illumination.

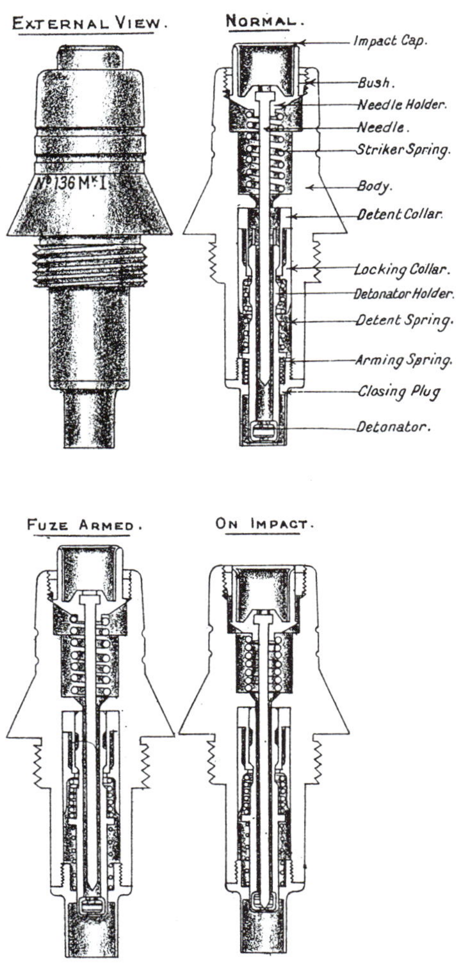

An early, simple impact fuze for mortars. Normally, the needle cannot touch the detonator. On firing, the detent collar slips down, grasps the detonator holder and lifts the detonator to the 'armed' position, ready for impact.

The performance of illuminating bombs varies according to their size, but is fairly constant within calibres; an average 81mm bomb will produce about 700,000 candle-power and will provide this for 50–60 seconds; a 120mm bomb will develop about one million candle-power for about the same time. Smaller bombs, in the 50–60mm bracket, produce about 250,000 candle-power for about 30 seconds; these are frequently used by anti-tank missile and gun crews to fire behind their target in order to silhouette it.

IMPACT FUZE Fuze for grenade or mortar bomb designed to operate the munition upon striking the target. See *Fuze* above.

INCENDIARY GRENADE Hand-grenade for setting fire to targets. They resemble smoke grenades in that they are filled with a solid substance which is ignited by the usual kind of delay fuze. The filling is usually thermite, a mixture of magnetic iron oxide and aluminium powder, which burns at an extremely high temperature, over 2,000°C, melting the metal casing of the grenade and igniting anything it touches and most things within a radius of several feet. Although occasionally used for throwing at targets, incendiary grenades are more usually used as demolition devices in order to destroy some piece of equipment; in the US Army they are carried in artillery gun tool kits to provide a method of destroying the gun, should the need arise.

For casual fire-raising on the battlefield, the standard white phosphorus smoke grenade has a useful incendiary effect, and for this reason special incendiary grenades are rarely encountered.

An American incendiary grenade. The US Army carry them in artillery tool kits, to destroy the gun if necessary.

L

LIVENS PROJECTOR A highly specialized form of mortar developed during World War I for discharging poison gas. Invented by Lieutenant W.H. Livens, it was first used experimentally at Beaumont Hamel in October 1916, and first used in quantity to support the

Loading a gas cylinder into a Livens projector. Notice the wires leading to the electrically-fired cartridges which have already been loaded.

Canadian attack on Vimy Ridge in April 1917; it was later adopted by the US Army and was copied by the Germans. It remained in reserve stocks after the war and quantities were prepared for service in 1939-40, but were never used. It was generally agreed to be the most efficient method of projecting gas against an enemy.

The Livens projector consisted of a 20cm (7.875in) wide barrel about 94cm (37in) long, attached to a flat baseplate. It had no bipod or other support; to emplace it a trench was dug, with the side toward the enemy sloped at 45°, and the barrel was lowered in so that the muzzle just cleared the ground. The earth was then shovelled back to hold the barrel steady. A propelling charge of guncotton was then lowered into the barrel; this was fitted with an electric detonator, and the wire leads were brought out through the muzzle of the projector. The projectile consisted of a cylinder weighing about 27kg (60lb) and containing about 13.6kg (30lb) of liquid chemical agent; down the centre of the cylinder was a tube containing a charge of TNT and a length of safety fuze to give a time delay. On top of this tube was a simple fuze consisting of a firing pin above a detonator, the two being held apart by a spring and a safety pin which was withdrawn when the cylinder was lowered into the projector.

The maximum range of the projector was about 1,660 metres (1,800 yards). Lesser ranges could be achieved by altering the size of the propelling charge, but it was customary to leave the charge alone and obtain the desired range by simply siting the projectors at the correct distance from the target.

Several hundred projectors would be buried in the ground, aligned with the selected target; the propelling charges would be wired up in parallel and connected to a number of electric generators. On being given the signal to fire several men, each with a generator, would press the firing switches to ignite the electric detonators of a group of projectors. The charges would detonate and fling the canisters into the air; as the canisters left the projector, the shock of discharge would cause the firing pin to drop and strike the detonator, so igniting the length of safety fuze. This would burn for 30-35 seconds and then detonate the explosive charge which would burst the cylinder and release the chemical agent.

As each cylinder contained 13.6kg (30lb) of gas, and the projectors were fired by the hundred, and often by the thousand, an immense amount of gas was suddenly released in the target area. The effect was to swamp the area with a lethal concentration so quickly that the victims were unable to put on gas masks or protective clothing before the gas had taken effect. The biggest Livens Projector operation was carried out by the British at St Quentin on 19 March, 1918; 5,649 projectors were fired simultaneously to pour 85 tons of phosgene gas into the German lines, causing 1,100 casualties of whom about 250 died.

M

MILLS BOMB General name applied to five British grenades which were derived from an invention of W. Mills of Birmingham, patented on 16 September, 1915. The original model had a serrated iron body with a central hole in which a spring-loaded striker was held up by a curved lever engaging in its notched head. The lever was of U-section and a safety pin passed through two lugs cast into the grenade body and through two holes in the lever, locking it securely and thus holding the striker against the pressure of its spring. At the bottom of the body was a removable plug which allowed the grenade to be opened and a cap, fuze and detonator assembly inserted so that the cap and fuze went into the central tube, underneath the striker, while the detonator slid into a second tube, to one side, where it was positioned alongside the bursting charge of Baratol high explosive. The plug was then replaced and screwed tight, and the grenade was ready for use.

The thrower gripped the grenade in his hand so as to keep the lever tightly against the grenade body, and withdrew the safety pin. When he threw it, the pressure of the striker spring overcame the lever, which flew off, the striker hit the cap, the cap ignited the fuze, and five seconds later the fuze fired the detonator and the grenade detonated.

Mills offered his design to the British Army in January 1915; development and testing was completed by the end of April, production began in June, and by the end of 1915, 800,000 per week were being made. It was introduced as the 'Grenade Hand No. 5' and by the end of 1917, when production ended, more than 33 million had been made.

In early 1916 it was remodelled, largely for manufacturing convenience but also to adapt it to rifle firing. The body now had larger lugs for the safety pin, a flat steel lever, a striker with a notch on one side only, and a hexagonal base plug with a threaded hole in the middle into which a rifle rod could be screwed. This became the 'Grenade, Hand or Rifle, No. 23' and was introduced early in 1917. By the end of the war more than 29 million had been made.

The next variant was the 'Grenade Hand or Rifle, No. 23M', which was filled with different explosives better suited to storage in hot climates; this grenade was for issue to troops in Mesopotamia, hence the 'M' in its title.

Finally, in the latter part of 1917, some small changes were made to the internal design in order to make production easier: the base plug was given milled edges and threaded for a discharger cup gas-check, and the number was changed to '36'; at the same time, a 'Mesopotamia' version, the 36M, was also

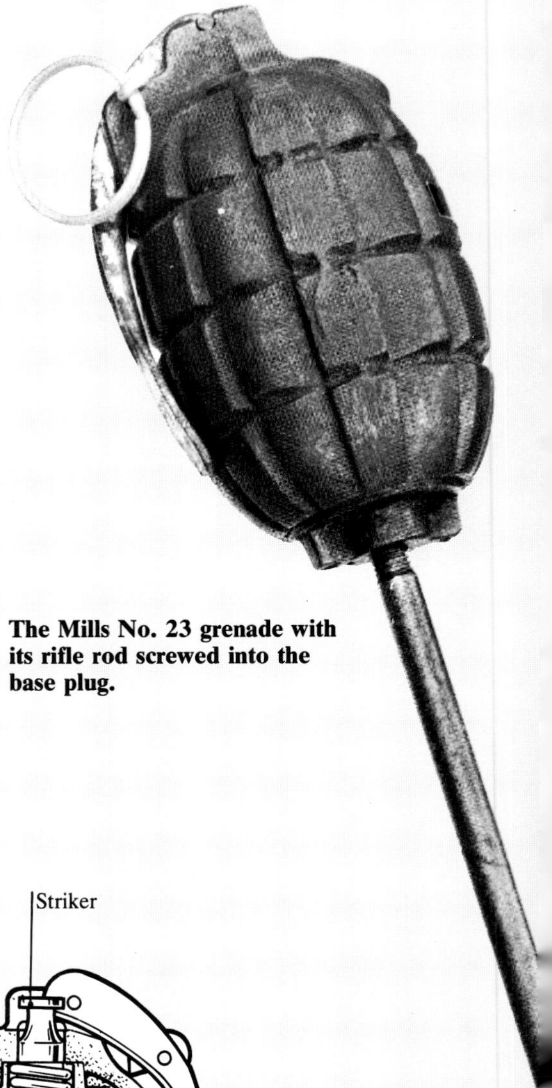

The Mills No. 23 grenade with its rifle rod screwed into the base plug.

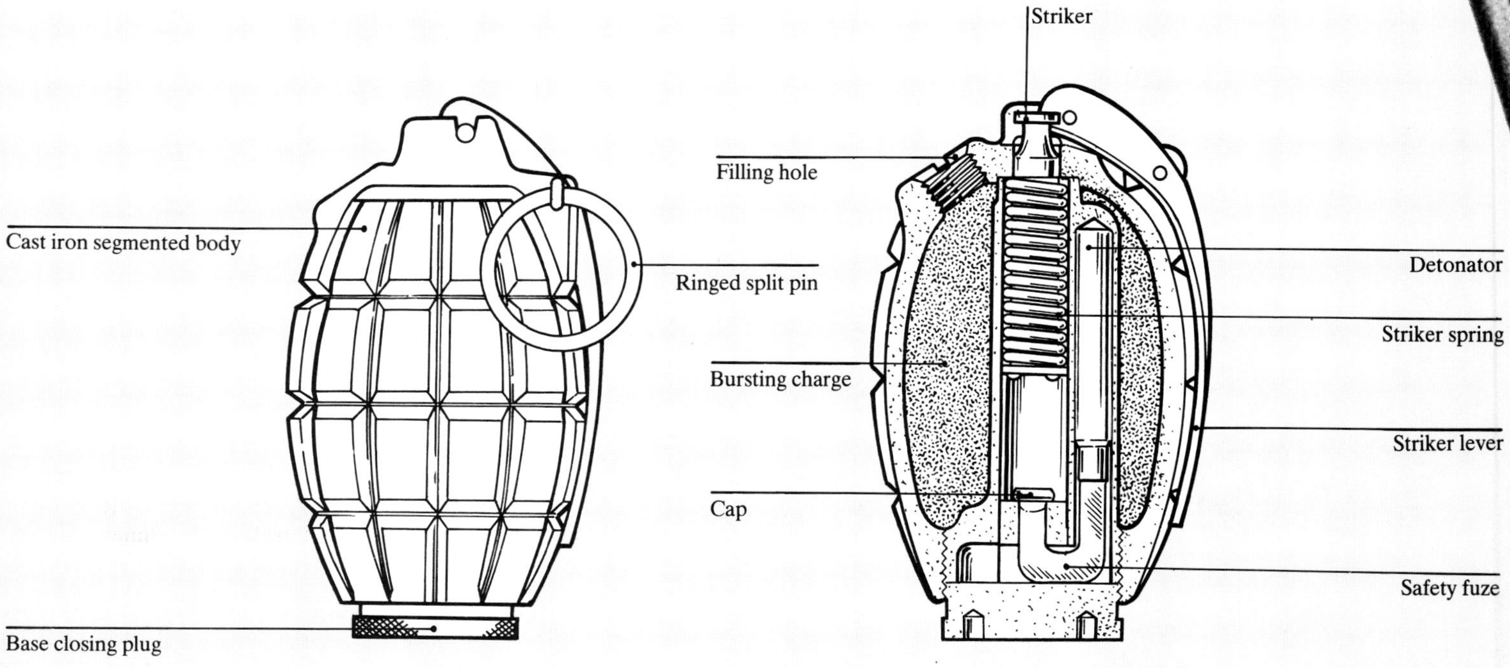

A diagram showing the operating mechanism of No. 5 hand-grenade, better known as a Mills bomb.

developed. A total of more than 5 million 36 and 36M were made before the end of the war.

After the war only the 36 and 36M survived, and the 36 was declared obsolete in 1932, leaving the 36M to continue as the British Army's standard hand- and rifle-grenade. It served throughout World War II and into the 1960s, its use as a rifle-grenade being officially abandoned in 1962, though it had not been used in that role since 1945.

One point about the Mills should be made; it is always supposed that the serrations in the cast-iron body were to assist fragmentation, but in fact the grenade never shattered along those lines. According to Mills' documentation, the idea was actually to allow a better grip with muddy hands.

The Mills No. 36 grenade with the gas check plate which allowed it to be fired from a rifle discharger cup.

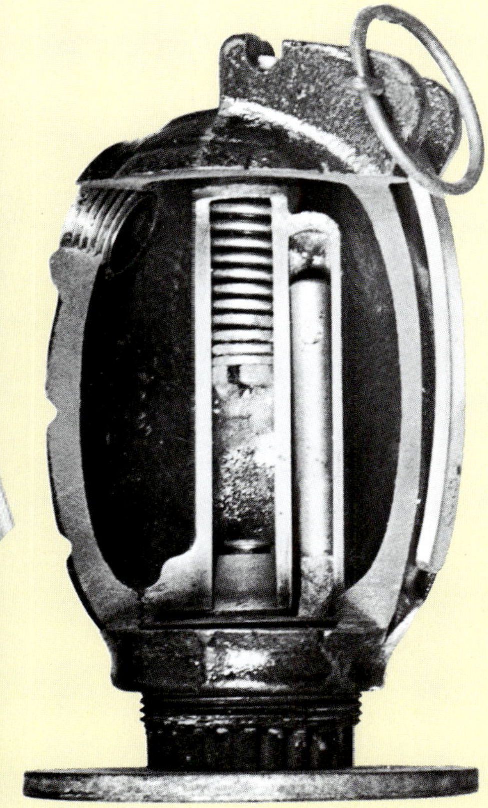

The Dutch mini-grenade showing how the casing was internally grooved to control fragmentation. The size of a golf ball, it could be thrown a considerable distance.

MINI-GRENADE Hand-grenade developed in the 1960s by the Netherlands Weapon & Munition Company (NWM). It was about the size of a golf-ball, had internal serrations in the cast steel body, was filled with RDX/TNT and had a small Bouchon igniter. It could be thrown for a considerable distance, but the fragments were lethal to a radius of only five metres, so that it was an offensive grenade. Despite its efficiency it does not appear to have been adopted by any major army and production ceased in the early 1970s.

MOLOTOV COCKTAIL General term for any home-made incendiary grenade manufactured by filling a glass bottle with petrol and attaching some form of burning fuze. When thrown, the bottle shatters, releasing the petrol, which is then ignited by the burning fuze.

This type of missile appears to have first been developed during the Spanish Civil War. It then re-appeared in Finland in 1939, during the 'Winter War' with Russia, where it acquired the 'Molotov' name. The origin of the name is doubtful, but it was probably coined by a newspaper reporter; at that time a type of multiple incendiary bomb was being dropped by Russian bombers, and this was called the 'Molotov Breadbasket' by the Finns. It is likely that the Molotov Cocktail was named in emulation.

Although the pure 'Molotov Cocktail' is always a home-made affair, there have been some official grenades which have been quite close to it. The British 'Grenade, Hand or Projector, No. 76' was perhaps the most famous; known also as the 'S.I.P. Bomb' (for self-igniting phosphorus) it was a clear glass bottle containing a mixture of white phosphorus, water and benzene together with a strip of crude rubber, 9cm (3½in) long. The rubber dissolved during storage and gave the mixture an adhesive property. The grenade was thrown or fired from a Northover Projector (a 2.5in smoothbore gun) and on impact the bottle broke, releasing the contents. On exposure to the air the white phosphorus immediately ignited and, in turn, ignited the benzene/rubber mixture. It was intended as an anti-tank grenade and was widely issued to Home Guard units from the autumn of 1940 onwards. In 1941 alone, some 5½ million were manufactured and issued.

The US 'Frangible Grenade' was a similar device, a glass bottle filled with either Napalm or with a mixture of gasolene and alcohol. The ignition system differed according to the filling: Napalm-filled bombs used a friction igniter which was pulled before throwing and remained alight so as to ignite the Napalm when the bottle broke, while the gasolene/alcohol filling had a capsule of chemical powder clipped to the bottle neck. On impact the powder mixed with the gasolene and, by chemical reaction, caused it to ignite. So far as is known very few Frangible Grenades were ever issued.

MOUSETRAP IGNITER Common term used to describe a Bouchon igniter, from the snap-over action of the firing pin which resembles the action of a common mousetrap.

The 'Grenade, Hand, No.76' or self-igniting phosphorus bomb.

N

NEWTON GRENADE World War I rifle-grenade invented by an officer named Newton; for obvious reasons it was often referred to by the troops as the 'Newton Pippin' or 'Pippin Grenade'. It consisted of a cast-iron segmented conical body with a flat front and with a rifle rod screwed into the rear. The body was filled with high explosive and had a central tube into which the detonator assembly was inserted. This consisted of a .303in rifle cartridge case, with its cap, into which a standard No. 8 mining detonator was pushed. Over the front end of the grenade fitted a spring cap with a central firing pin, which was held clear of the cartridge case cap by studs on the cap. The grenade was fired from a rifle, using a blank cartridge; when it struck the target the studs on the cap were overcome by the impact so that the firing pin was driven on to the cap. This flashed through and fired the detonator which then fired the explosive filling.

The design was manufactured in Second Army workshops at Hazebrouck, about 2 million being made there in late 1915 and early 1916. Manufacture then began in Britain, but in May 1916 it was suspended as a result of complaints that the grenade was detonating upon being fired. By that time other designs were appearing (notably the No. 23) and the Newton was retired. It was revived in late 1917, with a lighter cap which was less likely to set back on firing and give premature detonation, but it is not thought that many were made at that time.

O

OBTURATION Technical term meaning 'the sealing of a gun against unwanted escapes of propelling gas'. Obturation can refer to the sealing of a gun breech mechanism, or, as in the case of mortars, it can refer to the efficient sealing of the projectile in the bore so that the propelling gas is trapped behind it and put to work. Obturation in mortars is made difficult by the need to be able to drop the bomb down the barrel in order to load it, which necessitates a gap around the bomb to permit the air to escape as the bomb drops down. When the charge is fired,

Typical offensive grenades: the thin casing produces tiny fragments over a limited lethal area. This makes them suitable for use by advancing troops.

some gas will escape through this gap, thus wasting propellant power and also tending to upset the smooth travel of the bomb. Until the 1960s most designs relied upon the cutting of grooves at the largest diameter of the bomb in order that gas, trapped in the grooves, would form a barrier to prevent more gas leaking past, but this was barely efficient. In the 1960s the British 81mm mortar pioneered the use of an expanding plastic ring around the bomb which, when loaded was small enough to leave room for the air to pass, but under gas pressure would expand the seat tightly against the mortar barrel, so sealing the gases and centralizing the bomb. This improvement gave better range and accuracy than had been achieved by bombs without obturation control. The idea has since been adopted by several manufacturers.

OFFENSIVE GRENADE An anti-personnel, high-explosive grenade so designed that its fragmentation and lethal radius are considerably smaller than the distance to which the grenade can be thrown. It can therefore be thrown safely by an advancing soldier without his having to seek cover against the subsequent fragments from the explosion. Compare with *Defensive Grenade.*

The Polyvalent grenade in rifle and hand forms.

P

POLYVALENT GRENADE Specific name given to a grenade manufactured by Losfeld Industries of France, and so called because it can be configured as an offensive hand-grenade, a defensive hand-grenade, or as a rifle-grenade. It consists of three units; the basic offensive grenade which has a plastic body; a fragmentation sleeve which fits over the grenade and converts it to a defensive grenade; and a tail unit which can be attached to the base of the grenade and which permits it to be fired from a rifle. There is a multi-option fuze which can be set for impact action, time delay action (for when fired from the rifle) or impact with delay which allows reliable functioning when fired into soft targets such as sand or mud. Three tail units are provided: one is for use with blank cartridges; one is fitted with a bullet trap for 5.56mm rifles, and one has a bullet trap for use with 7.62mm rifles.

PRACTICE GRENADE A hand- or rifle-grenade used for training. Instead of having a steel casing and a filling of high explosive, a typical practice grenade has a plastic casing and a filling of harmless white powder. There is a very small gunpowder charge which is fired by the normal types of delay fuze; this cracks open the grenade body just enough to release the powder, which marks the point of impact, but without causing the grenade body to fragment or the fuze assembly to fly out and become a dangerous missile.

PRIMARY CARTRIDGE The ignition cartridge component of a mortar propelling charge; so called because it is the 'primary' stage in the operation of the charge. It invariably takes the form of a shotgun cartridge case filled with smokeless powder and inserted into the hollow tail boom of the bomb The tail boom is perforated, and the secondary or augmenting charges are arranged around it. When dropped down the barrel of the mortar, the cap of the primary cartridge strikes the firing pin at the bottom of the tube, and the impact fires the cap and thus the filling of smokeless powder. The flash passes through the perforations in the tail boom and ignites the augmenting charges, and the subsequent explosion lifts the bomb from the barrel. It is possible to fire some mortars with the primary cartridge only, in order to obtain the minimum range zone; the British 81mm, for example, can cover from 180 to 520 metres' range with the primary cartridge only, depending upon the elevation angle of the mortar.

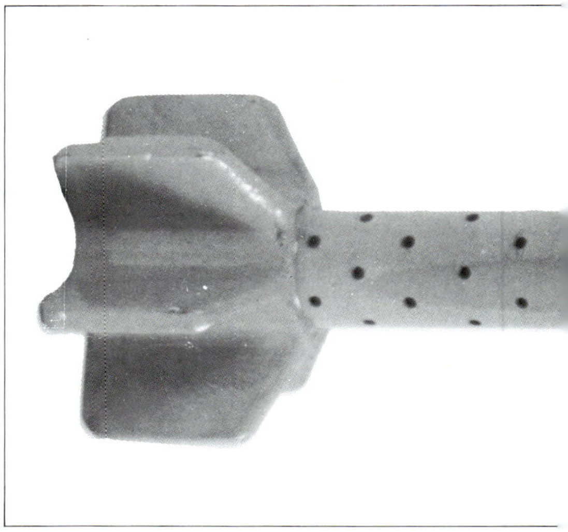

A practice grenade. A tiny charge splits the plastic body and ejects white powder.

PROPELLING CHARGE General term covering the charge of low explosive used to propel a projectile from any type of gun or mortar. In the case of mortars it is composed of two parts, the *primary* and *augmenting* or secondary cartridges. The number of augmenting cartridges can be varied so as to give different range zones; this is necessary because of the restricted range of elevations available with mortars. Each different combination of cartridges is known as a 'charge' and is numbered; the charge table for the British 81mm L16 mortar is as follows:

Charge	Muzzle velocity	Range
Primary	73 mps	180–520m
Charge 1	110 mps	390–1,120m
Charge 2	137 mps	580–1,710m
Charge 3	162 mps	780–2,265
Charge 4	195 mps	1,070–3,080m
Charge 5	224 mps	1,340–3,850m
Charge 6	250 mps	1,700–4,680m

(See *Muzzle Velocity* for fps conversion factor)

A practice bomb for the British 81mm mortar. Like a practice grenade, this type of bomb contains only a very small charge.

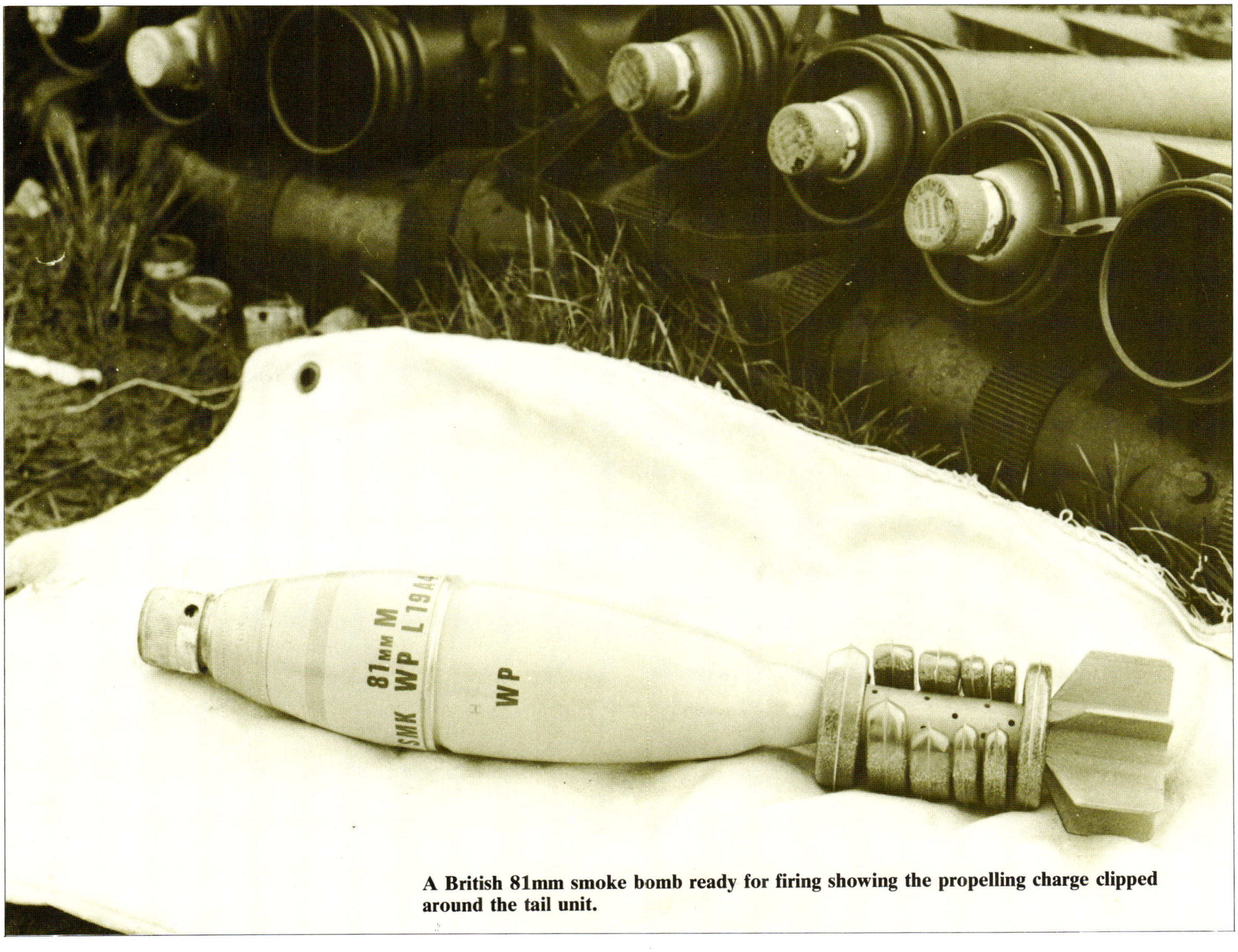

A British 81mm smoke bomb ready for firing showing the propelling charge clipped around the tail unit.

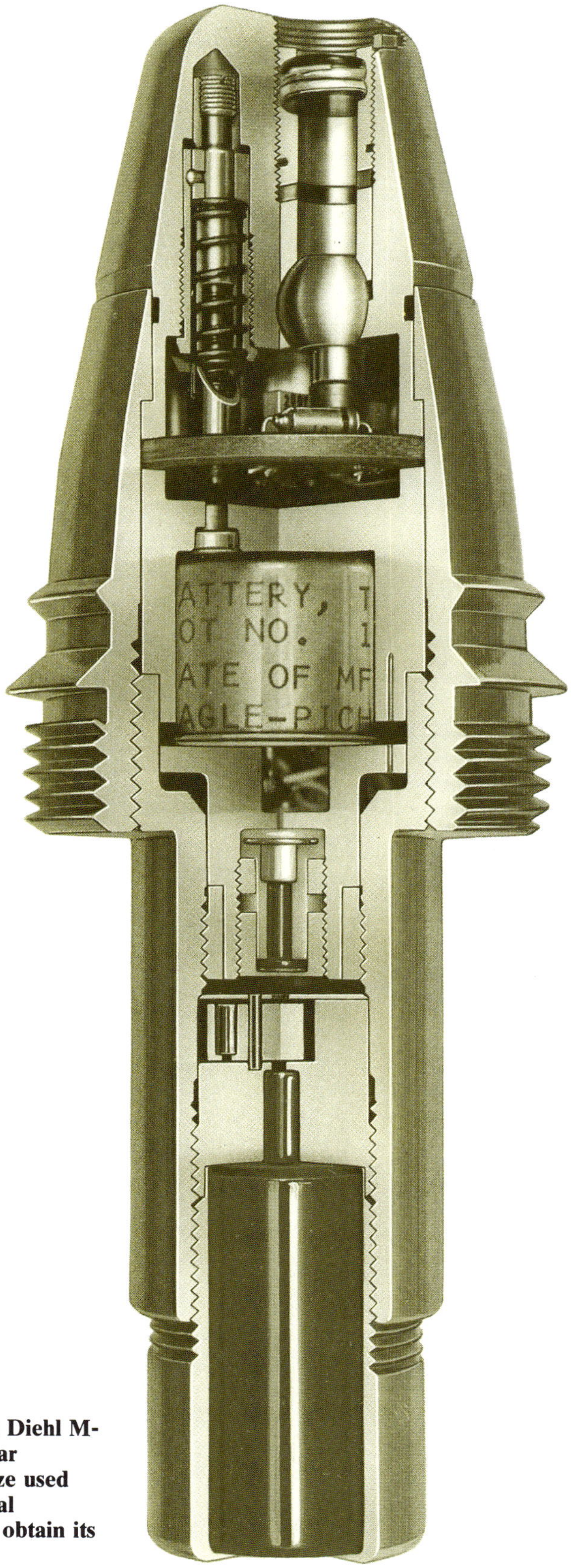

The German Diehl M-DN-14 mortar proximity fuze used electro-optical principles to obtain its firing signal.

PROXIMITY FUZE Type of mortar fuze which does not depend upon impact with the target nor upon being pre-set to a particular time before firing. Instead it relies upon a radio transmitter inside the fuze which sends out a short-range signal. This reflects from the target and is picked up by a receiver circuit in the fuze. The strength of the returning signal is monitored and when this indicates that the reflecting target is within the lethal radius of the bomb, the fuze fires a detonator and explodes the bomb.

Proximity fuzes for mortars are a fairly recent development, although the US Army had one in the late 1940s. The first proximity fuzes were designed for use in guns and, because of the state of electronic technology in the late 1940s, were somewhat bulky; moreover they relied upon a wet battery which, in turn, relied upon the spin of the shell to distribute the electrolyte across the battery plates to generate the necessary electric power. The US Army had developed a proximity bomb fuze for aerial bombs which used a wind-driven generator to provide the electricity (because the bomb did not spin) and these were used experimentally in 4.2in mortar bombs. However, the prime need for mortar fuzes is that they should be cheap, and it was not until micro-miniaturized circuits, micro-chips and similar electronic techniques were perfected that it became possible to make a mortar fuze which was an economic proposition.

The current fuzes are also called 'PPD' for 'Proximity/Point Detonating' since they can be adjusted by simply turning the cap of the fuze to give either impact functioning or proximity functioning. They are still comparatively expensive, but it has been estimated that a serviceable mortar proximity fuze can be manufactured for approximately $30, which is considered to be not unreasonable.

The purpose behind the proximity fuze is to burst the bomb at a height above the target so that the fragments can strike down, searching behind cover and into trenches, which would not be possible with a bomb bursting on impact. The proximity fuze is also necessary to the correct operation of some *smart bombs.*

R

RADAR ECHO BOMB A mortar bomb, usually of the base-ejection type, which carries a payload of finely shredded wire or metal foil; this material is sized to half the wave-length of some particular frequency-range of radar sets, eg, K-Band or S-Band. The bomb is fired and bursts in the air, releasing the cloud of metal

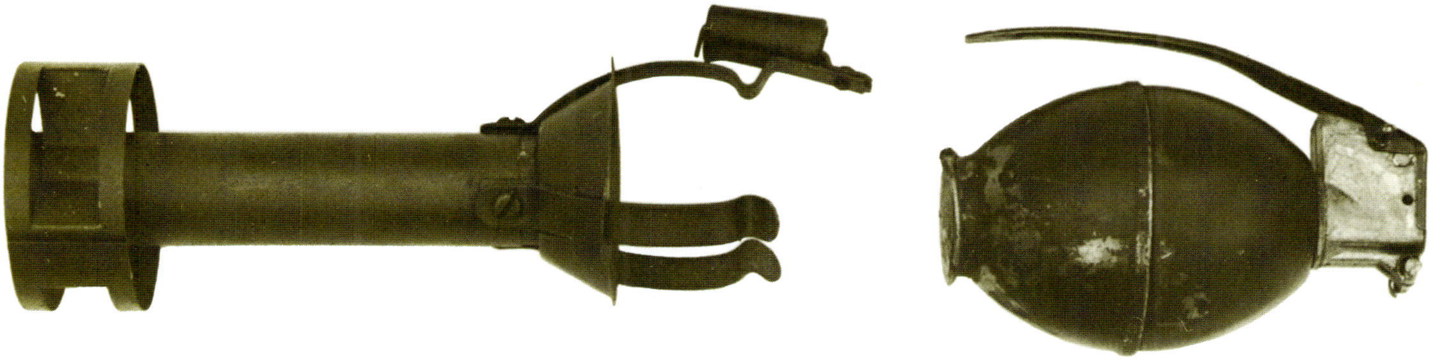

How a hand-grenade can be fitted with an adapter to allow it to be launched from a rifle.

strips which, because of their lightness, take a long time to fall to the ground. While in the air their precise sizing enables them to reflect any radar signals at considerable strength, and thus will totally blind a radar looking in their direction. Their use is thus analogous to a smoke-screen, in that they conceal activity from radar detection.

The radar echo bomb can also be used to locate a mortar in difficult terrain; if the mortar crew do not know exactly where they are but are in radio contact with their headquarters, they can fire a radar echo bomb at some carefully calculated elevation and fuze setting. The burst of the bomb can be pinpointed by a radar set and, the firing data being known, it is possible to calculate the position of the mortar with some degree of accuracy.

RED DEVIL GRENADES Series of hand-grenades used by the Italian Army during World War II. Though made by three different manufacturers they were all to a similar basic design. They were offensive grenades, with thin steel bodies and Allways fuzes, and frequently failed to operate when thrown. Because they were painted a vivid red and their action was uncertain particularly when found in an unexploded condition, they were nicknamed 'Red Devils' by the British troops during the 1941-42 desert campaigns.

RIFLE ADAPTER Metal tail unit which can be clipped to a standard hand-grenade to convert it to a rifle-grenade. It generally consists of a hollow tail boom with fins at one end and a set of claws at the other. The grenade is forced into the claws, where it is held securely. A sliding clip on one of the claws is slipped over the lever of the grenade and the safety pin is withdrawn. The tail unit is then slipped over the muzzle of the rifle and the missile is fired like any other rifle-grenade. As the assembly is launched, the restraining clip slips back and releases the fly-off lever so that the delay fuze is ignited.

The Italian Red Devil, an offensive grenade which frequently failed.

Firing a converted hand grenade from an assault rifle by adding a clip-on adapter.

RIFLE-GRENADE Any grenade launched from a rifle instead of being thrown by hand, the object being to propel the grenade farther than it could be thrown.

The idea of using the rifle (or a musket) to discharge a grenade is a very old one. The School of Infantry Museum at Warminster has a specimen weapon on display called the 'Tinker's Mortar', dating from 1681. It is a flintlock musket with its butt shaped into a cup, with a powder chamber. When required a strut could be unfolded from beneath the barrel and the muzzle of the musket was placed on the ground with the butt in the air. The chamber of the cup was charged with powder and a grenade was placed in the cup. A tap was turned to open a channel from the flintlock pan to the powder chamber, the pan was primed, and the trigger was pulled. The flash from the flint ignited the powder in the pan, and the flame ran to the powder in the chamber to discharge the grenade into the air, igniting the simple fuze as it did so. Other specimens, dating from about the same period, have a cup permanently attached to the muzzle of the musket. The charge of powder was poured into the barrel and the musket was fired using the normal flintlock.

The rifle-grenade concept then languished for many years, and was revived by Martin Hale in 1907 (see *Hale's Grenades*, above). These used rods which were inserted into the barrel of the rifle, and a blank cartridge was used to generate the propulsive power. A later refinement was to put a small copper cup on the base of the rod. This expanded to form a tight fit in the barrel and thus give the necessary obturation for more efficient discharge.

The drawback to the rod grenade was that the sudden pressure check, as the gas met the rod and began to lift the grenade's weight, led to ring-bulging of the rifle barrel, rendering it useless for normal shooting. In addition, the immense strain on the rifle caused by the weight of the grenade and rod, together with the short amount of space in the barrel in which the charge could develop its force, led to the rifles being shaken to pieces. As a result of this the cup discharger reappeared in 1917.

Above: Loading a grenade into the US M203 launcher unit which attaches beneath the M16A1 assault rifle.

Below: German troops using a rifle fitted with a grenade discharger cup.

As developed by the British and French it consisted of a cup or short wide barrel which could be clamped to the muzzle of a service rifle. In the case of the British design the cup was 6.5cm (2.5in) in diameter and was originally developed for use with the Mills No. 36 grenade. A 6.5cm round plate was screwed to the bottom of the grenade to act as a gas check, and the grenade was inserted into the cup, base first. The firing lever was inside the cup, against the inner surface, and the safety pin could be removed. A blank cartridge was placed in the rifle breech and fired, and the gas struck against the gas check plate and lifted the grenade from the cup. As it left, the lever was free to fly off and the striker went down, igniting the delay fuze system. A special 7-second delay fuze assembly was provided for use when the grenade was to be launched from a rifle.

Other grenades were later developed to be fired from the 6.5cm cup, and the system remained in use throughout World War II; it was not officially declared obsolete until 1962, though it is doubtful if discharger cups were ever used after about 1955. They have, though, recently begun to reappear for the purpose of discharging large-diameter, riot-control grenades from rifles or shotguns.

The French design was similar, but the operation of the grenade and the method of launching was much different; see under *VB Grenade* for further details. The German Army developed a small-calibre discharger cup which was internally rifled and fired a spin-stabilized, elongated grenade.

During World War II the US and German Armies, independently of each other, arrived at similar solutions. The grenade was designed with a hollow tail unit, on which stabilizing fins were carried, and the service rifle was provided with a clamp-on launcher, which was simply an extension to the barrel, its external diameter carefully machined to the inside diameter of the grenade tail unit. The rifle was loaded with the usual blank cartridge and the grenade tail was slipped over the barrel extension. This system had the advantage that the discharger unit could remain on the rifle at all times and did not interfere either with its normal use or with the line of sight — the large diameter cup dischargers had to be removed before firing the rifle since they obscured the sight line.

This system has now completely replaced any other system, to the extent that there is now an internationally-agreed standard dimension for the interior of a grenade tail unit and the exterior of a launcher. Today, however, separate launchers are uncommon; modern rifles have their barrels shaped and sized externally to provide two bearing surfaces of 22mm (about 1in) diameter so as to act as grenade launchers without further modification.

Loading a Brandt 120mm rifled mortar.

RIFLED MORTARS The conventional infantry mortar is smoothbored and fires a finned bomb; there are, however, some designs which use rifled barrels and rely upon the spinning of the projectile to stabilize the bomb in flight. For details of these designs see under *Bomb* above. The theoretical drawback to the rifled mortar is that unless the spin is carefully calculated there is a tendency for a spun bomb to retain its nose-up attitude after it has passed the maximum point of height (the 'vortex') of its trajectory, so that it tends to come to earth sideways. This objection seems to have been overcome in the designs in current use.

It is worth noting that the formal definition of a mortar is 'any weapon which fires only at elevations greater than 45°' and that it is common, outside Britain, to use the word mortar when speaking of quite heavy pieces of artillery which fit this definition. Thus the US Army coast artillery used 12in and 16in breech-loading rifled weapons which fired heavy armour-piercing shells against warships and, quite correctly, called them 'mortars'.

ROCKET ASSISTANCE Method of increasing the range of mortar bombs or rifle-grenades by incorporating a rocket drive. The grenade or bomb is fired in the normal way, but

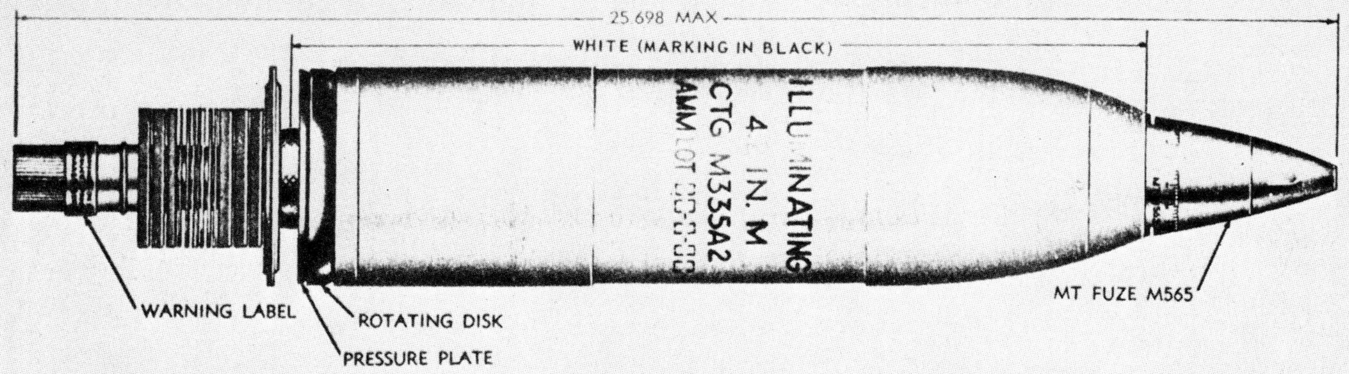

A projectile for the US 4.2in (107mm) rifled mortar. The explosion of the propelling charge drives the pressure plate forward and so expands the 'rotating disk' outwards into the rifling grooves.

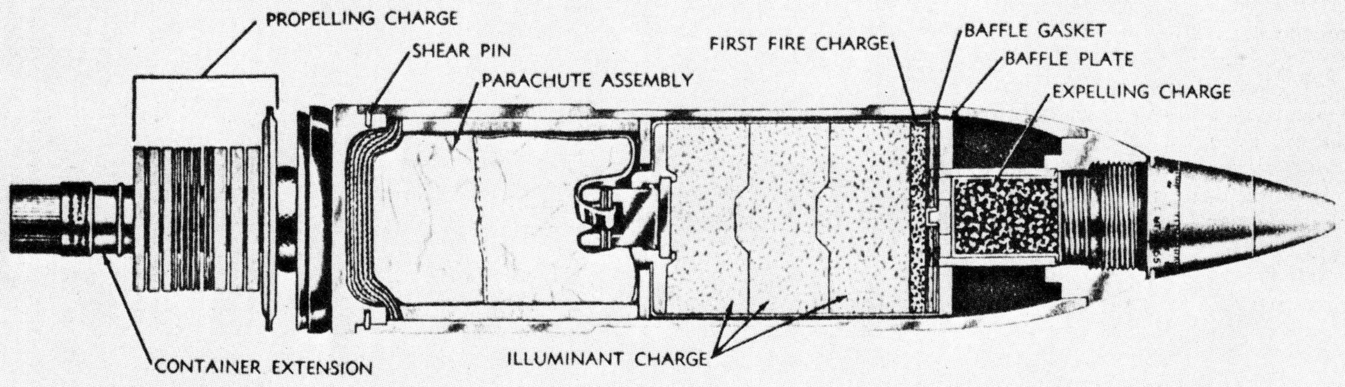

at some point on the trajectory the rocket motor is ignited and gives additional thrust, increasing the velocity and thus increasing the range and/or terminal velocity of the projectile.

The first such device appears to have been a rocket-boosted 'Energa' anti-tank rifle-grenade put forward in the early 1950s. The grenade tail boom was slightly longer than normal and carried a small charge of smokeless powder and a delay unit. The grenade was slipped over the end of the rifle and launched by means of the usual blank cartridges, and the flame from this ignited the delay unit as it launched the grenade. After one or two seconds of flight the delay burned through and ignited the rocket motor which exhausted down the tail boom and gave additional impulse to the grenade. The object was to extend the effective range to about 350–400 metres, but it does not appear to have been adopted by any army.

The Hotchkiss-Brandt company introduced rocket-boosted mortar bombs in the early 1960s for both smoothbore and rifled mortars. In their designs the rocket motor is installed in a central tube inside the body of the bomb, and the rocket venturi is sealed by the bomb tail unit which carries the normal propelling charge. There is a small channel leading from the primary cartridge compartment to a cavity inside the tail boom containing a small charge of powder, and the channel is packed with delay powder. In the case of the smoothbore bomb the tail unit carries the normal type of fin assembly, and there is a second fin assembly attached to the bomb, but folded up and locked in place by the tail unit.

The bombs are fired in the normal way, by dropping them into the mortar; the propelling charge fires and the bomb is launched on its trajectory. At the same time, the primary cartridge ignites the delay element in the tail unit; this burns for about ten seconds and then ignites the charge of powder which explodes, blows off the tail unit and ignites the rocket motor. As the tail unit falls away it unlocks the folded fins, which then spring out to stabilize the bomb during the remainder of the trajectory; these spring-out fins are much wider than the bomb calibre so that they protrude into the airstream for good control. The rocket unit is ignited while the bomb is still on the upward part of its trajectory, so that the height and range are increased. The maximum range of the smoothbore 120mm mortar is increased from 4,250 to 6,500 metres (4,650–7,110yards) by rocket assistance.

There are two drawbacks to rocket assistance. First, the incorporation of the rocket motor into the bomb means a reduction in the quantity of explosive carried, so the size of the rocket motor and the amount of increase in performance, must be balanced against the decrease in effectiveness of the bursting bomb. Secondly, any mis-alignment of the bomb on its trajectory at the time of rocket ignition is carried over to the rocket trajectory; ie if the bomb is yawing slightly to one side, the rocket portion of the trajectory will continue in this direction, and if the bomb is slightly 'nose-up' the bomb will achieve greater range than planned. As a result, rocket-assisted bombs are generally less accurate and consistent than unassisted ones.

A rocket-assisted rifle grenade.

An 81mm mortar bomb showing the plastic sealing ring and propellant charge.

Secondary or augmenting charges in cloth bags around the tail unit of a 120mm bomb.

S

SEALING RING The plastic obturating ring around the centre of a bomb is frequently referred to as the 'sealing ring' because it seals the propelling gas behind the bomb.

SECONDARY CARTRIDGE Another term for 'augmenting cartridge'. The major component of the mortar propelling charge, which is ignited by the primary cartridge and which can be altered in size in order to vary the range zone.

SKYTRAIL BOMB A coloured smoke bomb which ignites at a considerable height and emits smoke as it falls to the ground, leaving a coloured trail in the sky which points to the target. Used to guide air strikes against targets or to indicate a particular target to several observers at different places. Can also be used for signalling or for any other indicating task; eg, marking the boundaries between units during an advance, or marking the edges of an artillery barrage.

SMART BOMB A term which originated in the USA with reference to aircraft bombs which instead of falling freely are guided in their fall by remote control; the term has since been extended to cover other types of projectile, including mortar bombs.

The object in view is to be able to control the latter part of the bomb's trajectory so as to steer it to a specific target, and it is used particularly to guide anti-armour bombs on to tanks. At present there are four Smart bombs under development:

1. STRIX by Bofors of Sweden. This is for 120mm mortars and is a cylindrical projectile fitted with a discarding tail unit. It is fired from the mortar in the normal way and after about 20 seconds of flight the tail unit is thrown off by a delay charge. Folding fins spring out to stabilize the bomb for the remainder of its flight, and it is possible to fit a rocket boost to prolong the trajectory. In the nose of the bomb is an infra-red seeker which detects heat from a target and indicates its direction to an electronic control package. This activates jet thrusters in the afterbody to steer the bomb towards the target. The bomb contains a powerful hollow-charge warhead capable of penetrating the top armour of any main battle tank.

2. The German Diehl-Bussard bomb. This is fired in the usual way, after which wings fold out from the body to act as control surfaces. It

Above: A sectioned view of the STRIX terminally-guided mortar bomb. The shaped-charge warhead is at the rear, with a central channel for the jet, around which is the guidance system.

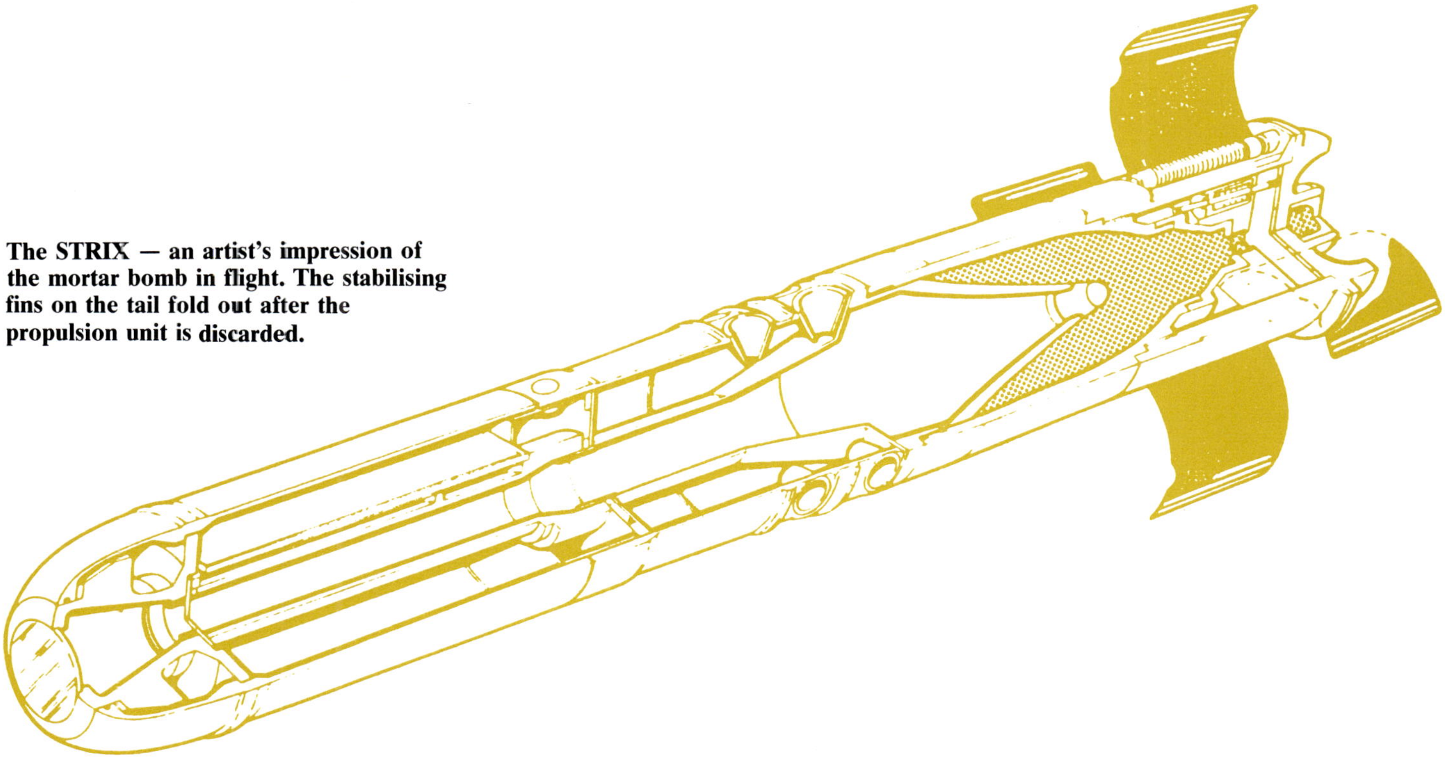

The STRIX — an artist's impression of the mortar bomb in flight. The stabilising fins on the tail fold out after the propulsion unit is discarded.

contains a laser seeker head which reacts to reflections of laser illumination provided by a ground observer; he calls for the bomb to be fired and then aims a laser illuminator at the target. The laser beam is reflected into the air and the bomb picks up this reflection and, by means of the wings, steers towards the target. The warhead is a hollow-charge type.

3. XM 899 GAMP (Guided Anti-Armor Mortar Projectile) being developed in the USA, is another hollow-charge weapon for the top attack of tanks. It is fired from the 4.2in (107mm) rifled mortar and folds out a set of wings shortly after leaving the barrel. Two forms of guidance are under development; an infra-red detector by Raytheon and a millimetric-wave radar seeker by Martin Marietta. A third option, under development by General Dynamics, is an infra-red device which scans in a circular pattern while the bomb's flight is slowed down by parachute. The US Army hope to have selected one system, completed development and placed the chosen munition in production by 1989.

4. MERLIN, developed by Marconi in Britain, is an 81mm bomb for the standard British mortar. It uses a millimetric-wave radar seeker which controls a set of spring-out fins to steer the bomb to the target.

All these devices reflect the concern felt over the considerable imbalance in armour of the Warsaw Pact armies compared to the NATO armies, and they are all intended to attack tanks from the top, where the armour is thinnest, giving the infantry an anti-armour capability at greater range than currently can be achieved by wire-guided missiles.

SMOKE BOMB A mortar bomb designed to emit smoke for signalling or screening purposes. Screening smoke is always white as this gives the greatest obscuring power, while signalling smoke can be of any desired colour, though the most usual are red, green and yellow; blue smoke is generally available, but it is not so easily distinguished at long range as are the other colours.

White smoke can be produced by releasing white phosphorus (WP), titanium tetrachloride (FM), chlorsulphonic acid mixture (FS) or other chemicals into the air where they combine with the water vapour to produce smoke. Or a composition of zinc and hexachloroethane (HC) can be ignited and burned to provide a white smoke. Coloured smokes are usually mixtures of chlorate salts and sugar with dyes of the requisite colour and have to be burned so that the smoke carries the dye particles into the air.

Smoke bombs are of two types, bursting or base-ejection, and their operation is detailed under those headings.

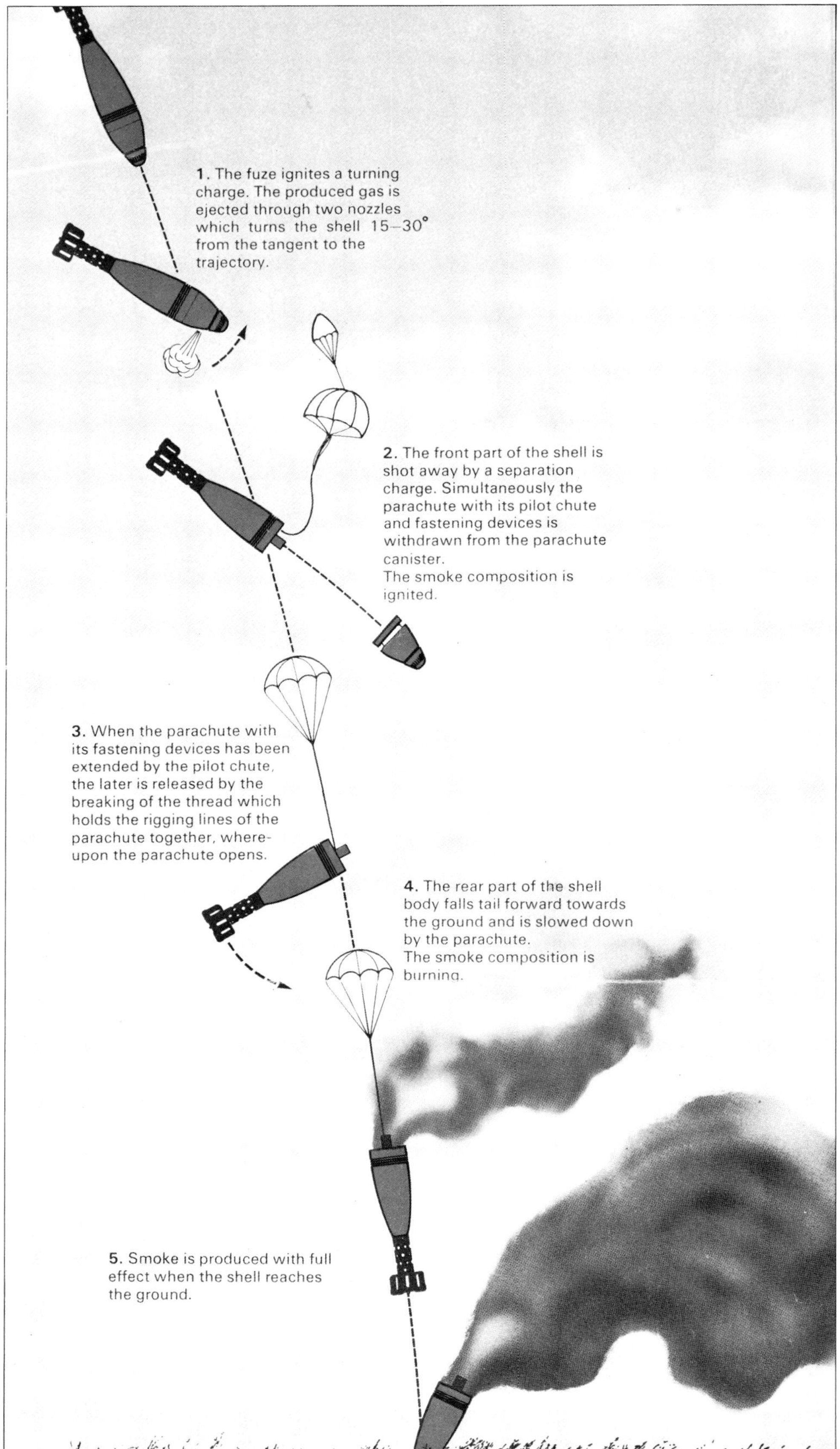

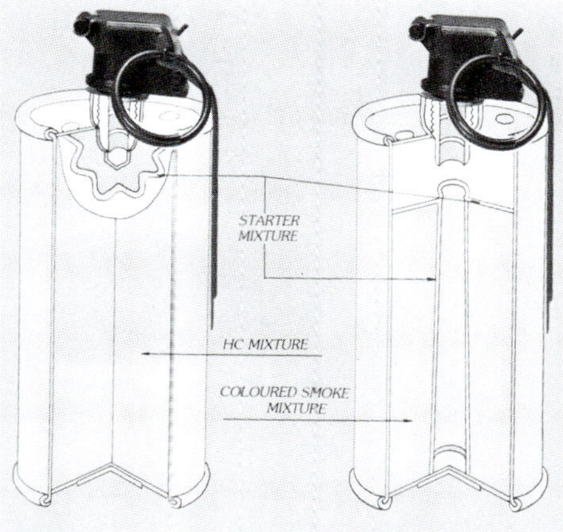

Above: Sections of two smoke grenades; left, white screening smoke and, right, coloured signalling smoke.

Left: An unusual design of nose ejecting bomb from Sweden. The smoke composition is in the rear section of the bomb, and a parachute ensures that it lands in the correct position to allow free emission of the smoke.

Right: Typical coloured signal smoke grenades.

SMOKE GRENADE A smoke grenade, like a smoke bomb, can be used to produce white smoke for screening or coloured smoke for signalling, and the chemical compositions used are the same.

Screening smoke grenades may use WP or HC; other compositions are rarely found today. When filled with WP the grenade requires a small explosive charge to split the casing open and allow the WP to escape and burn. When filled with HC mixture the fuze will light an igniter which, in turn, lights the smoke mixture which is then allowed to burn inside the grenade body, the smoke passing out through a series of holes in the casing. The method of fuzing smoke grenades is exactly the same as that used for anti-personnel explosive grenades, time fuzes being standard. Coloured smoke grenades operate in the same way as HC screening smoke types, burning the composition inside the grenade body and allowing it to escape from holes.

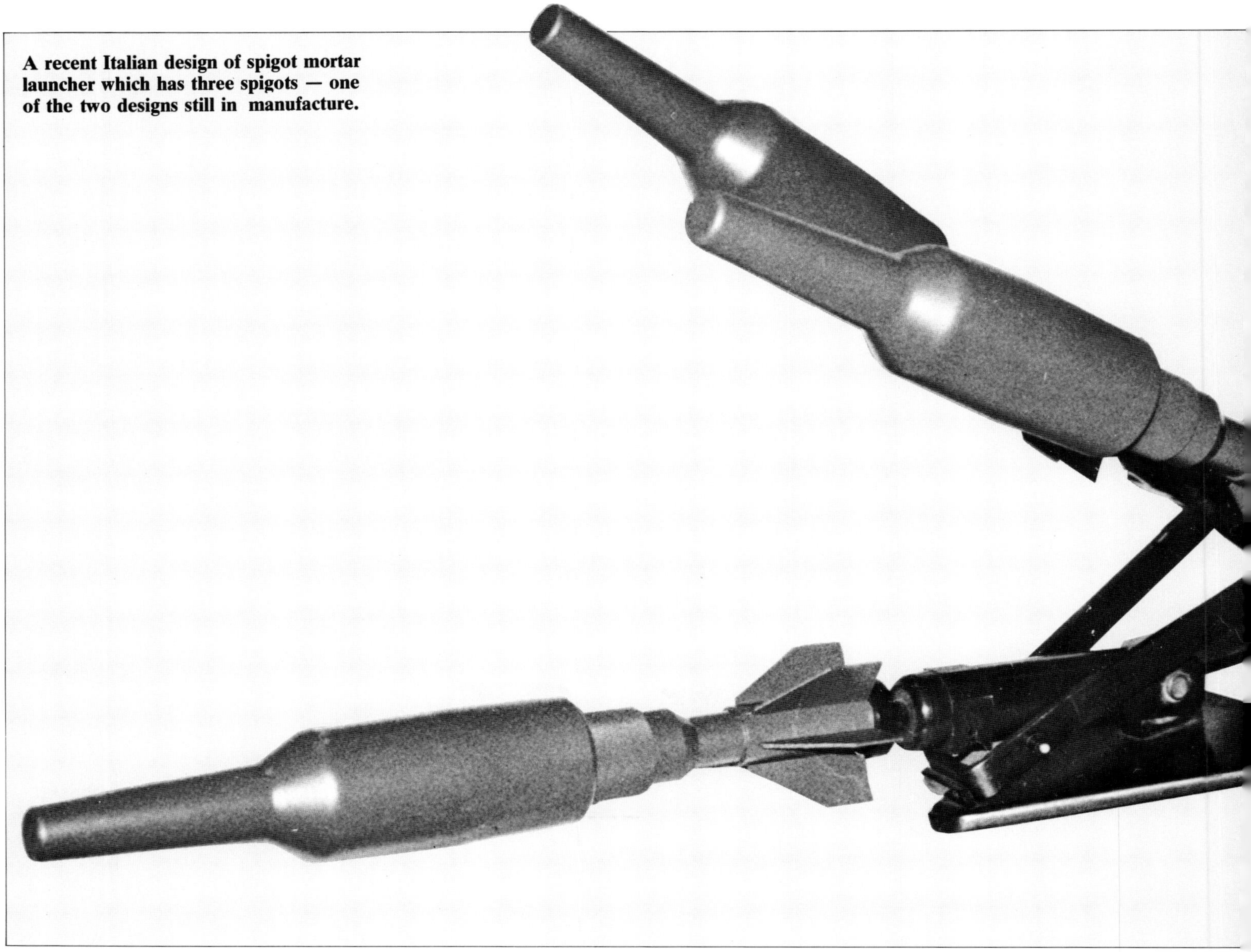

A recent Italian design of spigot mortar launcher which has three spigots — one of the two designs still in manufacture.

SPIGOT MORTARS Mortars which, in effect, reverse the normal method of discharging a bomb. Instead of having a hollow barrel and firing the bomb from inside it, they use a solid steel rod (the 'Spigot') as the 'barrel' and the bomb has a hollow tail which slides over the rod. The propelling cartridge is inside the tail of the bomb, so that when it is fired (by a firing pin concealed inside the spigot) the explosion blows the bomb off the spigot and into the air. Direction and elevation are given solely by the short travel of the bomb leaving the spigot.

Examples of spigot mortars include the British *Blacker Bombard* used by the Home Guard during World War II; the Japanese 'barrage mortar' of the same era which fired a massive 306kg (675lb) bomb to about 1,000 metres range; and a light German design which fired a 27 kg (59.52 lb) bomb to a range of 750 metres (820 yards). At the present the only spigot mortars in existence are the Belgian 'FLY-K' mortar, described above, and the Italian AV/700 three-spigot multiple launcher.

The principal advantage of the spigot mortar is that the weapon itself is cheap and easy to manufacture, since it lacks a barrel, the most difficult part of the conventional mortar to make. On the other hand the ammunition is only slightly more complicated to make than a conventional bomb.

STICK GRENADE Any grenade, having a metal head with explosive or other content, attached to a stick for ease of throwing. Often specifically used to refer to the standard German grenade of the two world wars, when this was the most common stick grenade.

The first 'modern' grenade ever issued to the British Army was a stick grenade; the Grenade Hand No. 1 appeared in 1908 and consisted of a tubular brass body with a cast-iron segmented ring around it, attached to a 39.64cm (16in) cane handle which had a 91.44cm (36in) canvas streamer attached. The object of the handle and streamer was to stabilize the grenade in flight so that it landed nose-first; the fuze was a very simple impact arrangement. In the flat head of the grenade body was a hole, which ran centrally into the explosive filling. Into this hole a standard mining detonator was inserted and a loose cap was then placed over the head; this cap had a fixed firing pin centrally mounted. When the nose of the grenade hit the ground, the cap was driven in, the pin struck the detonator, and the grenade exploded.

When this grenade was put to its first serious use in the trenches in 1915 the major design defect appeared: when the thrower took hold of the cane handle and swung the grenade back to throw it, he usually struck the head against the

Below: A German stick grenade with the cap removed and the pull-cord for the friction igniter exposed. The marking reads: 'Before using, insert a detonator'. Bottom: One of the few stick grenades still in service is this Hungarian M42 model.

rear wall of the trench and blew himself (and his companions) up. The Grenade Hand No. 2 was therefore developed, having a cane handle 17.78cm (7in) long, but this was little more popular and as soon as other designs appeared the stick grenades were abandoned.

The German stick grenade was smaller and far more efficient. It consisted of a steel canister attached to the top of a short hollow handle. The canister was filled with explosive, and into the bottom a detonator was inserted. This was connected by a short length of delay fuze to a friction igniter. The friction igniter works on the same principle as a common match; a jagged steel peg is embedded into a sensitive material, and when this peg is suddenly pulled out, the friction ignites the composition. On the German stick grenade the friction igniter was operated by a length of string, which had a porcelain bead attached, and which was concealed inside the handle and retained by a screw cap at the bottom of the handle. To use the grenade the screw cap was removed, allowing the porcelain bead to drop out and reveal the string. The thrower now took the handle in his throwing hand and the bead and string in the other, so that as he drew back his arm to throw, he instinctively pulled on the string and so lit the friction igniter. He then threw the grenade; after four or five seconds the delay burned through and the grenade exploded.

Similar designs were used by the Chinese, Japanese and other armies, and the Germans had variant models with impact fuzes, automatic-igniting systems, and with heads filled with smoke mixtures. The only countries using stick grenades today are China and Hungary.

STICKY BOMB Term used to describe any hand-grenade which is coated with adhesive material in order to make it stick to its target. Specifically applied to the British Grenade No. 74, developed early in 1940 by a clandestine Intelligence department. The original intention was to use them as sabotage devices on barges on the Danube, but this project fell through and the grenade was then offered to the Army, but refused as being too dangerous. It eventually did appear in service, being issued mainly to the Home Guard, though it is believed a few Army units had them. There is no record of its having been used in combat.

The 74 Grenade consisted of a short handle screwed into a spherical glass flask. The handle contained a striker and release lever, similar in principle to the fuzing system of the Mills grenade, and the flask was filled with 6 hectograms (20 ounces) of pure nitro-glycerine — which was why the Army originally refused it. The exterior of the flask was covered in cloth impregnated with a very tenacious adhesive, and two hemispheres of sheet metal were then clamped around the flask to prevent the adhesive surface sticking to anything while the grenade was in store.

To use, the thrower pulled a first safety pin which allowed the hemispheres to spring open and fall off, exposing the adhesive surface; he then removed a second safety pin from the handle, holding down the firing lever as he did so. He then threw the grenade or, preferably, ran up and jammed it against the target, in either case so that the adhesive held the grenade tightly against the surface of the target. As the grenade was thrown, or as the handle was released, the lever flew off and initiated a 5-second delay, after which the nitro-glycerine detonated. It was intended as an anti-tank grenade, and there is no doubt that this quantity of nitro-glycerine in contact with any wartime tank would have put it out of action.

There were two drawbacks; first, nitro-glycerine is a highly sensitive explosive and could well detonate from a sharp blow when handled; secondly there was a constant danger that when throwing the grenade the sticky surface might come into contact with the thrower's clothing and stick to it. To put it mildly, the Sticky Bomb was not popular.

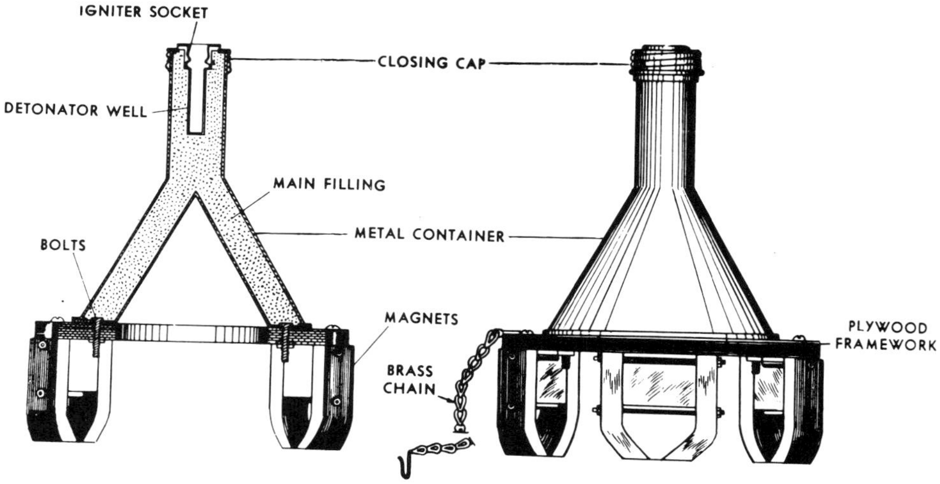

The German equivalent of the British Sticky Bomb used three magnets and a hollow charge.

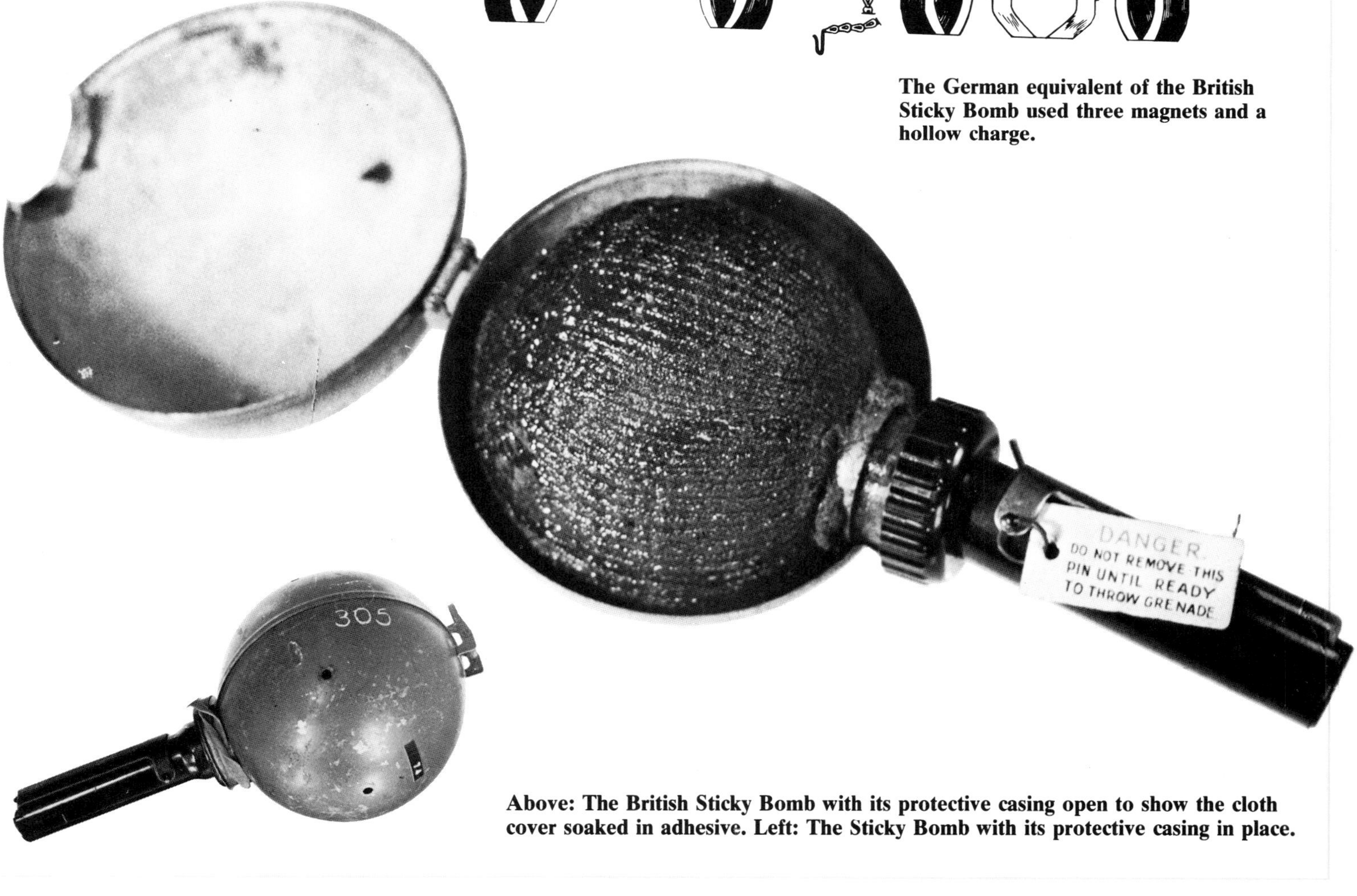

Above: The British Sticky Bomb with its protective casing open to show the cloth cover soaked in adhesive. Left: The Sticky Bomb with its protective casing in place.

The Stokes 4in mortar together with its un-stabilized bomb of World War I vintage.

STOKES BOMB Projectile used with the original Stokes trench mortar of 1915, though the term was subsequently used loosely to describe any mortar bomb. The Stokes bomb was cylindrical and had a tubular cartridge container at one end and a fuze at the other. The cartridge container held the primary cartridge, and the augmenting charges were wrapped around it. The fuze was a tube protruding from the flat nose of the bomb, which held a striker, cap, fuze and detonator resembling the fuze unit of the Mills grenade. The striker was held by a

The original un-stabilized Stokes bomb for the 3in trench mortar, with its early Allways fuze.

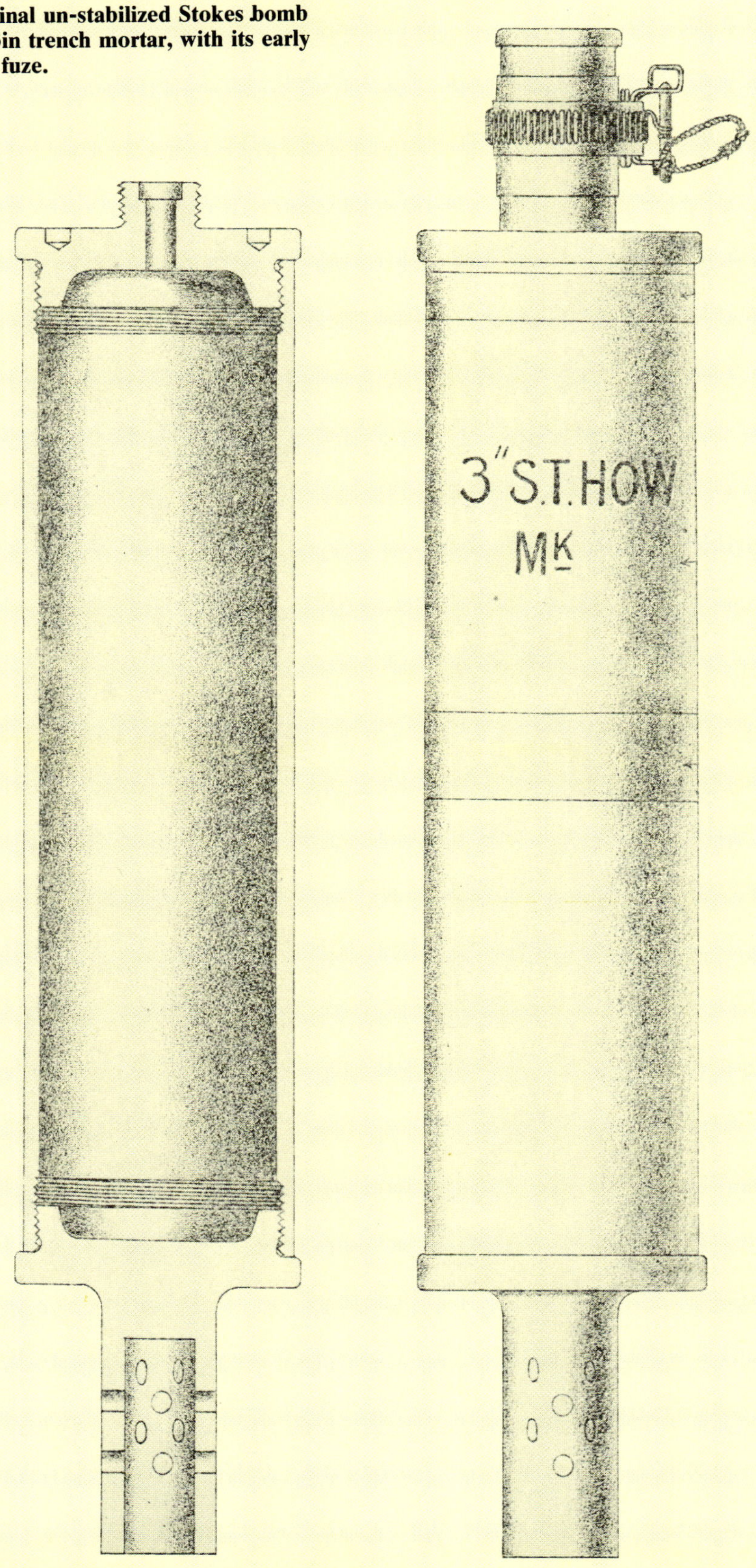

fly-off lever locked by a collar which ran around the fuze body and the lever.

The bomb was dropped base-first into the mortar and slid down the barrel; the cap of the primary cartridge struck the fixed firing pin, the propelling charge exploded, and the bomb was launched from the barrel. The shock of acceleration caused the locking ring to slip down and release the fly-off lever which, however, struck against the inside of the mortar barrel and did not therefore release the striker until the bomb actually left the muzzle. At the end of the delay set by the fuze (about 25 seconds) the bomb exploded.

The principal drawback to the Stokes design was that there was no stabilization and thus the bomb turned over and over as it flew through the air; it was for this reason that a time fuze was first used, though an 'Allways' impact fuze was later developed. Stokes did design a finned bomb, but it was considered too difficult to manufacture at the time and since the cylindrical bomb appeared to be perfectly satisfactory — and was a good deal more accurate than might be expected — it continued in use until 1918.

STUN GRENADE A hand-grenade which explodes with a vivid flash and loud report, but which does not discharge any fragments and is non-lethal. It is used in situations where it is desired to shock the target personnel without injuring them, and is specifically intended for use against terrorists holding hostages. The grenade will temporarily disorient everyone in the room, allowing the rescue party to enter and take the terrorists, leaving the hostages unharmed (though shaken). The precise nature of the filling is a trade secret (these grenades are all manufactured by commercial companies), but it is based on known firework compositions.

An American stun grenade which has a rubber body to avoid fragments.

SUB-CALIBRE TRAINING DEVICES A conversion system which allows the firing of very small projectiles from a mortar in order to enable training and practice to be conducted in restricted areas or within barracks where it would be impossible to fire a service mortar.

The systems in use are broadly similar, involving the use of dummy full-sized bombs which have a central hole. Into this hole a small 25mm calibre bomb and propelling charge is loaded. The entire assembly is then handled just like a full-sized service bomb and is drop-loaded into the mortar. The propelling charge fires when it strikes the firing pin; the small bomb is launched from the tube in the dummy bomb and follows a conventional trajectory to a maximum range of about 500 metres. It has a small powder charge in its head which gives a puff of smoke on impact in order that the point of fall can be easily seen. The large dummy bomb is also ejected by the propelling charge, but this falls about ten metres in front of the mortar and can be picked up and reloaded for further use. By varying the charge, ranges as short as 75 metres can be achieved, and the trajectory is accurate and consistent so that training in ranging and the observation of fire can be carried out. Such devices are manufactured in Germany by Dynamit-Nobel and Nico-Pyrotechnik and in Spain by Esperanza y Cia.

Some examples of Spanish sub-calibre training bombs (right) and the sub-projectile and propelling charge (left).

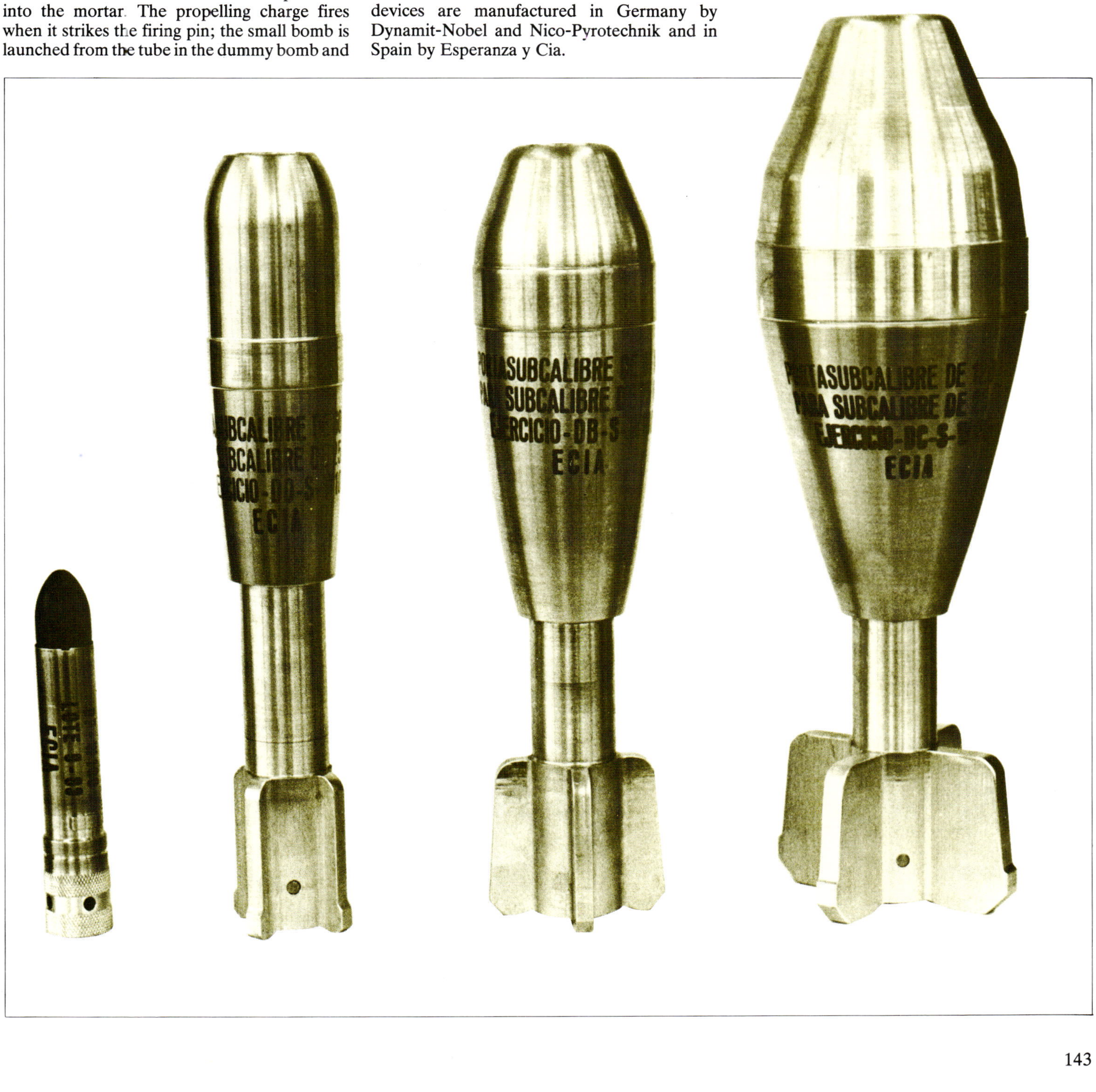

A modern tail unit of extruded alloy with multiple flash holes to ensure adequate ignition of the augmenting cartridges.

T

TAIL UNIT The rear section of a mortar bomb comprising the tubular tail boom, the fins, and the cartridge-holding assemblies. Tail units were originally put together in a rather rudimentary fashion, for cheapness and speed of manufacture, but experience and tests showed that a more refined design and construction would improve accuracy. Today's tail units are frequently made of solid metal, extruded and machined to the required shape, rather than stamped fins spot-welded to the tail tube.

TEAR GAS GRENADE Hand- or rifle-grenade containing a filling which generates a gas or smoke with lachrymatory properties, causing sneezing, tears and nausea in the victims. Used entirely for riot control.

The fillings used today are either CS or DM compositions; these are both solid materials which are burned to produce a smoke which conveys the active agent. CS is the more mild of the two, DM being more likely to cause nausea and vomiting. The grenade construction is quite similar to a smoke grenade, using the normal type of Bouchon igniter with delay and permitting the smoke to leave the grenade body through a series of holes.

Unfortunately this simple grenade is easily picked up and thrown back at the sender, and a number of more sophisticated designs have appeared. These burst open and scatter the chemical agent so that it is impossible to gather it up and return it; other designs have a small explosive charge with an additional delay so that shortly after landing and beginning to burn the grenade explodes, scattering the contents and injuring anyone attempting to pick it up. Another pattern ejects a number of burning pellets of chemical agent in various directions — yet another way of preventing the material being thrown back.

Poison gas grenades were proposed during World War I, but were not used in any number because the short range of the grenade rendered the thrower likely to be caught by his own gas. Tear gas grenades, though, have been used since 1912 when the French police first adopted them for dealing with criminals. They were extensively used as offensive grenades during World War I, but since then have been confined entirely to riot control.

TIME FUZE Any fuze which can be set by the operator to function after a specific period of time has elapsed. In mortar ammunition, time

A typical tear gas grenade.

An illuminating bomb for the British 81mm mortar.

fuzes are found only on illuminating bombs, and are so arranged as to cause the flare to eject at the optimum point so that it will have the maximum fall and illuminating time. Time fuzes may be operated by a train of burning gunpowder or by a clockwork mechanism; the former is most usual, because great precision of timing is not required, and the expense of a clockwork fuze is rarely justifiable in this role.

A typical powder-burning fuze has a rotatable ring which contains a filling of compressed powder extending around the ring for less than the complete circle; this moves below a second, fixed ring which also contains a filling of powder. On setting the fuze, the lower ring is moved until an index mark on its outer surface registers against the desired time on a scale. When the bomb is fired, a detonator lights the top powder ring at some point along its length; it then burns to the end and ignites the lower ring, again at some point along its length. This burns until its end and then fires the expelling charge in the bomb. The turning of the lower ring adjusts the length of powder which must be burned between ignition of the top ring and ignition of the expelling charge.

The French Vivien-Bessiere grenade. Note the striker, which is in the path of the bullet as it emerges from the central channel.

V

VB GRENADE French rifle-grenade adopted during World War I and which remained in use with the French and US Armies until the late 1930s. VB stands for Vivien-Bessiere, the inventor, and the grenade was quite unique.

The VB grenade was cylindrical and had a hole running through its axis. To one side of this hole was a well containing a detonator and the usual length of delay fuze at the top of which was a percussion cap with its sensitive end facing towards the central hole. Riveted to the side of the hole was a leaf spring with a small firing pin, aligned with a cap on the fuze assembly.

The grenade was launched from a cup discharger which fitted on the muzzle of the French Lebel service rifle. The soldier loaded a ball cartridge into the chamber of the rifle, inserted the grenade, and fired; the bullet went up the rifle barrel and into the cup discharger, where it then passed through the hole in the grenade and out into the air. As it passed through the grenade, it struck the leaf spring and drove it sideways so that the firing pin struck the percussion cap and ignited the delay fuze. The propelling gas behind the bullet lifted the grenade and launched it from the cup. The grenade had a range of about 150 metres.

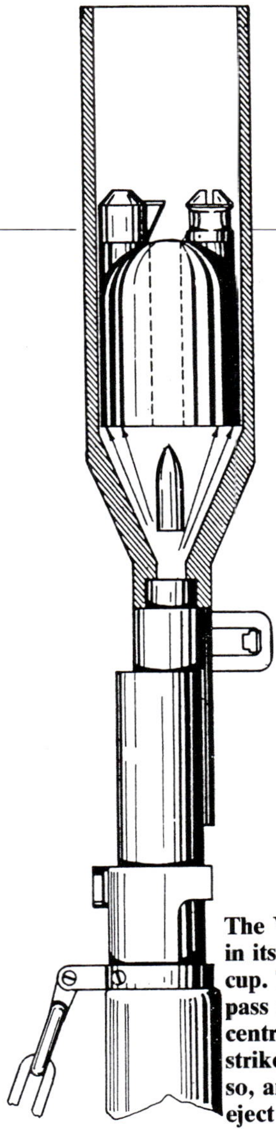

The VB Grenade in its discharger cup. The bullet will pass through the centre, hitting the striker as it does so, and the gas will eject the grenade.

The only drawbacks to the VB grenade were that the grenade had to be quite precisely manufactured in order to ensure that the hole lined up with the rifle barrel, and practice in its use demanded a very large area of ground because of the flying bullet, which left the rifle at an elevation of about 45° and thus could travel for three or four miles before reaching the ground, where it still had sufficient energy to be lethal.

W

WINDAGE The difference in diameter between the exterior of the mortar bomb and the interior of the mortar barrel. A small space is necessary so that air can escape as the bomb slides down the barrel when loading; insufficient space means a slow escape of air, and thus the bomb not falling with sufficient force to fire the primary cartridge. Too much space means that the propellant gases will escape and the bomb will wobble from side to side as it goes up the bore, leading to inaccuracy. The modern development of expanding obturating rings has made this rather less of a problem.

A British 105mm light gun in action. Note the loaders holding a shell and cartridge, and the breech operator with a rammer in his hands.

ARTILLERY

A

AIRBURST Type of artillery fire in which the shell is caused to burst in the air above the target so that fragments strike down and thus reach behind protection where more direct attack would not reach. Performed with *high-explosive* shells, using *time* or *proximity* fuzes.

ALUMINIZED EXPLOSIVES High explosives to which a quantity of powdered aluminium has been added. This increases the severity of the blast and raises the temperature of the detonation, so that a projectile filled with aluminized explosive is much more damaging to matériel and has enhanced incendiary effect. First introduced in mines and torpedoes for the increased blast benefit, their principal artillery application is to small-calibre *mine shells* for anti-aircraft shooting so as to have the greatest damage potential and an incendiary effect against fuel.

AMATOL High explosive made by mixing ammonium nitrate into molten TNT and allowing it to cool. The nitrate, being an oxygen carrier, improves the efficiency of combustion of the TNT though it reduces the rate of detonation. The principal purpose in adopting amatol as a shell-filling explosive is economy, since ammonium nitrate is a great deal cheaper than TNT and is used in proportions as high as 80 per cent of the mixture. Widely used during World War II as an artillery shell filling and to fill the 1-ton warhead of the V1 missile, it is less common today; its wartime use was largely based on the concurrent use of low-grade steel for shell bodies. When this type of steel is used, amatol gives satisfactory fragmentation, whereas RDX/TNT and similar more powerful explosives blow the shell casing into fine fragments which are less effective. Modern shell designs are built around the use of higher-grade steels which allow thinner shell walls and thus a higher capacity for explosive, and filling with amatol would give poor fragmentation.

AP Abbreviation for 'Armour Piercing'; see below.

APC Abbreviation for 'Armour Piercing, Capped', meaning an armour-piercing projectile which is fitted with a soft steel cap over the point. The principle was discovered in the late 1880s by the Russian Admiral Makarov and is intended to prevent the sudden shock of impact breaking the relatively brittle tip of the shot or shell. The cap rests on the shoulders of the projectile and is internally recessed so that it does not bear on the tip. On striking, the cap takes the initial shock and transmits it to the shoulders of the projectile, relieving the tip of stress. The cap then deforms and acts as a support and, to some extent, as a lubricant to the projectile as it begins to penetrate the armour. Capped shot is particularly effective against face-hardened armour.

APCBC Abbreviation for 'Armour Piercing, Capped, Ballistic Capped'. The cap of an APC shot (see above) is designed to assist penetration, and its exterior shape is not compatible with optimum flight characteristics. As a result, it became common to place a second cap on top of the piercing cap; this was the 'ballistic cap' and was no more than a light metal cover of pointed form such as would give the projectile the best head shape for flight. It is so light that on impact it collapses immediately and makes no difference to the subsequent action of the piercing cap, but its use means that the projectile arrives at the target with a greater velocity and thus has a better chance of success.

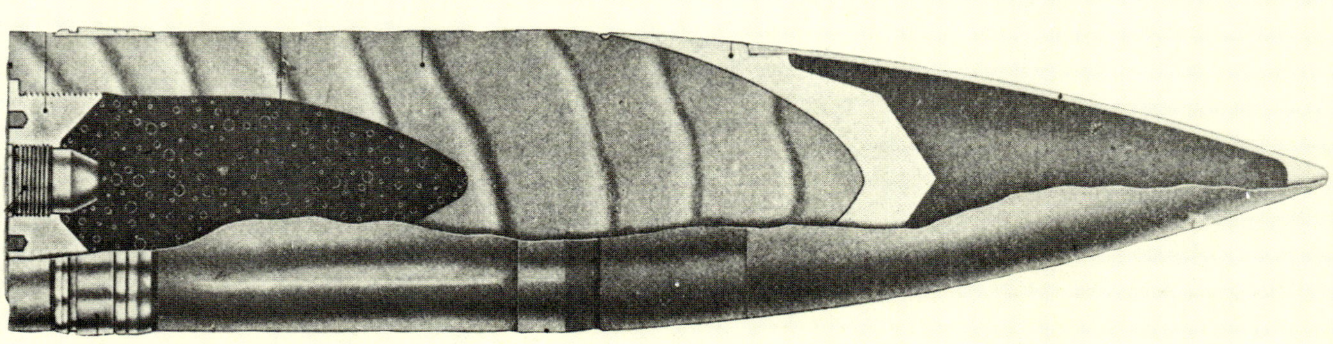

An American APCBC shell for a coastal defence gun, showing the explosive content, fuze and arrangement of piercing and ballistic caps.

A variety of armour-piercing projectiles. Opposite: An AP shot with its caps. Left: Cut-away APCR showing its solid core. Below: An APDS with its sabot intact and cut away to reveal the sub-projectile.

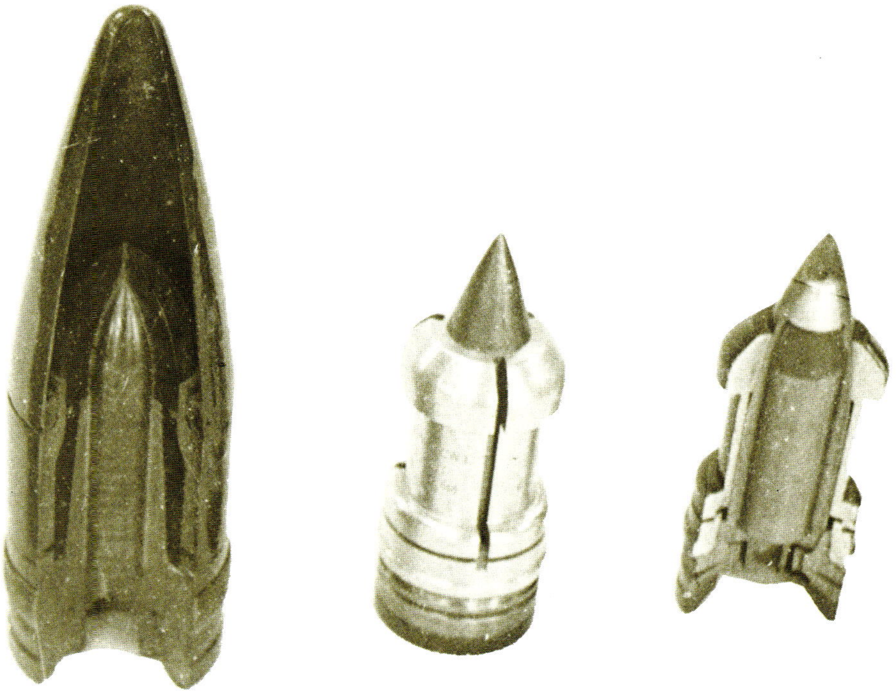

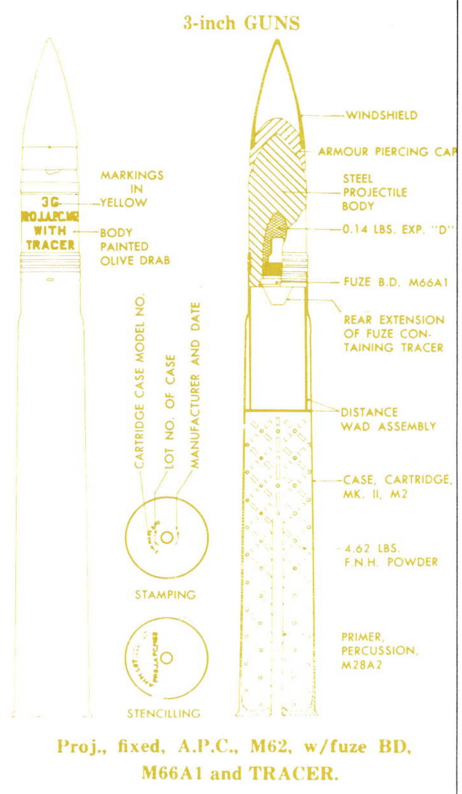

The construction of a typical APC shot for use in an anti-tank gun.

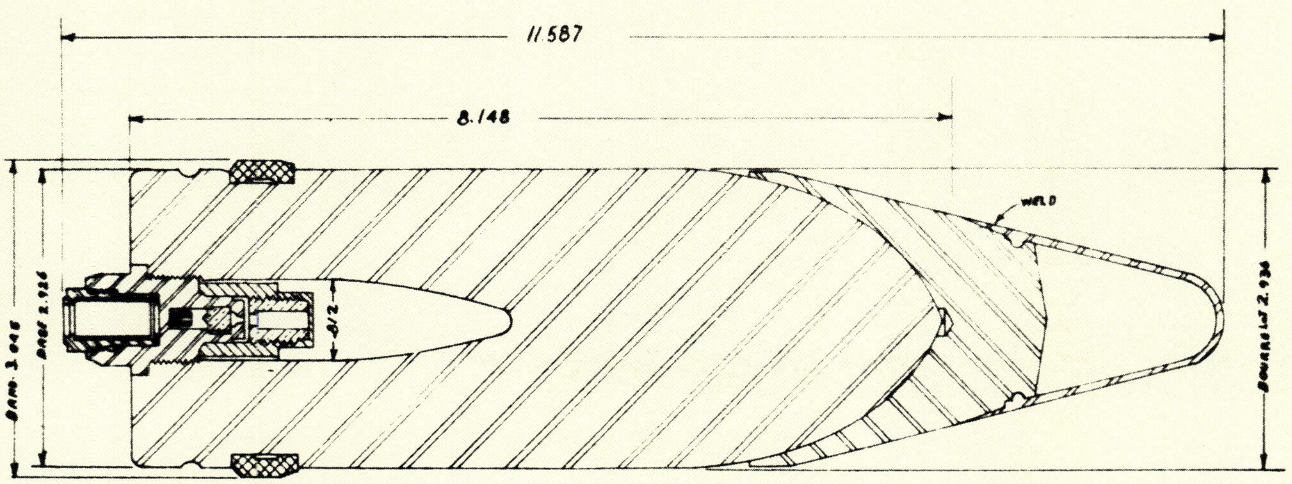

A typical APCBC shell showing the poor ballistic shape of the piercing cap and the better shape of the ballistic cap which covers it.

A British APCNR shot for a two-pounder anti-tank gun with the adapter, which was screwed into the gun muzzle to squeeze the shot down during its passage. Note that the shot in the muzzle of the adapter is much slimmer than the unfired shot alongside.

A typical armour piercing composite rigid shot: 1. Ballistic cap. 2. Alloy body. 3. Tungsten core. 4. Driving band. 5. Tracer. 6. Alloy base.

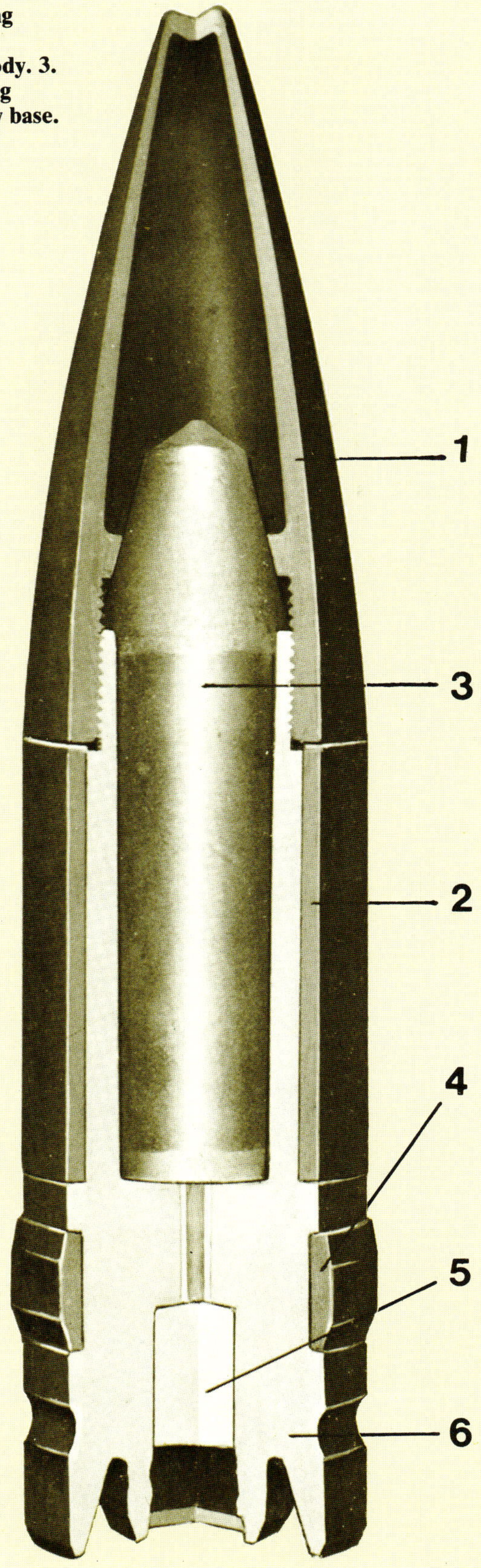

APCNR Abbreviation for 'Armour Piercing, Composite, Non-Rigid', a British term indicating an armour-piercing shot using a tungsten carbide penetrative core surrounded by a light alloy body of full calibre size which was designed to be used with a gun having a bore tapering in calibre. The light alloy surround had, therefore, to be capable of diminishing in diameter as it passed down the bore. The idea was first used by the German designer Gerlich in sporting rifles in the 1920s, and it was then adopted by the German Army for a number of anti-tank weapons. The British adopted a design developed by Janáček, a Czechoslovakian refugee and weapons designer, in which the gun barrel was conventional and the taper was added by a bolt-on adapter. The object is to develop high velocity: first by making the shot of lesser diameter in relation to its weight once outside the gun, and secondly by developing greater unit pressure on the base during its travel because while the gas pressure remains constant, the area of the base is decreasing. The principle was successfully applied, but the idea was dropped in favour of *discarding sabot* ammunition which does not involve making a tapering gun barrel.

APCR Abbreviation for 'Armour Piercing, Composite, Rigid', a British term signifying an armour-piercing shot which uses a core of dense and strong material (usually tungsten carbide) surrounded by a light alloy body to bring the projectile up to the requisite bore size. The object is to have a projectile strong enough to penetrate armour, but light enough to attain the high velocity necessary to achieve penetration. Known in US service as 'HVAP' for High Velocity Armor Piercing, in German service as 'AP40'.

APDS Abbreviation for 'Armour Piercing, Discarding Sabot', a type of armour-piercing projectile in which a heavy metal sub-projectile is surrounded by a light alloy sabot of bore diameter. The sabot is designed so that it splits open and leaves the sub-projectile just outside the muzzle of the gun.

APDS was developed in Britain by Permutter and Coppock in 1941-43 and was first introduced for the 6 pdr anti-tank gun in 1943. The object is to reconcile two conflicting requirements. In order to pierce armour successfully it is necessary to strike at high velocity; steel shot striking at velocities higher than about 854 metres per second (2,800fps) shatters instead of piercing, so tungsten carbide was adopted as the penetrative agent. This, being more dense than steel, cannot be used as a full-calibre shot, since it would be too heavy to be able to reach a worthwhile velocity. By using a sub-projectile inside a light alloy casing (as in APCR — see above) the weight of the complete projectile is less than the weight of a steel shot, and thus the propelling charge will give the composite shot a greater muzzle velocity.

Outside the muzzle, however, a light shot of full diameter has poor 'carrying power' and will soon lose velocity; for flight purposes it is better to have a small and heavy shot. The APDS shot reconciles these two demands by having a light shot of full diameter in the bore, to give maximum velocity, and then, after discarding the sabot, having a small-calibre shot of high density outside the bore for good flight characteristics.

The penetrative core is now often made from depleted uranium; this has no radioactive properties, but is dense and strong, giving good penetration performance.

A sectioned APDS shot for the 120mm tank gun used on the British Chieftain tank, showing the tungsten penetrator surrounded by the light alloy sabot. This sabot splits open and leaves the sub-projectile just outside the muzzle of the gun.

APFSDS Abbreviation for 'Armour Piercing, Fin Stabilized, Discarding Sabot', a type of APDS shot which uses a finned dart-like sub-projectile and was originally intended to be fired from smoothbore guns. It was found to have superior performance, as a result of compressing all its momentum into a very small area, and its use has been extended to rifled guns. In order to obtain the benefits of fin stabilization it is necessary to fit the sabot with loose 'driving bands' which transmit relatively little spin to the sub-projectile.

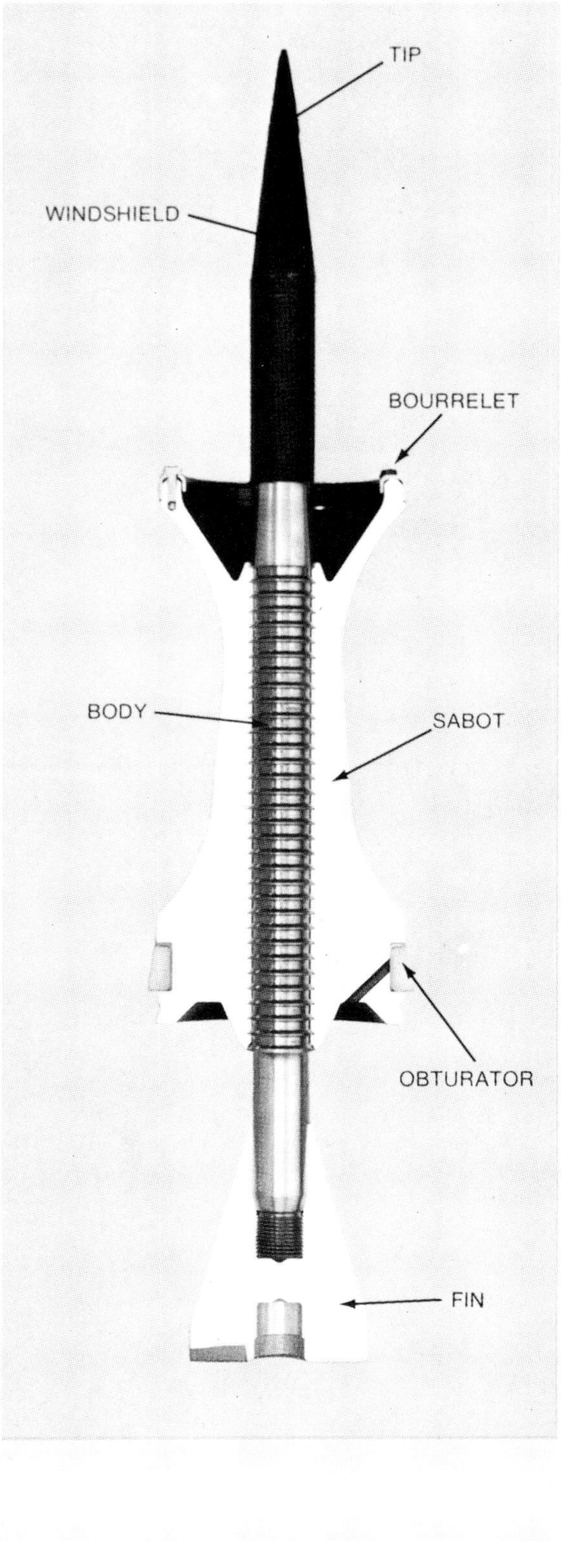

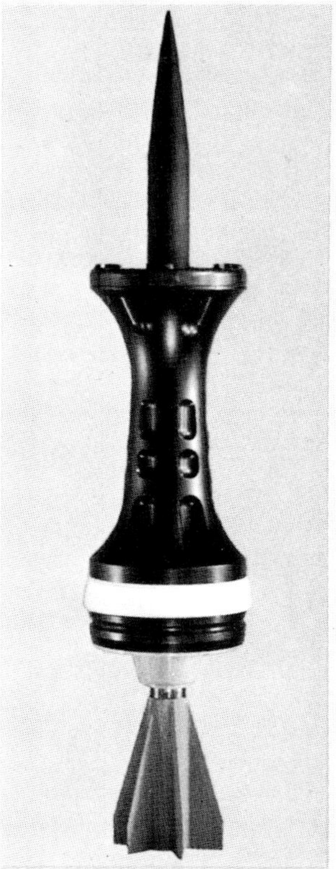

An American 105mm XM833 APFSDS shot, intact and cut away to show the method of construction.

Left: A French APFSDS round for the 105mm gun on the AMX-30 tank. It is known in France as 'Munition Fleche'.

A selection of armour-piercing ammunition of British, German, American and Russian manufacture.

APSV Abbreviation for 'Armour Piercing, Super Velocity', another term for the type of shot described under *APCR*.

ARMOUR PIERCING Any projectile designed for the penetration of hard armour. May be solid shot, composite shot (with a hard core and light alloy surround), discarding sabot shot, or piercing shell, which has a small explosive charge intended to detonate after penetrating the armour.

ARROWHEAD SHOT General name for *APCR* shot in which the centre section is cut away to reduce weight, thus leaving the projectile with a sharp tip resembling the point of an arrow. First developed by the Germans during World War II, the design is still used in some older Soviet ammunition.

A 76mm arrowhead shot for use with the Soviet Army's outdated 76mm regimental gun.

B

BACK-FIRING TARGET SHELL A type of shell developed in Britain during World War II to act as a target for light anti-aircraft weapons. The projectile was a carrier shell which contained a small rocket; at the selected height a time fuze blew off the base of the shell and ignited the rocket, which was then launched back towards the gun which fired the shell. This rocket was then to be shot at by other, lighter, weapons for training. Used only by the Royal Navy for practice with shipboard armament.

A Bofors 25.4cm AP shell, after piercing 2.5cm of Krupp cemented armour in 1932.

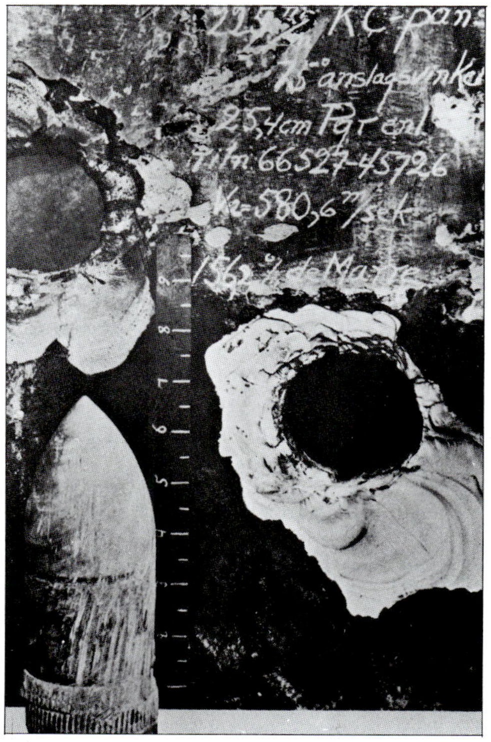

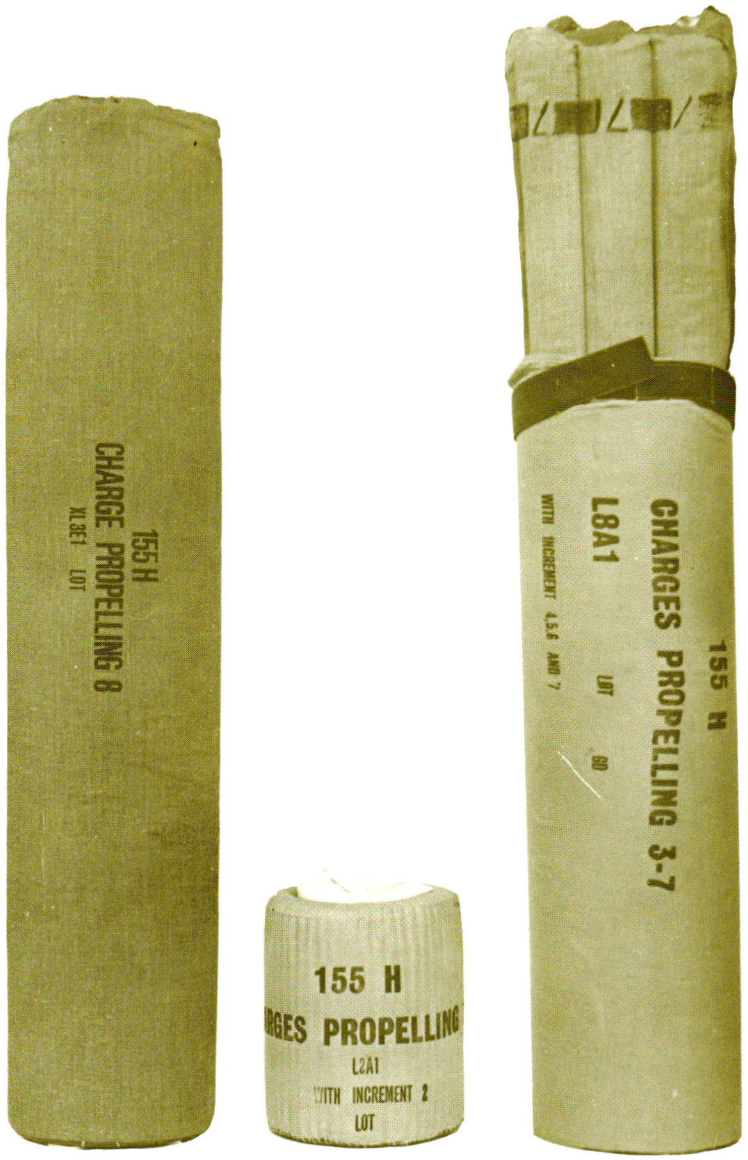

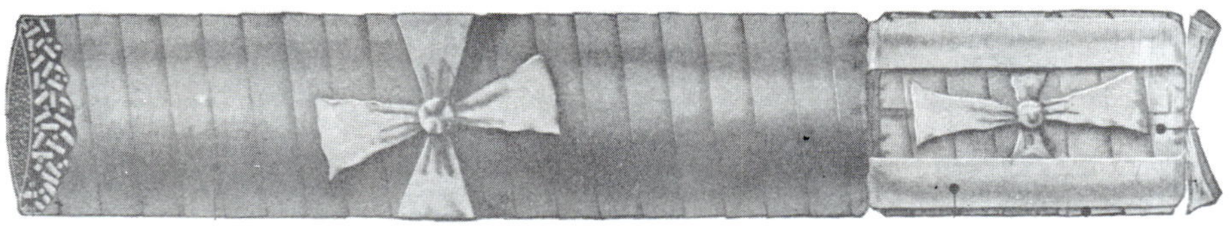

Top: Cartridges for the 155mm FH70 howitzer which are made up by combining bags.
Above: An American bag charge for use with a 155mm gun, detailing its construction.

BAG CHARGE Propelling charge for artillery. It is contained in a cloth bag instead of a metal cartridge case, which means that the sealing of the breech must be performed by the gun and not by the ammunition. The bag is of shalloon, cotton or silk cloth, chemically treated to be waterproof, but also to be totally consumed during the explosion of the charge and not leave any smouldering residue in the gun after firing. Depending upon the size of the charge it may be in one or in a number of bags, and the charge may be adjustable, that is it is possible to alter the size of the charge for different ranges by adding or subtracting bags.

If cordite propellant was used the bag was filled with strips or sticks and thus formed a rigid and easily handled package; if granular propellant was used, the bag was likely to be less rigid and more difficult to handle in large sizes. American heavy bag charges had the propellant grains 'stacked' or carefully set out in layers, which helped the rigidity but made the charge more difficult to manufacture.

BALLISTIC CAP A cap attached to the nose of a shell in order to give it a better shape for flight. Some types of shell, particularly armour- or concrete-piercing types, have blunt noses for the best penetrative performance, but these mean poor flight performance. The attachment of a thin metal ballistic cap makes the nose more tapered and thus gives better flight, increasing the range and terminal velocity.

BASE BLEED A method of increasing the range of a shell by reducing base drag. In the normal shell there is an area of low pressure behind the shell in flight, caused by its passage through the air, which sets up drag and reduces its speed and hence its range. This can be partly overcome by streamlining, but in recent years a system of burning a small quantity of propellant in a chamber in the base of the shell has been developed. The gas from this propellant is allowed to 'bleed out' into the base area, so filling up the low pressure area and removing the drag. The adoption of base bleed can add as much as 20 per cent to the maximum range of a 155mm shell. Work on it has been done in several countries, but the only successful and combat-tested application so far has been by the South African artillery.

BASE EJECTION SHELL Type of carrier shell in which the payload is ejected from the rear by the action of a time fuze. First developed in Britain in the middle 1930s for the 25-pounder gun smoke shell, it has since been widely adopted.

The shell is a cylinder and the payload is in the form of a canister or several canisters which fit inside. At the head of the shell is a compartment containing a small, black-powder expelling charge, and beneath this is a loose 'pusher

An APCBC shot for the abortive 32-pounder anti-tank gun of 1945. The ballistic cap is painted blue.

A base bleed 155mm shell, showing the base cavity which contains the propellant gas- producing powder.

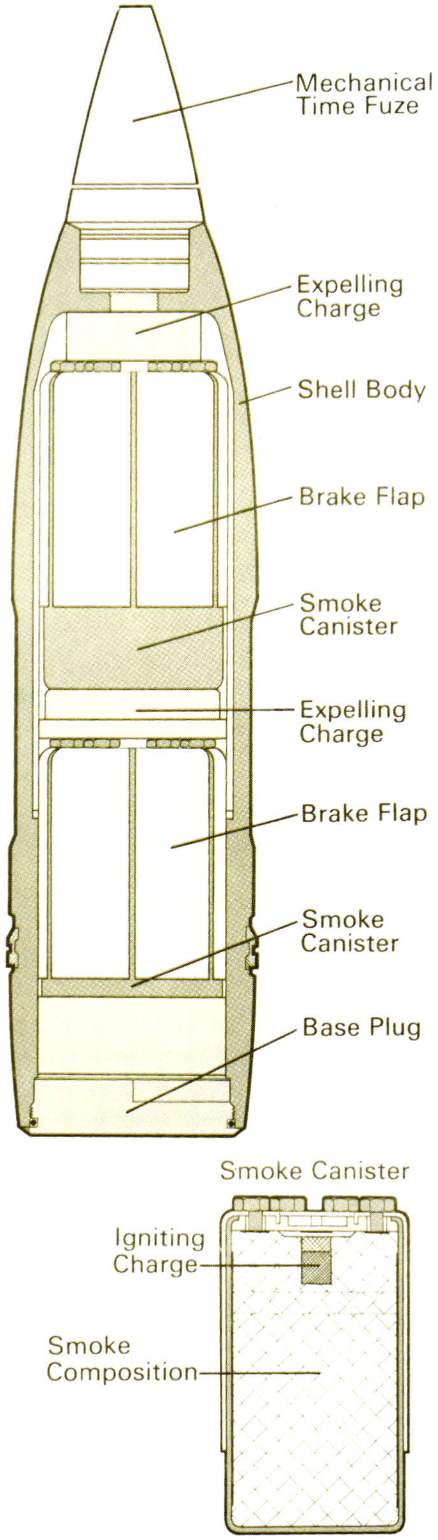

The Swedish 105mm smoke shell uses two canisters which have brakes to control their fall and placement.

plate' with a small hole in its centre. The canisters are secured in place by the base-plate of the shell, which may be screwed into the shell body or riveted into place. A time fuze is screwed into the nose of the shell and when this operates during flight it ignites the black-powder charge which explodes. The flame from the explosion passes through the hole in the pusher plate and into the body of the shell, where it ignites the payload if this is necessary — as it is in smoke, incendiary and illuminating shells. At the same time the gas generated by the explosion of the charge forces the pusher plate down, against the payload, and this thrust is transferred to the base-plate of the shell. The thrust shears the securing threads or rivets and the base-plate is thrown clear, the pressure then forcing the payload out to fall to the ground beneath.

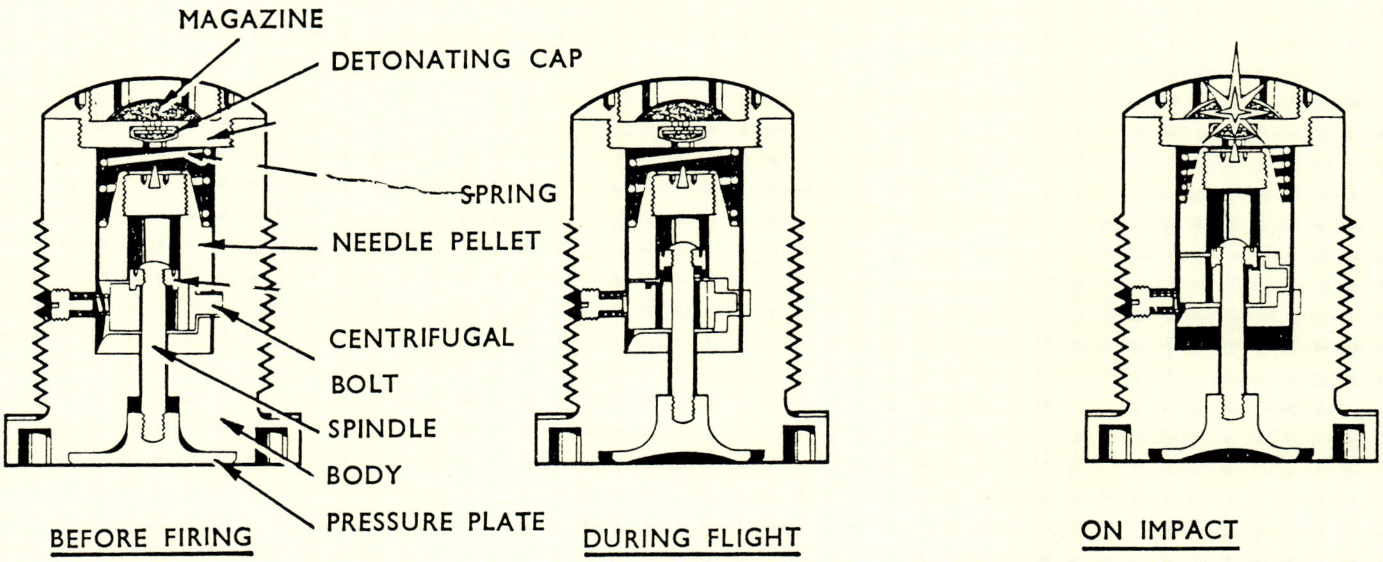

A typical base fuze. Pressure from propellant gas drives in the pressure plate and spindle. This frees the centrifugal bolt, which spins outwards and unlocks the needle pellet. On impact, the pellet flies forward and strikes the detonator cap which ignites the powder.

BASE FUZE Type of fuze fitted into shells where it is necessary that the nose be unobstructed on impact. Originally used on armour-piercing shells where the nose had to be of solid steel. The base fuze operates on the graze principle, a heavy inertia pellet being normally restrained by a spring, but flying forward under its own momentum when the shell strikes. This causes a firing pin to strike a detonator and thus begin the initiation of the shell filling. The position of the base fuze inevitably causes delay between the impact of the shell and the functioning of the fuze, because of the movement of the graze pellet, and this serves to delay the detonation of the filling until the shell has passed through the armour, so that the detonation takes effect against the personnel or equipment protected by the armour. In the larger calibres of shell, intended to be fired against greater thicknesses of armour on warships, the inherent delay is insufficient and it is necessary to interpose additional delay mechanisms into the initiation process, usually by having the detonator ignite a train of black powder which must burn for a short period before firing the fuze magazine and thus the shell filling. Some base fuzes were fitted with adjustable delays which could be set before loading to give delay times of, for example, 0.1, 0.15 or 0.25 seconds.

In order to eliminate the need to calculate delay requirements, either in the design stage or when loading the gun, the 'thinking fuze' has often been proposed. This incorporates a mechanism which senses the passage of the shell through the armour; it arms the fuze on impact, ready for it to be fired, and then fires the fuze when it senses the shell's slight acceleration as it pulls free from the restraint of the armour. However this type of mechanism is somewhat delicate and has rarely been successfully applied.

With the development of *HEAT* and *HESH* shells base fuzes were adopted for these, since the nose of these shells needs to be clear of unnecessary clutter which would interfere with the operation. However, the inherent delay of a base fuze is generally too long for reliable functioning of a HEAT shell unless an inordinately long ballistic cap is fitted, and this type of fuze is now rarely used. In the case of a HESH shell the inherent delay is ideal, allowing the plastic filling to smear on to the target before being detonated.

BEEHIVE SHELL Shell developed by the US Army in the late 1950s and filled with several thousand flechettes, small dart-like sub-projectiles about one inch long. The shell was operated by a time fuze which fired a charge to split the shell walls and liberate the flechettes, allowing them to spread outwards by centrifugal force (derived from the spinning of the shell) and then fly forward in a continuation of the shell's trajectory in an ever-widening cone. The flechettes were lethal to a considerable distance from the point of burst and were a potent anti-personnel weapon. The fuze could be set to operate at any distance from the muzzle of the gun to the maximum range, so that in an emergency it could be used for local defence of the gun position. The name was derived from the noise of the flechettes as they flew through the air. Originally for the 105mm howitzer, shells in other calibres were developed and some are still in service with the US Army. The idea has also been taken up elsewhere. The Soviets have 122mm and 152mm flechette-filled shells.

BINARY SHELL Type of chemical shell in which the dangerous aspects of storage are minimized by using two chemical solutions which are packed separately and are not mixed until the shell has left the gun. Used only with nerve gas agents, and developed in the USA in the 1960s, the binary shell contains two canisters of liquids which by themselves are relatively harmless and thus make filling and storing the shell less hazardous. On firing the shell a delay mechanism functions which, after allowing the shell some seconds of flight, opens the two containers and allows the two liquids to mix. These react chemically to produce a lethal nerve agent which is dispersed at the target when the shell lands.

BLANK CARTRIDGE Cartridge which provides a loud report, for use in training or for firing salutes. Generally filled with black powder, since smokeless powders cannot burn sufficiently fast in an empty gun to provide the rapid explosion of gas which generates noise. Blank cartridges are always prominently marked, since they can be extremely dangerous if inadvertently fired behind a projectile; the rapid development of gas is too fast, the projectile fails to move quickly enough, and the result is usually a burst gun or damaged breech mechanism. See also *Saluting Charges.*

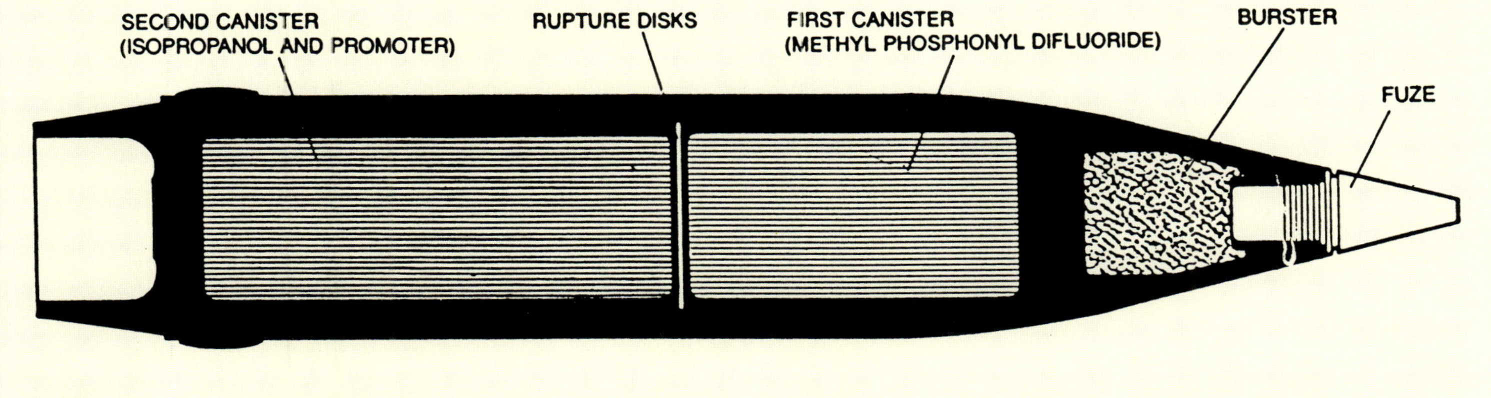

A 155mm binary shell, which contains two chemical containers. The dividing disc ruptures in flight and the chemicals mix.

The design of a typical blank cartridge — in this case it is for use with the American 105mm howitzer.

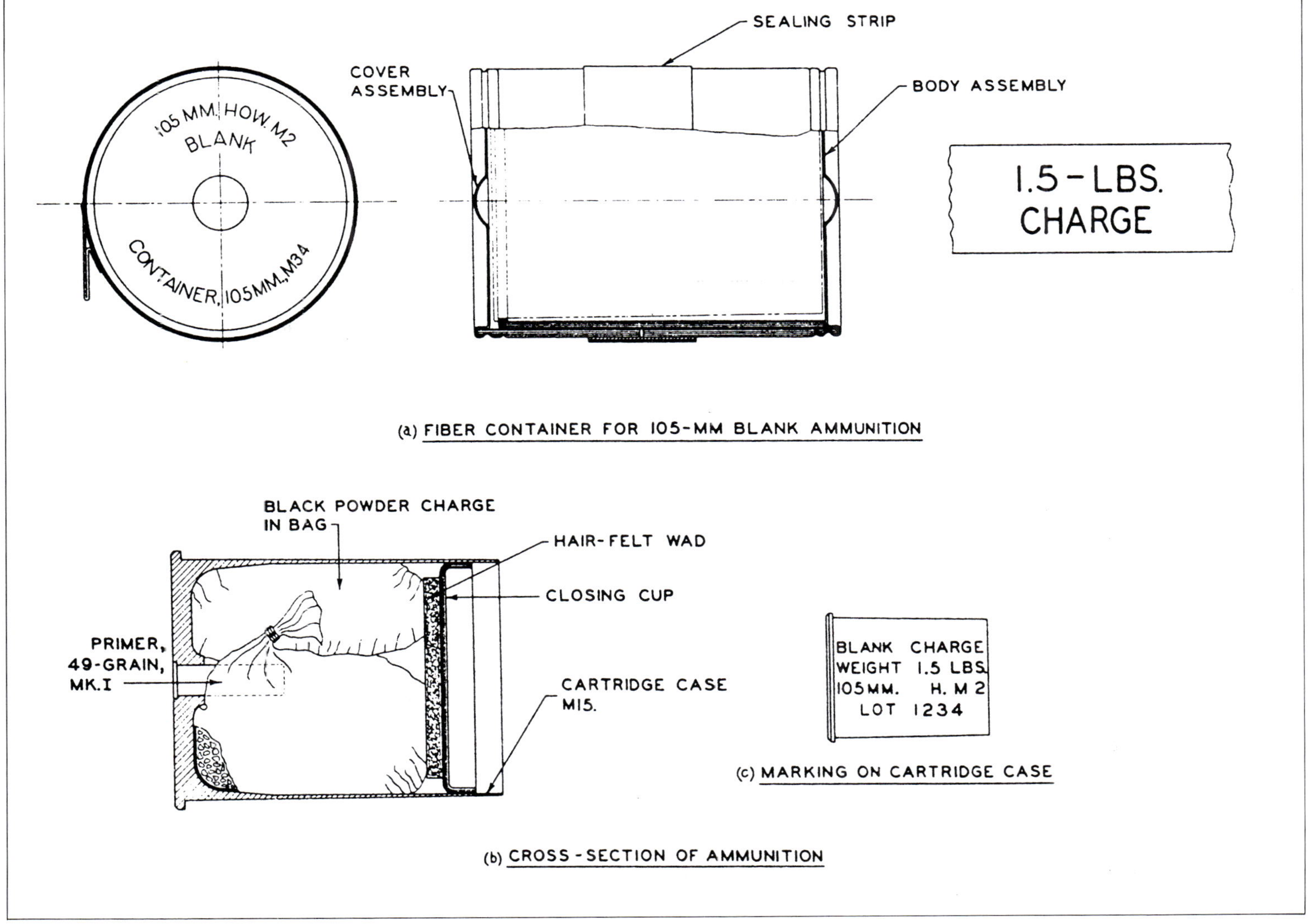

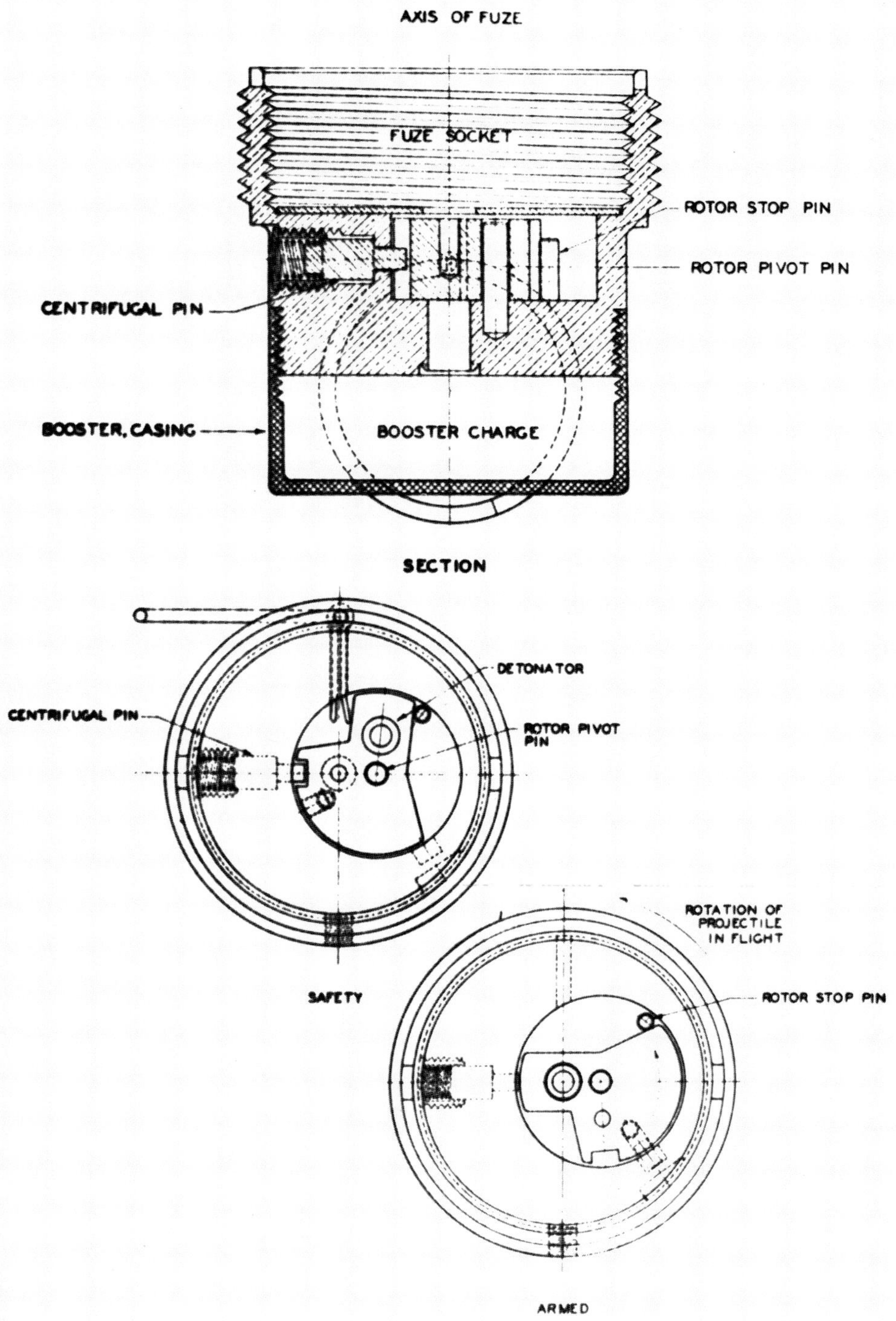

The American M21 booster which was used with almost all US fuzes during the period from 1930 to 1955. It screws to the base of the fuze and converts the flash from the fuze into a detonation suitable to initiate a high-explosive shell. The shutter mechanism provides bore safety.

BOAT-TAIL The tapering portion of a shell behind the driving band. The taper is intended to smooth the passage of air into the low-pressure area behind the shell and thus reduce the drag. Choice of the optimum taper can improve the shell's range by as much as 25 per cent over that of a square-based shell, but the benefit is only significant in the sub-sonic portion of the shell's flight. During the super-sonic part of the trajectory the nose drag, resulting from the Mach Wave on the shell nose, exceeds the base drag.

BOOSTER American term to describe a device attached to the base of a fuze which provides bore-safety and converts the flash from the fuze into a detonation suitable for initiating a high-explosive shell. The M21 Booster, the standard pattern, contained a centrifugally-actuated 'shutter' which contained a detonator filled with lead azide. The booster was screwed into the nose of the shell and the fuze was then screwed into the booster. When the shell was fired, centrifugal force resulting from spin caused a locking detent (or catch) to pull clear of the shutter, allowing it to move round, by centrifugal force, until the detonator was aligned centrally with the fuze and with a magazine containing a charge of tetryl in the lower part of the booster. When the fuze operated, it delivered a flash which struck the lead azide detonator and ignited it; the lead azide burned at an increasing rate until it detonated, and this, in turn, detonated the tetryl, so initiating the shell filling.

In later years it became normal to assemble the fuze and booster and lock them together, so that they could be inserted into the shell as one unit; in recent years it has become the practice to incorporate the booster mechanism in the construction of the fuze and cease to consider it as a separate item.

It should be noted that boosters, as such, have rarely been used outside the USA, other countries preferring to include the bore-safety mechanism in the fuze and to arrange for the fuze to deliver a detonation without requiring a conversion system. Where such a device has been used in Europe it has generally been known as a 'gaine'.

BORE-SAFE An artillery fuze operates by igniting a detonator and then passing this detonation through some relay train until it initiates the shell filling. If the mechanism of the fuze incorporates some physical break in this system, such that the detonation sequence is positively interrupted, and if this break is arranged so that it remains until the shell has left the gun and is in free flight, the fuze is said to be 'bore-safe'.

In the early days of fuze design, when the standard shell was shrapnel or filled with black powder, the danger to the gun and its crew of a fuze malfunctioning and igniting the shell in the bore was relatively small; the shell would shatter, but the damage would be contained within the gun barrel. When high-explosive shells were adopted, a malfunction of the fuze which caused a 'bore premature' could be fatal, since the high explosive would shatter the gun barrel. As a result, bore-safe fuzes began to be introduced just before the 1914-18 war as high explosives came into service. Today it is rare to find a fuze which is not bore-safe.

The most usual bore-safety device is a centrifugally-actuated 'shutter', a solid block of metal which, at rest, forms a positive barrier to the passage of flash or detonation. Once the shell begins to spin, the centrifugal force gradually overcomes a spring, holding the shutter closed, and it slides or revolves until the metal barrier is removed and, usually, a detonator or channel of explosive is put in its place so as to form part of the detonation train. The speed of the shutter's opening can be controlled either by springs or, more accurately, by a ratchet and escapement similar to the escapement of a clock. In the case of fuzes for smooth-bored weapons, where centrifugal force is not available for the purpose, the acceleration of the fuze can be used to operate a rearward moving ratchet and escapement device. In some modern fuzes, electronic circuitry using a form of resistor-condenser delay, has been used to provide bore-safety.

BOURRELET That part of a shell body which is carefully machined to the correct calibre diameter so that it rides on the lands of the gun barrel (the lands being the raised portions between the rifling grooves). There are usually two bourrelets, one at the shoulder and one behind the driving band, so that the shell is well supported in its passage up the gun barrel and does not wobble. Some designs of shell, notably British designs below about 100mm calibre, do not use bourrelets, but machine the entire parallel section of the shell to give a bearing surface. Whether or not to make such shells with bourrelets is entirely a question of production convenience, since in small calibres it is often more easy to machine the whole shell in one pass than to move it from machine to machine in order to finish the surface piecemeal. In major calibres machining the whole surface would be slow and wasteful, and only the bourrelets are turned to calibre, the remainder of the shell being below that diameter and left more or less as forged.

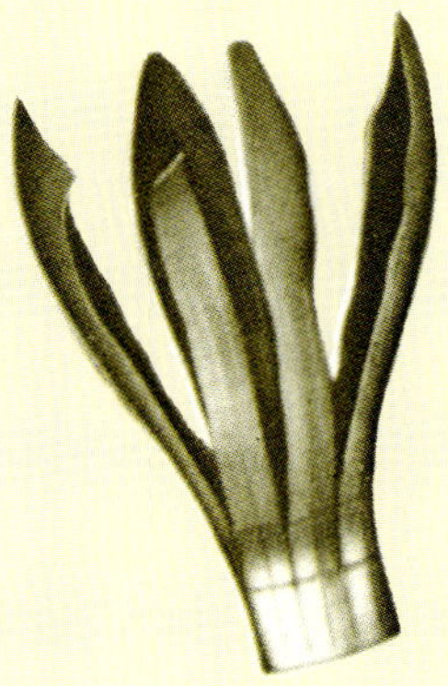
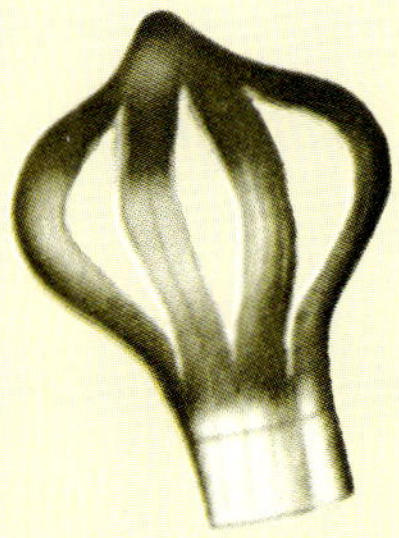
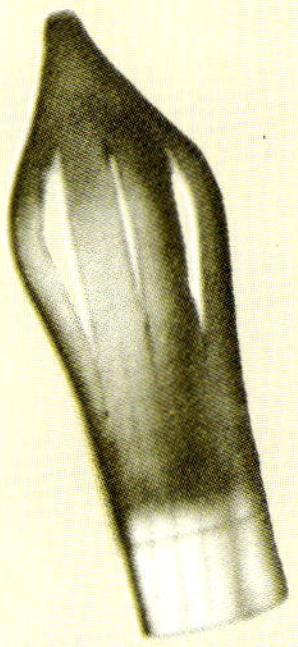
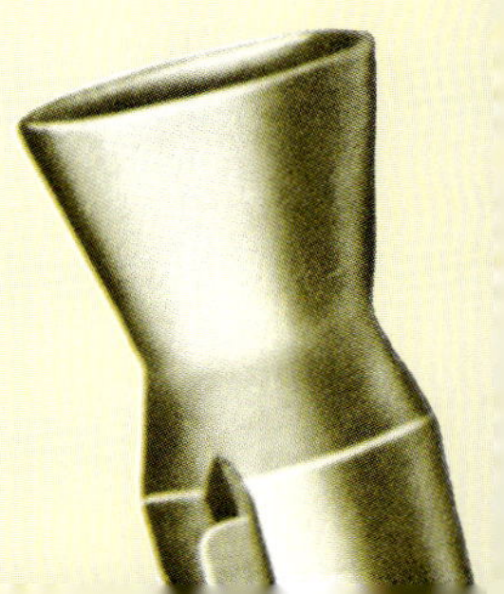

The action of Break-up Shot as it leaves the gun. Centrifugal force drives the filling of heavy dust or liquid outwards, splitting the plastic projectile along pre-formed lines of weakness.

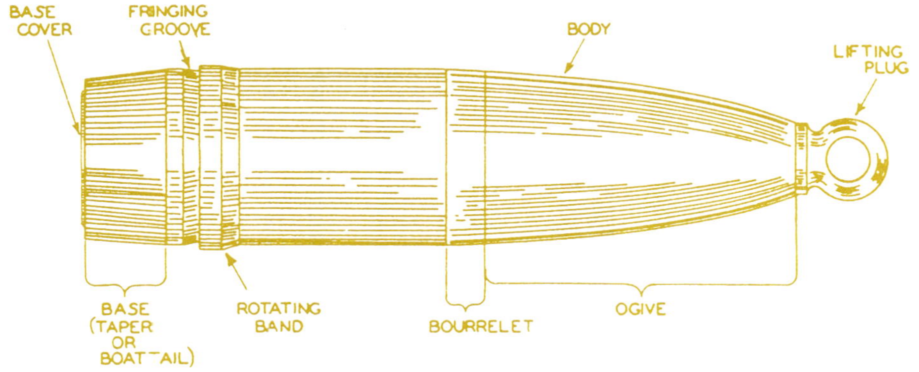

The structure of a conventional shell showing the position of the Bourrelet. This is machined to the diameter of the gun barrel's lands.

BREAK-UP SHOT A projectile which, when fired, sets up the necessary resistance in the bore which will cause the gun to recoil and any automatic mechanism to function, but which then disintegrates either inside the bore or shortly after leaving the muzzle and disperses harmlessly. The object is to operate the gun for test or drill purposes, but not to discharge a projectile which will fly any great distance or cause any damage. Originally they were cylinders of waxed paper filled with sand or water provided for coast artillery guns in locations where it was not possible to fire a service projectile in peacetime because of danger to shipping. The weight would give rise to recoil, so ensuring that the gun mechanism was functioning correctly, but the shot would split in the bore and the water or sand would be ejected from the muzzle and would evaporate or fall to the ground within a few feet.

The principle was then adapted to light anti-aircraft guns in order to test their automatic functioning without the need to take them to a proper firing range; by using break-up shot it was possible to test them within a workshop area. For this application, a plastic projectile filled with either fine dust or a thick innocuous liquid is used; the shot passes up the gun barrel and the gun mechanism functions correctly, but centrifugal force, imparted by the spin of the

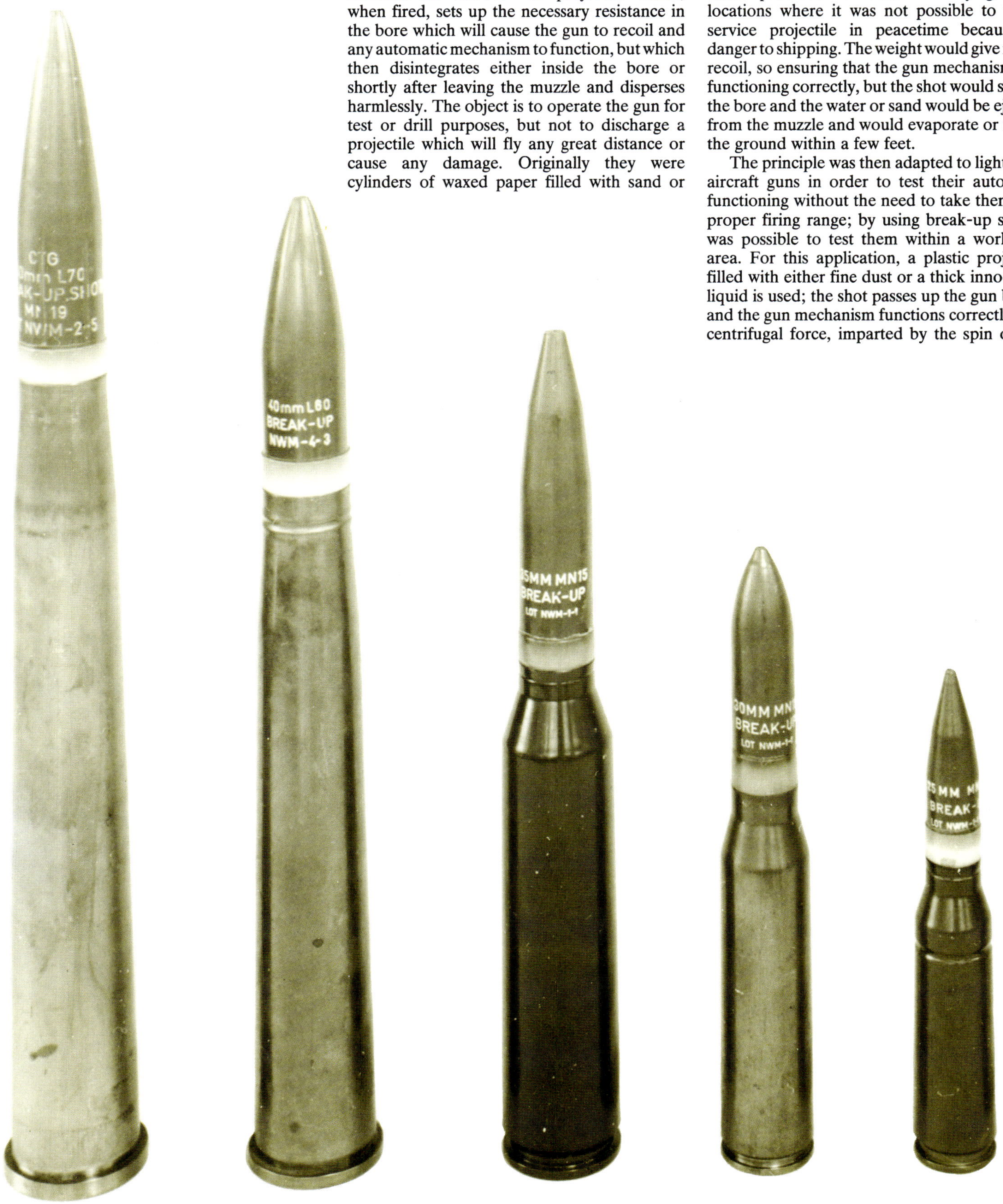

shot, causes the heavy contents to split the shell and disperse within a few feet of the muzzle. Break-up shot has also been adopted as a training ammunition, since it can be fired on manoeuvres by light automatic cannon such as are used on armoured personnel carriers.

Break-up shot is not well-liked by the people who use it since, depending upon the direction of the wind, it often coats the gun and its crew in fine dust or sticky liquid and necessitates a great deal of additional cleaning.

BURSTER The charge of high explosive used to break open a bursting-type carrier shell, eg, a white phosphorus smoke shell. Usually in the form of a thin column of explosive carried in a tube running down the centre of the shell, the payload filling being packed around the burster.

A sectioned chemical shell showing the tubular burster running down the centre. The burster contains explosive which is fired by the fuze screwed into the nose of the shell.

Rheinmetall break-up shot in a range of calibres from 20mm to 40mm. These are used to check automatic recoil mechanisms and the projectile breaks up harmlessly after leaving the muzzle.

BURSTING SMOKE A smoke shell designed to operate on striking the ground and there burst open to release its contents. Generally used with white phosphorus as a filling, since this substance is self-igniting on contact with air and therefore does not require any involved ignition system; other fillings such as sulphur trioxide, titanium tetrachloride or chlorsulphonic acid can also be used since they react with the water vapour in the atmosphere to develop smoke and, again, do not require ignition.

The principal advantages of bursting smoke shells are: they provide smoke very rapidly; they do not require a time fuze to be set, an impact fuze being sufficient; and since they are of the same size, weight and balance as a high-explosive shell they do not demand any adjustment of the weapon sights.

The usual design of a bursting smoke shell is a body of the same design as the standard high-explosive shell but with a central tube running down the shell's axis. This contains a high explosive burster, and the impact fuze screwed into the nose of the shell detonates this burster on striking the target. The smoke material is filled into the shell body around the central burster, either through the shell nose before the burster tube is inserted or through a hole on the side which is plugged and sealed after filling. The quantity of high explosive in the burster is sufficient to shatter the shell and so liberate the smoke mixture, but not so large as to scatter the mixture over a very wide area, so that the smoke composition is concentrated in one place and generates a compact volume of smoke.

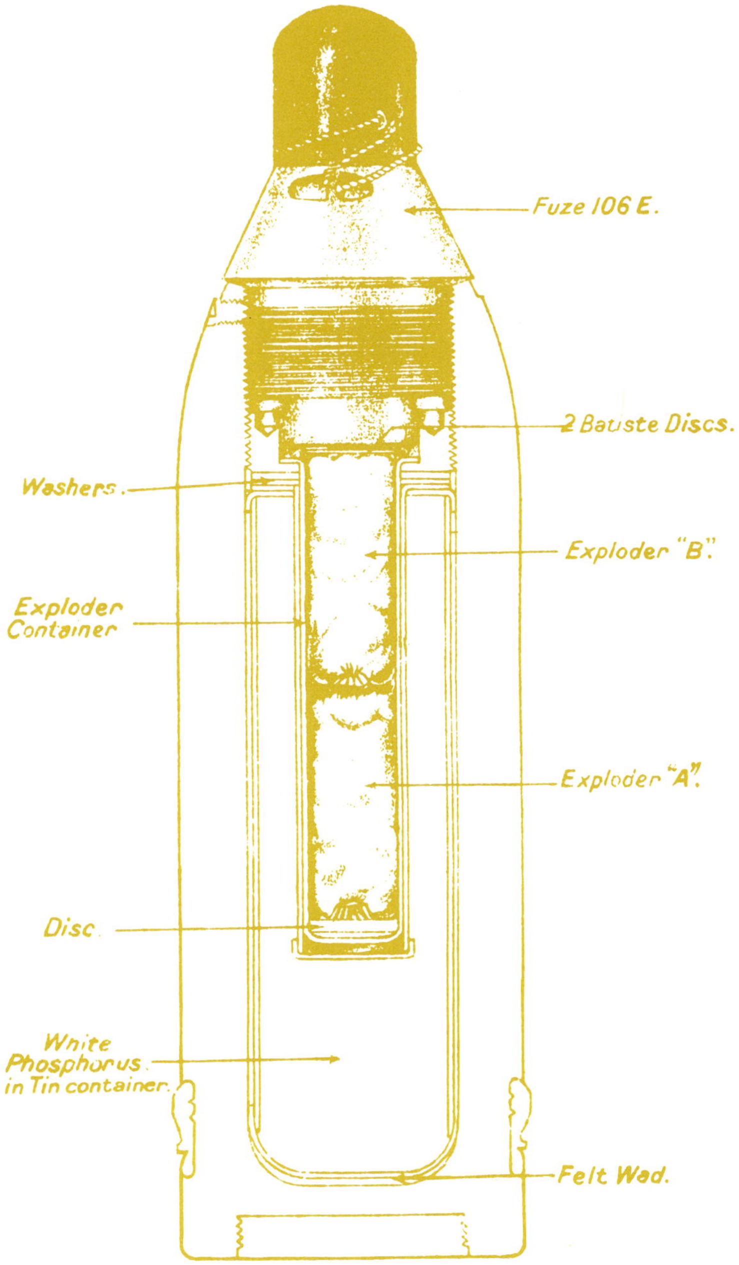

An early design of a bursting smoke shell which used a short burster tube.

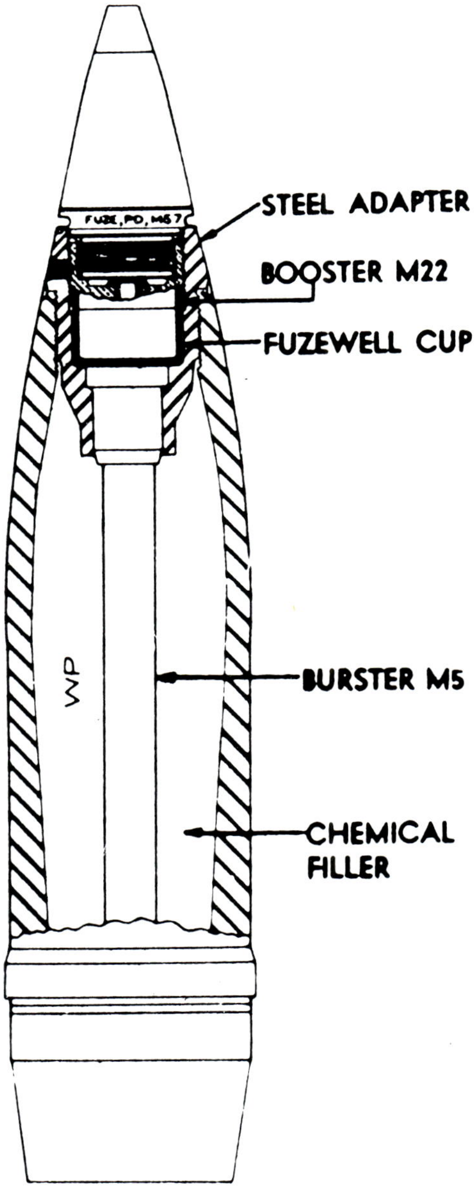

An American M60 white phosphorus smoke shell which is widely used throughout the world.

The head of a projectile is given a gently curving contour whose radius is a multiple of its calibre. The shape of the head is specified by the 'crh' — or calibre/radius/head.

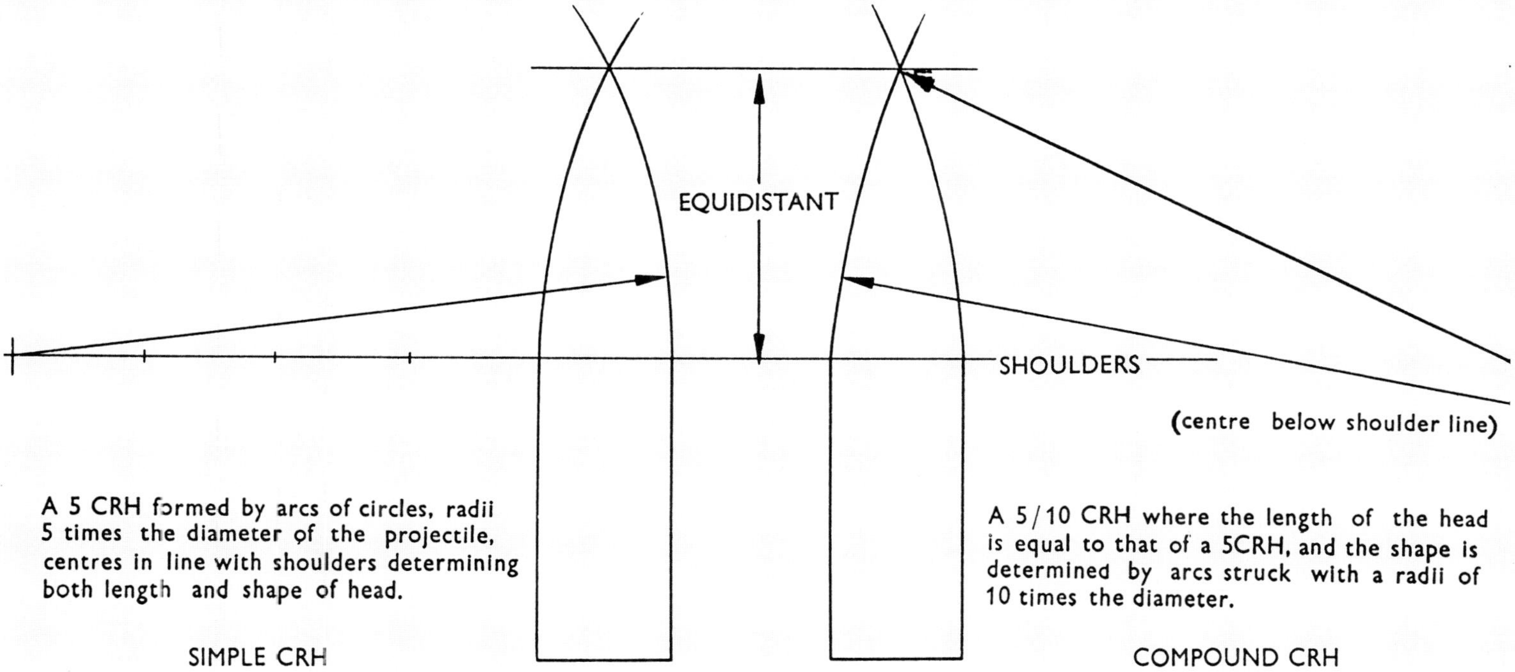

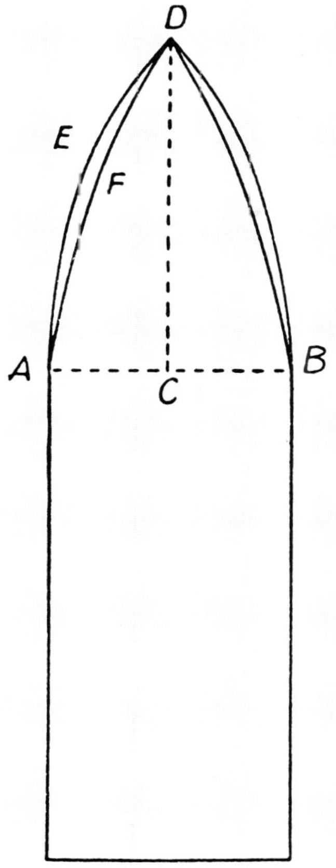

Left: In the design of a modern shell the ballistic length CD is of primary importance. The crh is given by the ratio of the radii of the arcs AED and AFD.

C

CALIBRE/RADIUS/HEAD Description of the head shape of a projectile, frequently abbreviated to 'crh'.

The head of a projectile is usually ogival, that is to say a gently curving contour which blends the parallel sides into a curved nose of good aerodynamic shape. The curve is drawn as a radius which is a multiple of the shell's calibre; thus, a 150mm shell with a head curvature struck on the radius of 600mm would be said to have a head shape of 4crh. The locus of curve is in line with the shoulder of the shell — that point where the straight wall ends and the curve begins — and such a construction is called 'simple crh'.

A more slender head, with better flight characteristics, can be achieved in the same overall length by striking the curve at a greater diameter, but from a point behind the shell shoulder; thus an arc from a radius of 8crh can be struck so as to fit the curve into the same head length as a simple 4crh curve. This is known as 'compound crh' and would be described as having a head shape of 4/8crh.

The amount of curvature for a given projectile depends upon several factors, notably its employment. A piercing shell requires a very short, blunt head shape, possibly 2 or 2.5crh, in order to achieve optimum penetration into hard surfaces. A shell operating entirely in the subsonic velocity range does not require a particularly sharp or tapering nose; but a shell for a high-velocity weapon needs a tapering head in order to reduce the drag produced by the standing Mach Wave set up by air resistance.

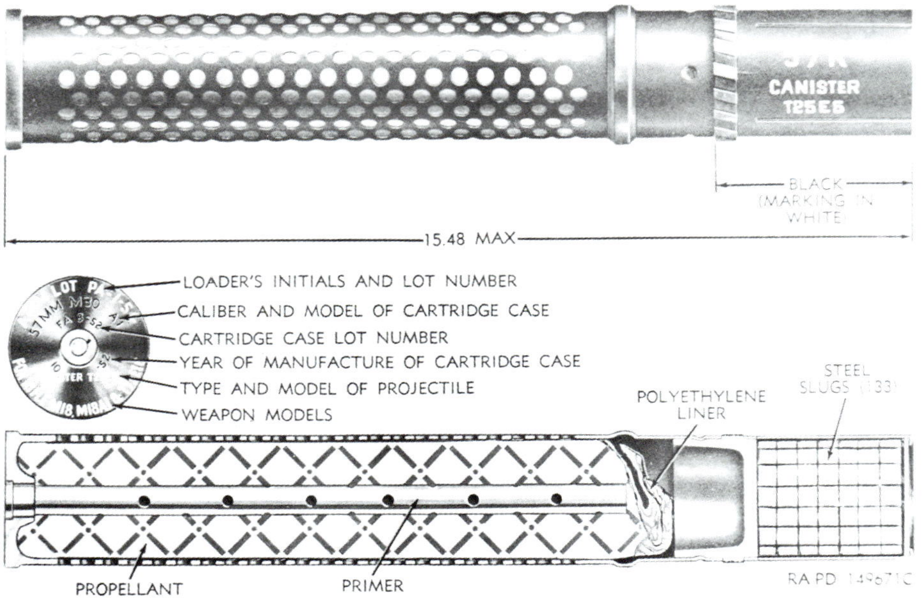

A canister shot round for the American 57mm recoilless gun. It uses cylindrical steel pellets instead of balls.

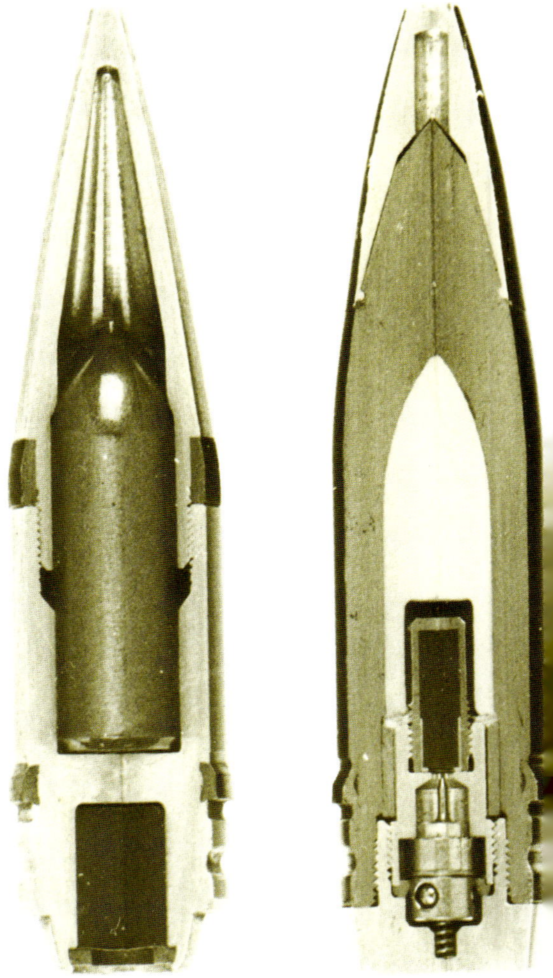

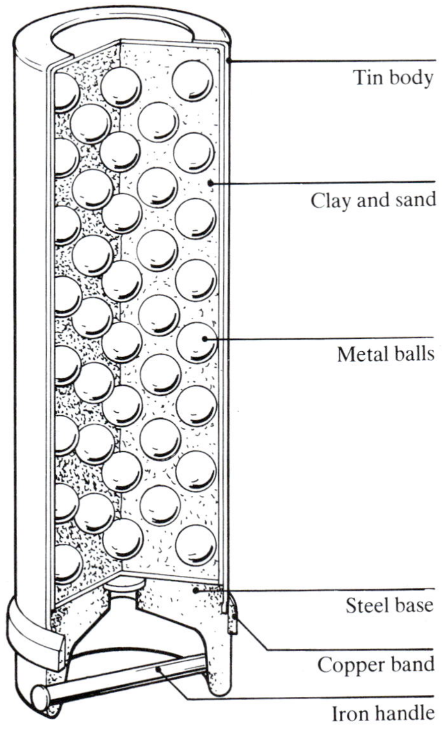

An older type of canister shot. The iron handle is used when loading. The copper band locates the canister.

CANISTER SHOT An inert anti-personnel projectile which consists of a large number of balls, slugs or pellets inside a simple metal casing. The canister is loaded into the gun and fired, and splits open inside the bore so that the remains of the canister and the contents are ejected from the muzzle in the manner of a charge of shot from a shotgun.

Canister was originally called 'case shot' and has been in use for centuries as a close-quarter defensive weapon. It was standard for field artillery for defending the gun position against assault and was also issued to major-calibre coast defence weapons to rake landing-beaches and destroy parties landing from small boats. In the late 19th century it was largely superseded by improved designs of shrapnel shell with fuzes which allowed the shell to function close to the gun muzzle. It was briefly revived during World War I for the early tanks, to be fired as a close-quarter weapon when clearing trenches. It fell into disuse after the war and was not revived until the 1950s, when British armour in Korea was provided with a 20-pounder canister to allow them to shoot at friendly tanks in order to rid them of enemy infantry swarming over them.

A typical modern canister shot is that for the British 76mm gun used on armoured vehicles: it consists of a tinplate cylinder packed with 800 steel slugs, 9mm in diameter and 9mm long. On firing, the slugs are ejected in a cone and have lethal effect to some 150 metres from the gun muzzle.

CANNON AMMUNITION 'Cannon' is the term currently used to classify those weapons which fall between small arms on the one hand and artillery on the other; ie, the group of weapons with calibres from 20mm to 30mm. In broad terms they are automatic weapons resembling overgrown machine-guns, their principal attraction being that they fire explosive projectiles and not inert bullets.

The first 20mm cannon was invented by Becker, in Germany during World War I, and was intended for the armament of aircraft. In postwar years his designs and patents changed hands and wound up with the Oerlikon Machine Tool Company of Switzerland who perfected the design and sold 20mm cannon around the world. Other designs appeared, and during World War II the 20mm cannon became a major aircraft and anti-aircraft weapon. Since 1945 other calibres — 23mm, 25mm, 27mm, 30mm — have been adopted, and at present there is a revival of interest in these weapons as light anti-aircraft defence, due largely to the perfection of small radar and electronic fire-control equipment, and as armament for light armoured infantry fighting vehicles.

The ammunition used with cannon is always a fixed round, and the projectiles cover a range of high-explosive, incendiary, and armour-piercing types, usually complete with tracer. High-explosive shells, because of the small calibre, need to be efficient in order to do damage, and cannon shells are usually filled with

Above: Oerlikon ammunition for the 20mm cannon (left to right): AP tracer, semi-AP/explosive/incendiary, semi-AP/HE/I/tracer, HE/I, HE/I/tracer, target practice, target practice/tracer.

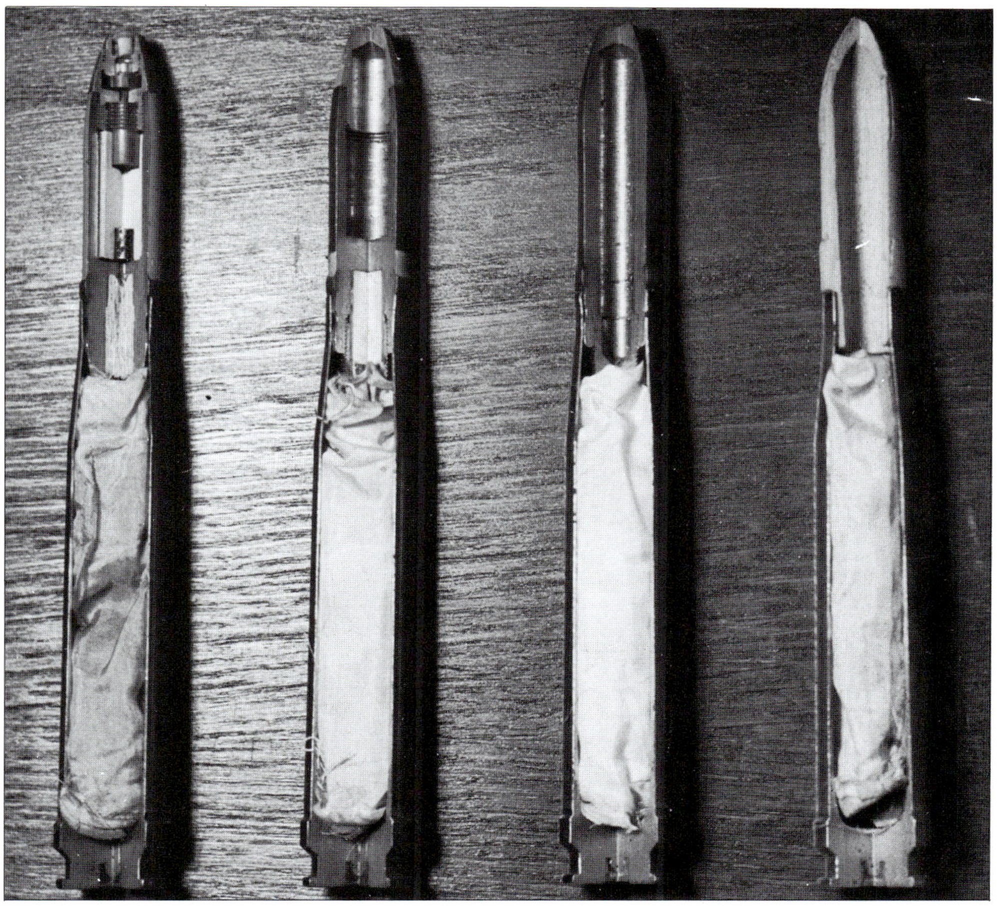

Right: German World War II 20mm cartridges (left to right): HE/tracer, HE/practice/tracer, inert practice round, wooden-bulleted drill cartridge.

the more powerful types of high explosive such as *Composition B*, *Hexogen* or *PETN*. In order to achieve the greatest possible payload, the *mine shell* was developed for cannon use and has been widely adopted.

Fuzing for cannon shell poses some problems, the main requirement being that the fuze should be sensitive enough to function on striking the rather thin skin of an aircraft. Another vital requirement is that the ammunition should be *self-destroying*, since otherwise it would be likely to fall on friendly locations while still in a dangerous condition.

Armour-piercing ammunition for cannon is usually of the *APCR* type, though *APDS* is now being produced in several calibres despite considerable problems in design and manufacture in these small sizes. The expense of this type of ammunition has led to the development of cannon having dual feed; ie, with mechanisms which can be primed with two belts of ammunition, one containing standard high-explosive ammunition and the other loaded with armour-piercing ammunition. The weapon is operated in, say, the anti-aircraft role, with the explosive ammunition belt feeding; should an enemy armoured vehicle appear, the feed can be switched to the armour-piercing ammunition and the weapon can be used against the tank. This system allows a quick change of ammunition without having to unload the gun and thread a new belt, and allows the minimum of the more expensive ammunition to be used.

CAPPED SHOT General term covering any armour-piercing shot fitted with a piercing cap. See *APC* above.

CARRIER SHELL Any shell in which the shell body acts merely as a vehicle to carry the payload to the target, the payload having the desired tactical effect. Thus a high-explosive shell is not a carrier shell since its effect depends upon shattering the shell body for fragmentation, but a bursting smoke shell is a carrier since the body is merely opened by the burster so as to release the smoke mixture which then provides the required tactical effect. Broadly speaking any projectile other than a high-explosive shell or an armour-defeating shot or shell is likely to be a carrier shell of some sort.

Carrier shells fall into two main groups, *bursting* or *base ejection*, though there are one or two which do not fit these categories, such as *Beehive shell* and some of the recently-developed *remotely delivered munitions*.

CASE SHOT Old term for *Canister Shot*.

A 17-pounder anti-tank gun capped APC shot, showing how the piercing cap fits on top of the steel shot to protect the tip during the initial impact.

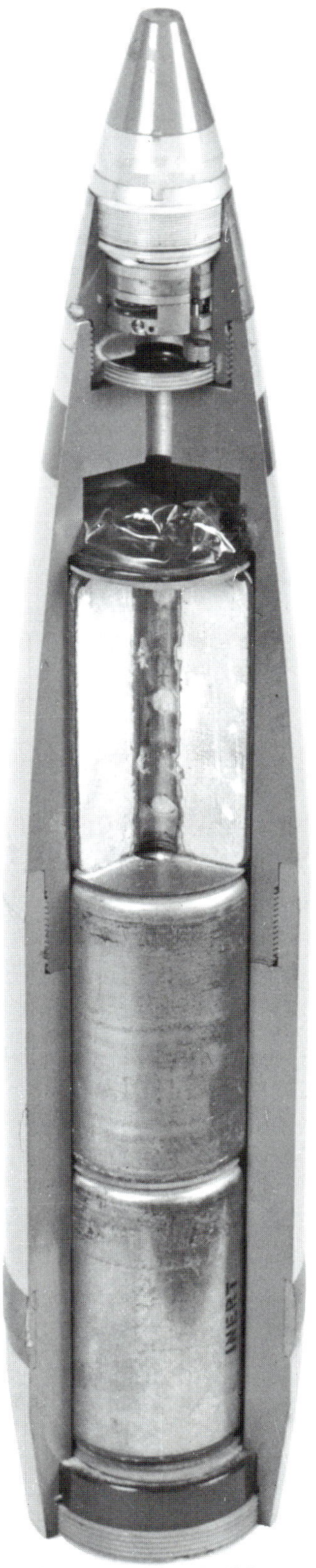

A base ejection carrier shell for a 105mm light gun containing three smoke canisters.

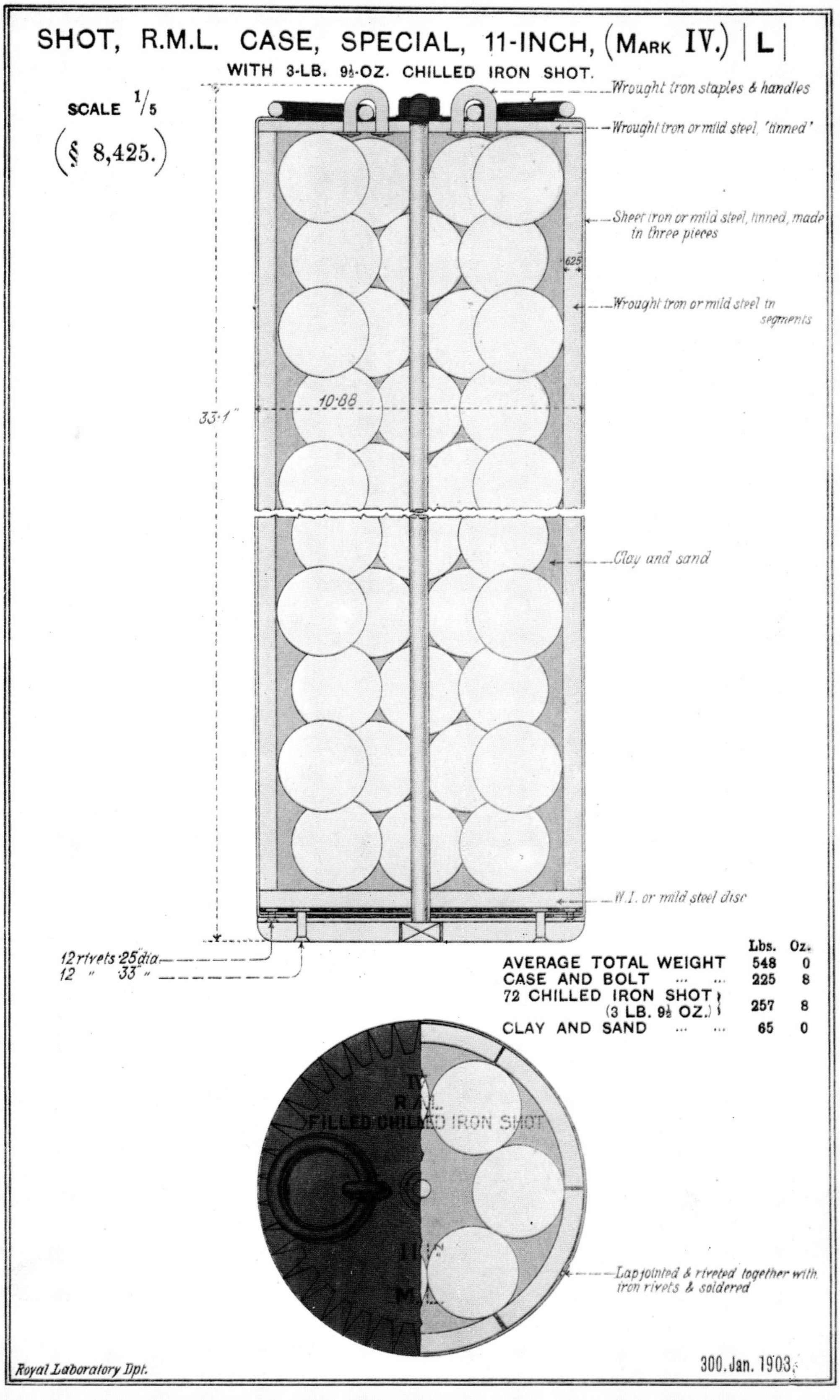

A heavy case shot designed for use by coastal defence guns against landing parties. In effect, it launched 72 separate 3lb cannon balls at the target.

CASED CHARGE A propelling charge contained in a metal cartridge case; as opposed to a *bag charge* which is contained in a cloth bag.

The propellant may be loosely poured into the case, if granular in form, or it may be composed of sticks which are bundled together and tied. Ignition is achieved by a primer screwed or otherwise fixed into the base of the case and may be assisted by an igniter placed around or inside the propellant. In *fixed* rounds of ammunition, where the projectile is fixed into the mouth of the case, it is sometimes desirable to put a cardboard distance piece between the base of the shell and the propellant charge so as to keep the charge in the lower part of the case and in contact with the primer and igniter. Without such a distance piece the act of loading could easily cause the propellant to surge forward in the case and away from the primer, so that ignition of the propellant would become difficult and inefficient.

In *semi-fixed* and *separate-loading* ammunition the propellant is likely to be packed into small cloth bags and inserted into the case, so that the charge can be adjusted for power by removing one or more of the bags prior to loading. In separate-loading cases, the charge is retained in the case by some form of closing cap or lid, which can be removed for adjustment of the charge and then replaced prior to loading; as with the distance piece in fixed ammunition, the presence of the cap ensures that loading the cartridge does not allow the charge to part company with the primer.

A World War II 25-pounder chemical shell. Stripes indicate the filling and the band of reactive paint close to the fuze joint detects leakage.

CHEMICAL SHELL Euphemistic term generally used to describe shells filled with chemical agents — ie, poison gas — though some countries apply it also to shells filled with smoke or incendiary mixtures.

Chemical agents are generally liquid, and the chemical shell usually resembles the bursting smoke shell, having a central burster and an impact fuze, since the requirement is much the same — that the shell should break open and release the agent with minimum force. The greatest problem with chemical shell design is to get the liquid filling to conform with what the shell body is doing in flight; it is not desirable to fill the shell cavity completely, and a space of about 5 per cent of the volume is left to allow for expansion. This means that the liquid can move inside the shell, so that it congregates at the rear during acceleration, intermittently sloshes forward during flight, and must be forced to spin with the shell by a series of internal baffles.

Liquid fillings can also be discharged by base-ejection action; a small ejecting charge, similar to that of a base-ejection smoke shell, is carried in a short tube at the top of the shell. When fired by the fuze, this tube is sheared off and expanded and then driven down the cylindrical interior of the shell, forcing the liquid in front of it. This puts pressure on the base-plate, which is blown out, and the liquid is then ejected. The rear part of the shell cavity is sometimes formed into a choke or venturi in order to ensure the agent leaves in a fine spray rather than simply slopping out in one splash. This system is not common,

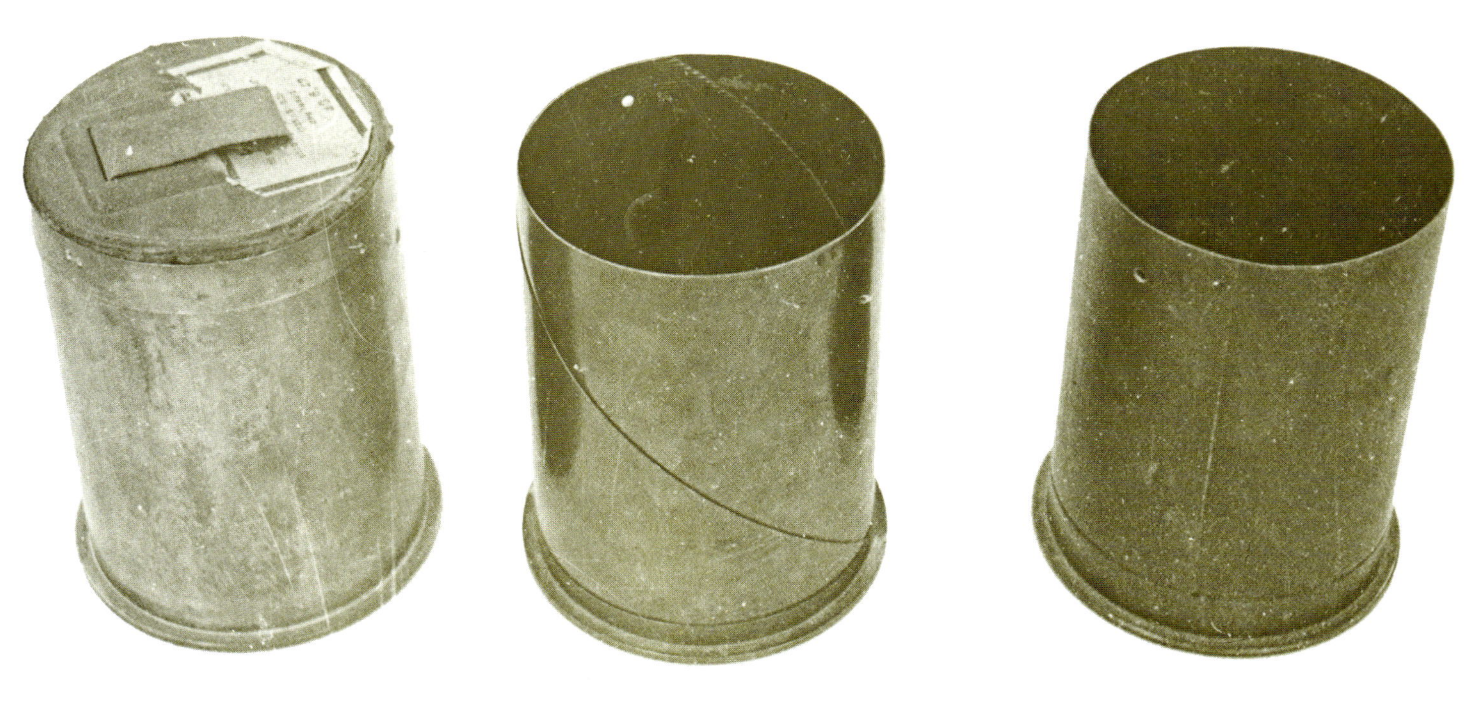

These are cartridge cases for the German 105mm howitzer model 18. The charge is contained in a brass, wrapped steel or drawn steel case.

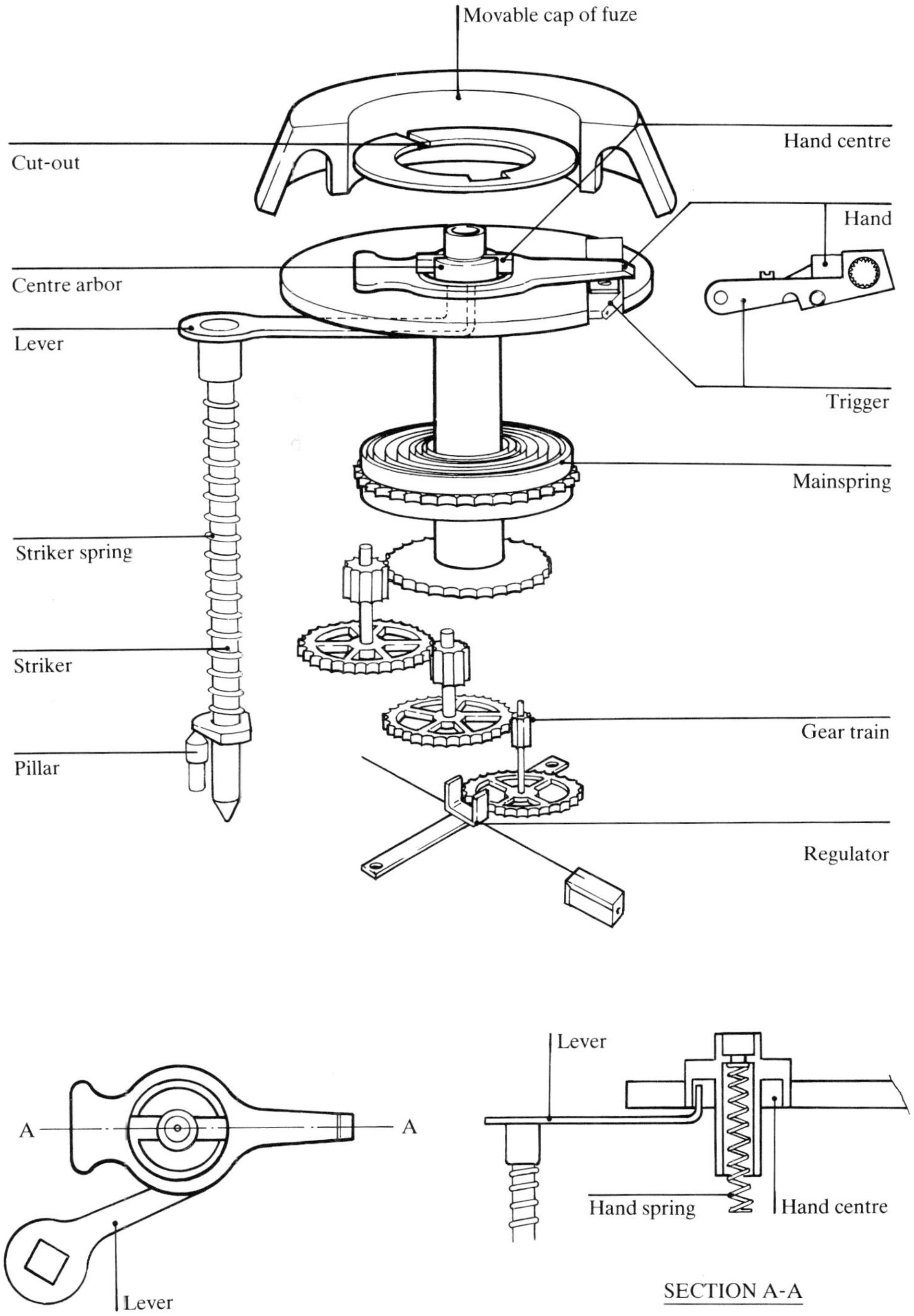

The mechanism of the Krupp-Thiel clockwork time fuze. The clockwork is wound when the fuze is assembled in the factory.

since it is difficult to fit efficient baffles to make the liquid spin. In the British BE design a metal baffle was fixed to the base-plate and protruded a short distance into the cavity, but it was probably not very effective in ensuring all the liquid took up spin.

Those few chemical agents which are solids can be dealt with in easier fashion, either by filling into canisters and ejecting from a base-ejection shell or by simply allowing the substance to be dispersed from a bursting shell.

Most chemical agents are corrosive, and all are either lethal or extremely injurious, and therefore construction of the projectile requires special precautions. The shell cavity needs to be lined with some impervious material so that the agent does not attack the metal, and all joints must be sealed against leaks. It is customary, too, to paint a ring of detector paint on to the outside of the shell close to any major joint so that should the agent leak the detector paint will change colour and thus warn of danger.

CLOCKWORK FUZE Mechanical time fuze relying upon a pre-wound clockwork mechanism for its timing function. There have been two principal types: the Krupp-Thiel, in which the clockwork is wound when the fuze is assembled in the factory, and the Tavaro, which is wound during the setting operation prior to loading. The Krupp-Thiel is the easier mechanism to understand and is shown in the attached drawing.

The mechanism contains a mainspring and a regulator which allows the mechanism to run at a pre-determined rate. The main shaft carries a 'hand', beneath which is a spring forcing it upwards. Under the rim of the hand is a lever controlling a striker which is located above a detonator in the fuze body. The upper part of the fuze body can be rotated and carries an index arrow which is set against a time scale engraved on the lower part of the body. The bottom of the movable portion is highly polished and has a cut-out exactly the same shape as the hand on the mechanism; the spring beneath the hand keeps it pressed up against this polished surface.

Setting the fuze by rotating the upper section displaces the cut-out portion. When the shell carrying the fuze is fired, the force of acceleration allows a trigger to drop and release the hand and mechanism so that the running of the mechanism by its mainspring causes the hand to rotate. At the end of the time set, the hand aligns with the cut-out and the spring pushes it up into the upper section. This raising of the hand releases the firing lever which then allows the striker to hit the detonator and fire the fuze magazine and thus detonate the shell.

COMBUSTIBLE CASE CHARGE Type of bag charge in which the bag is actually of rigid combustible material which is entirely consumed in the explosion of the charge and which adds to the propellant effect. The only advantage of this type of charge over a conventional bag charge is that the rigidity of the casing permits the charge to be used to ram the projectile ahead of it into the chamber of the gun, thus making it possible to adopt mechanical ramming systems with bag charges. It can be used with some types of tank gun and with modern 155mm howitzers having mechanical loading systems.

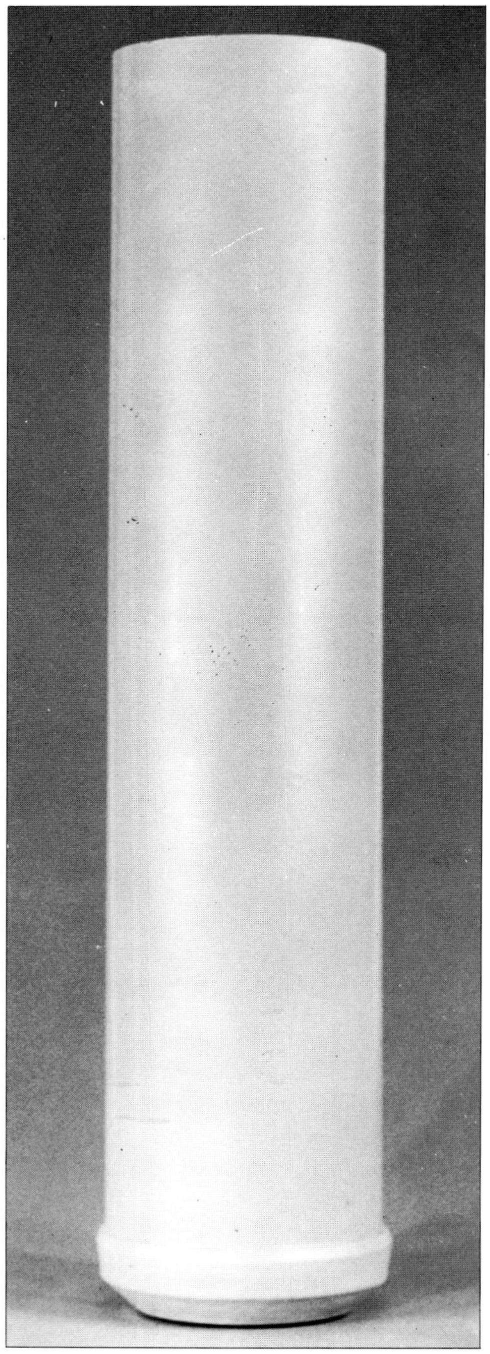

A French combustible case charge for a 155mm howitzer.

COMBUSTION TIME FUZE Time fuze in which the timing is performed by burning a pre-set length of gunpowder or similar combustible substance.

A typical combustion time fuze consists of a central pillar and base unit, two metal rings each containing a powder train for about 320° of their circumference, and a nose cap which holds everything together. The central pillar contains the ignition system — a detonator supported by a spring above a firing pin. The base contains a charge of black powder. The two rings surround the central pillar and are kept in place by the nose cap. The top ring is fixed, while the lower ring can be revolved; it carries an index mark which is set against a time scale engraved on the base unit.

When the fuze is fired from a gun the detonator sets back, impales itself on the firing pin, and generates a flash. This is channelled to the top time ring and begins burning the powder filling at a regular speed. The lower time ring has a channel leading from its upper surface (which is adjacent to the top ring's powder train) to its own powder train, and when the top ring burns to this point it ignites the lower train. This, in turn, burns until it meets a port leading to the powder charge in the bottom of the fuze which is then ignited and which, in turn, ignites the main charge — the contents of the shell itself.

Since the top ring is fixed, as is the channel leading to the fuze magazine, the only variable is the port which leads from the top ring to the bottom ring. Setting the fuze to a specific time moves this port so that the top ring must burn longer in order to reach it and the bottom ring must then burn longer to reach the magazine channel.

In order to achieve regularity the consistency and compactness of the powder rings must be within certain limits, and this, in turn, imposes some limits on the operation of the fuze; with two time rings, 30 seconds is about the maximum time which can be achieved. To obtain longer times fuzes with three and even four rings have been developed, but the advent of mechanical fuzes removed the need of such complication. Combustion fuzes are cheap and sufficiently accurate for use with most carrier shell; their accuracy of timing is about 4 per cent of the time of flight of the shell, so that a shell having a time of flight of 30 seconds to the target can be expected to burst within about 1.2 seconds of the set time. Their principal defect is a certain sensitivity to atmospheric pressure, though this was mainly seen when firing at high altitudes in the anti-aircraft role and is no longer of much importance.

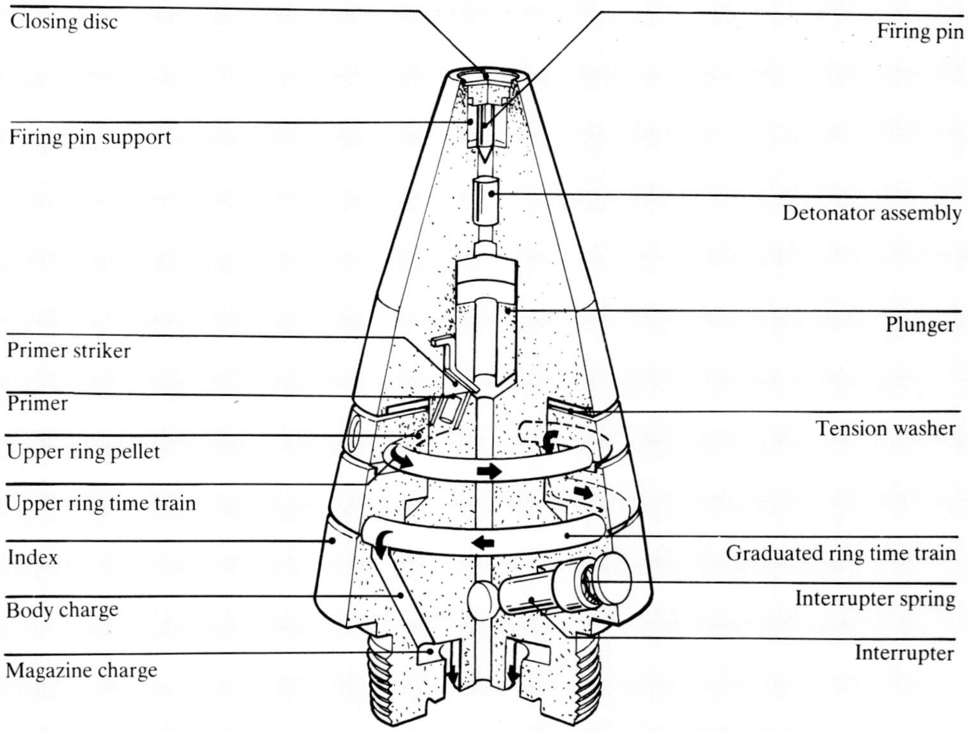

A typical combustion time fuze with arrows indicating the path of burning from the primer to the magazine charge.

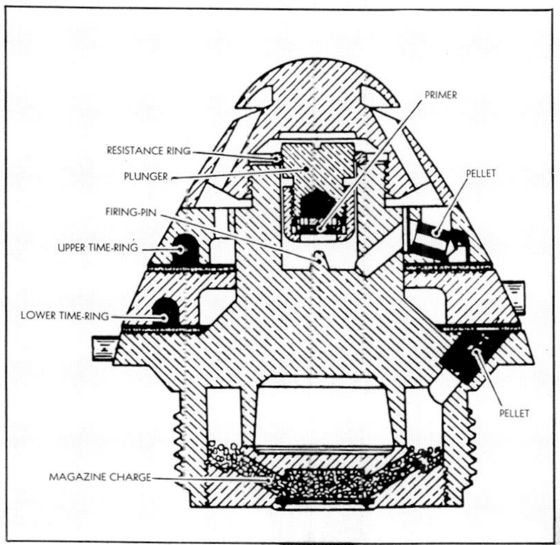

An American combustion time fuze. On firing, the plunger pulls free from the resistance ring and the primer strikes the firing pin. This ignites the pellet, which ignites the upper time ring. The lower time ring ignites the second pellet which fires the magazine charge.

A Japanese combustion time fuze with four time rings. The grooves in the rings were filled with powder and the holes allowed the flash to pass from ring to ring and eventually into the fuze magazine

COMPOSITION A High explosive consisting of 90–97 per cent RDX and 3–10 per cent montan wax; known in British service simply as 'RDX/Wax'. Too sensitive to be used in large quantities as a shell filling, it is used in fuze magazines, boosters and exploder systems.

COMPOSITION B High explosive consisting of RDX and TNT in varying proportions, the most usual being 60 per cent RDX and 40 per cent TNT. RDX is too sensitive to be used alone as a shell filling, but mixed with TNT in this way it produces an extremely powerful mixture having a detonation rate of 8,000 metres per second (26,240 feet per second) and yet which is sufficiently robust to withstand being fired from the gun.

The American M7A1 concrete-piercing fuze: A. Shell mouth holding the fuze and booster. S1. Solid steel head. S2. Spring. S3. Detonator and delay. D1. Booster detonator. D2./D3. Tetryl filling. D4. Centrifugal shutter.

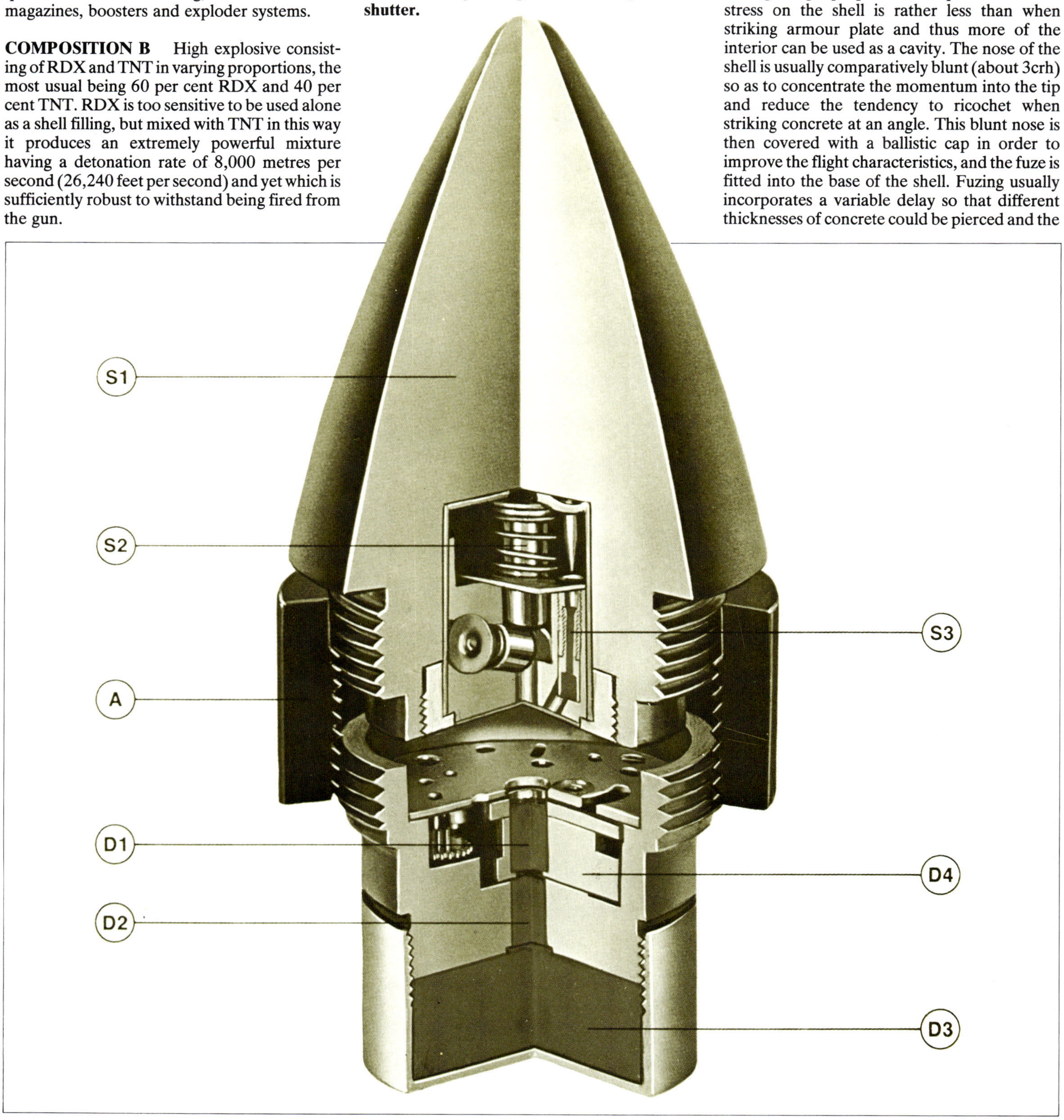

CONCRETE-PIERCING SHELL Special-purpose shell developed, largely by European powers, for the attack of fortifications. Generally similar to armour-piercing shell, but having a larger proportion of explosive since the stress on the shell is rather less than when striking armour plate and thus more of the interior can be used as a cavity. The nose of the shell is usually comparatively blunt (about 3crh) so as to concentrate the momentum into the tip and reduce the tendency to ricochet when striking concrete at an angle. This blunt nose is then covered with a ballistic cap in order to improve the flight characteristics, and the fuze is fitted into the base of the shell. Fuzing usually incorporates a variable delay so that different thicknesses of concrete could be pierced and the shell detonated on the interior side. Since fortification is virtually non-existent today, concrete-piercing shells have all but vanished; it is understood that some Soviet artillery weapons still carry CP shells in their inventories, but these are hangovers from the last war.

An interesting variation developed by the US Army during World War II was the concrete-piercing fuze. This was a delay fuze set into an exceptionally hard steel false point which could be screwed into the nose of any standard artillery shell in place of the standard fuze, so converting the shell into a piercing shell. The 'Fuze Concrete-Piercing M78' appears to have been successful on the few occasions it was used against field fortifications, but was relatively useless against properly built permanent fortifications such as those of the Fortress of Metz.

CONED BORE A gun using an add-on barrel extension which reduced the calibre, or one which had a reducing-calibre section in a barrel otherwise parallel. An example of the former was the 'Littlejohn Adapter' fitted to the muzzle of the British 2pdr and US 37mm armoured car guns. An example of the latter was the German 75mm PaK44 anti-tank gun in which the barrel had a sharply sloped conical section immediately in front of the chamber and was thereafter conventionally rifled at the reduced calibre of 55mm. Both designs were for use with *APCNR* projectiles which were reduced in calibre as they passed through the conical section. Also known as 'squeeze-bore'. Compare with *Taper Bore.*

CORDITE British gun propellant developed in the late 1880s and used until the 1950s. It consisted of nitro-cellulose, nitro-glycerine and mineral jelly (Vaseline) incorporated and gelatinized by the action of acetone. By gelatinizing, the fibrous nature of the nitro-cellulose was destroyed and the substance was converted to a colloidal form which could then be worked into whatever shape or size was convenient. It was usually extruded in the form of long cords — hence its name — or sticks.

The original cordite had 58 per cent nitroglycerine and this gave it a high explosion temperature and was responsible for excessive wear in the guns, leading to inaccurate shooting. It was therefore changed to 30 per cent nitroglycerine which reduced the temperature and thus the erosion of the guns, but also reduced the power so that charges had to be rather larger to obtain the same performance.

CYCLONITE Name frequently used in European countries for *RDX.*

Right: A round of ammunition showing the de-coppering foil on top of the charge propellant.

D

DE-COPPERING CHARGE When a shell with a copper *driving band* is fired, friction causes the band to deposit a layer of copper on the interior of the gun. Subsequent shots add to this, and unless something is done to remove this build-up, the gun becomes a victim of 'copper choke' and the shell is unable to pass up the bore. The solution is to add a quantity of metallic lead, in the form of wire or foil sheets, to the propelling charge. This is vaporized in the explosion of the charge and deposited on to the bore of the gun, where it amalgamates with the copper to form a brittle alloy. The next shell to be fired sweeps this alloy away, thus removing the copper build-up. It is customary to have a quantity of lead foil or wire in all propelling charges so that the de-coppering action is constant, but in guns and howitzers with adjustable charges the de-coppering material is usually in the bag containing the highest charge. If this bag is discarded, as is done when firing a lower charge, no de-coppering material is used and copper starts to build up. In such cases a special de-coppering charge, a normal charge to which a larger amount of lead was added, would be fired periodically to remove the accumulated fouling.

DELAY FUZE A fuze used in order to delay the detonation of the shell after impact. This may be done in order to (1) permit the shell to ricochet; (2) permit the shell to break through light obstructions such as a house roof before detonating inside the house; or (3) to permit the shell to break through a major protective obstacle such as armour plate or concrete before detonating on the far side of the protection. Case (1) is covered under the heading *ricochet fuze.* Case (2) is generally satisfied by having a delay setting on an otherwise normal nose impact fuze. Case (3) is covered by using a *base fuze* with inherent or additional delay built in.

DETONATION The rapid molecular disintegration which takes place when a high explosive is initiated. The detonation wave moves at a speed between 3,000-10,000mps (9,840-32,800fps) and is capable of crossing a metal barrier, though it can be stopped by an air gap. In munitions which require to detonate, therefore, the train of explosive must be continuous.

An explosive which detonates (a high explosive) must be initiated by a detonator, ie, an initiating device which produces a detonation; some detonators can be made to function by being struck with a firing pin, others can be of an explosive which is first ignited and the burning of which accelerates until it converts to a detonation. The sensitivity of explosives to detonation varies considerably, and those explosives which are difficult to initiate require a graduated train of explosive (an *exploder system*) to amplify the relatively small impulse from the detonator.

DISCARDING SABOT SHELL Similar to a discarding sabot shot (see *APDS*) but where the sub-projectile is a high-explosive shell of more or less conventional form.

The principle was first explored by Edgar Brandt, a French ordnance engineer, in the 1930s; his object was to develop long-range shells for conventional field artillery by using, say, a 75mm sub-projectile in a 105mm gun. The lower weight and smaller size of the sub-projectile would allow it to be fired at a much higher velocity than a standard 105mm shell, and therefore, since it started with more velocity, it would have a greater range. The whole idea, of course, revolves around the question: is the increase in range worth the decrease in projectile size and effect at the target?

Brandt's researches were interrupted by the German occupation of France in 1940 and his ideas were taken by the Germans, who began several lines of research. Their principal interest was in developing anti-aircraft projectiles which, due to their higher velocity, would have a shorter

time of flight and thus simplify the entire anti-aircraft fire-control problem. (Put simply, this problem was that the longer the shell took to reach the aircraft's altitude, the further the aircraft would have flown, and therefore predicting its position *vis-à-vis* the arriving shell became more difficult.) Many designs were developed, particularly for the 10.5cm and 8.8cm air defence guns, but none were put into manufacture.

One advantage of the discarding sabot principle is that the sub-projectile can be designed for the best flight performance, without having to conform to any in-bore requirements; for example it is no longer necessary to have a parallel-walled section or bourrelets to ride on the bore because the sabot elements are taking care of that. Consequently the shell can be designed with a long ogive and long rear taper or whatever other shape the designer wants; this feature is obvious in several of the German wartime designs.

Discarding sabot shells were proposed for one or two post-war anti-aircraft guns, particularly the British 5in calibre 'Green Mace', but when guns were displaced by the universal adoption of guided missiles interest in this type of projectile waned. There appears to have been no post-war development in the sphere of ground artillery.

Various types of German experimental discarding sabot shells allowing an 88mm sub-projectile to be used with a 105mm anti-aircraft gun. The advantage of this arrangement is that the smaller shell will reach a higher velocity in the barrel of the gun — giving it greater range.

DISTANCE FUZE An early type of time fuze, insofar as the object was to cause the shell to function at some particular point on the trajectory, but differed because it relied not upon measuring the time of flight but upon measuring the distance travelled by the shell. The method adopted was always to measure the spin, counting the number of revolutions. Since the distance travelled during one revolution is known precisely from the gun's twist of rifling, counting the revolutions would give a very accurate measure of the shell's range. The counting was invariably performed by having a pendant weight inside the fuze or a vane which would be held stationary by the air-flow over the fuze. While the weight or vane remained still, the rest of the fuze would revolve around it, so causing gears to actuate the counting mechanism.

Designs of distance fuze were experimented with by most countries during the period 1885 to 1900, but none was ever put into service. While perfectly sound in theory, most designers overlooked the fact that the acceleration of the shell on being fired generates enormous rearward stresses which effectively lock any machinery solid, so that the commencement of operation of the fuze becomes a variable point, and the subsequent spin generates a powerful centrifugal force which frequently causes the pendant weight or vane to rotate with the fuze, entirely negating the mechanical action. Some early models, in low-velocity, low-spin guns, worked well enough to encourage further research, but as guns increased in power the drawbacks became more obvious and the idea was gradually abandoned.

Two experimental distance fuzes dating from the 1890s. Both have wind-vanes which were supposed to remain still in the airflow while the shell revolved — thus counting the revolutions. Neither was successful.

Right: The Mieg and Gathmann distance fuze of 1904 as outlined in *Modern Guns and Gunnery* by Colonel H.A. Bethell, 1910.

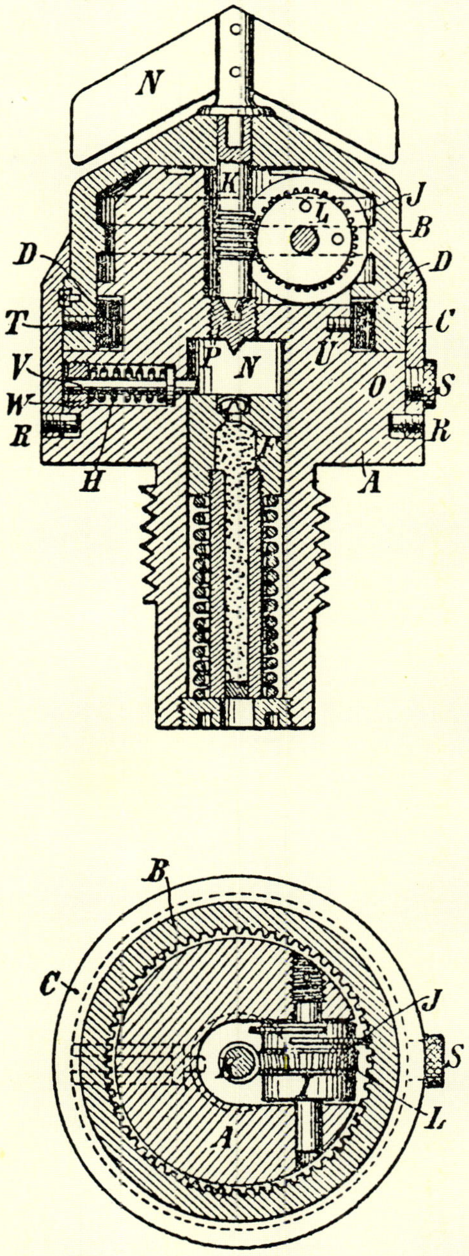

FIG. 75.

K is the spindle, stepped upon a plug P which carries the needle N. The endless screw on the spindle engages with the worm wheel L. The axle of this wheel carries the worm J carrying one turn of thread. The worm J engages with teeth inside the dome, which can be turned to set the fuze. A coiled spring D is fixed at one end to the fuze, at the other to the dome, and tends to revolve the dome on its seat. The dome is held down, yet so as to be free to turn, by the setting ring C. The ring has ratchet teeth inside so that it can only be turned in one direction, namely that tending to wind up the coil spring.

The striker F, which carries the cap, is in the stem of the fuze; it is actuated by the spiral spring surrounding it. It is held down by the locking-bolt V, actuated by the spring W, which tends to withdraw it outwards. The locking-bolt is prevented from moving outwards and unlocking the striker by the wall of the setting-ring, against which it abuts; when the hole O comes opposite the locking bolt the latter is free to fly out, allowing the striker to fire the fuze.

Before loading, the hole O is closed by the milled-headed screw S.

The spring constantly tends to unwind and revolve the dome, but is prevented from doing so by the worm J which engages with the latter. It is only as this worm turns under the action of the vane that the dome is allowed to revolve.

Thus in this fuze the action of the mechanism is assisted by the coiled spring.

To set the fuze, the dome and setting-ring are turned in the direction allowed by the ratchet till the hole O is at a given distance (corresponding to a given number of turns of the shell in flight) from the locking bolt.

DRAG STABILIZATION Method of stabilizing a projectile in flight, so that it arrives nose-first, by placing the centre of balance well back; analogous to stabilizing a spearhead by the drag of the shaft. Rarely used, and then only on short-range smooth-bore weapons. The only near-application of drag stabilization in current use is the 84mm shell used with the Carl Gustav recoilless anti-tank gun; this has an extended skirt for some distance behind the shell which adds an element of drag stability to the projectile which is primarily stabilized by spin.

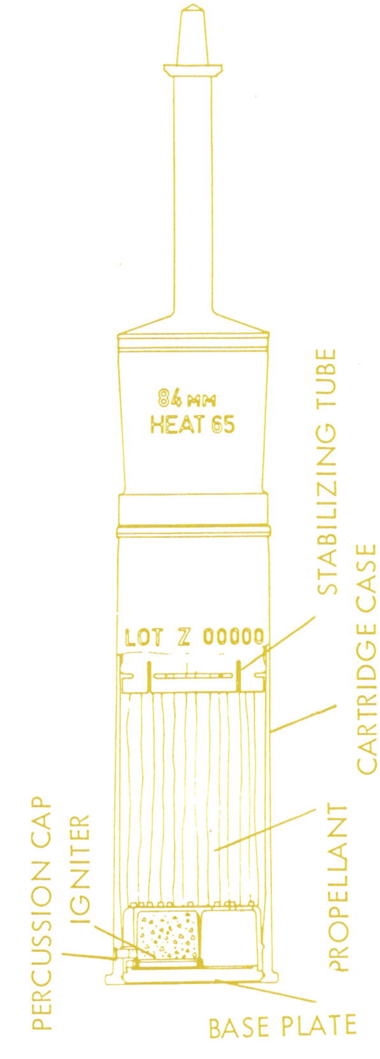

The 84m Carl Gustav recoilless projectile which uses a drag stabilization tube at the rear end.

**Below: A Victorian solid bronze drill shell for a 12-pounder coastal defence gun.
Right: A wood and brass drill shell for a 25-pounder field gun.**

DRILL AMMUNITION Ammunition which is completely inert, ie without any explosive components, and which is provided for the training of troops in loading and operating guns. The ammunition resembles the live round in shape, size and weight, but is usually of wood, bronze or other solid material so as to withstand the rough and tumble of repetitious drill use. The cartridge case may be of the service pattern, but the charge bags, if used, will be of canvas or leather and there will be no primer in the case so that the firing pin of the gun cannot be damaged.

DRIVING BAND Soft metal band pressed into the circumference of a shell towards its rear end and which is engraved into the gun's rifling grooves by the force of the cartridge explosion. In following the curve of the grooves the shell is spun to stabilize it in flight. Known in the USA as the 'rotating band'.

The driving band as it is known today was developed by Vavasseur, a British engineer, in the 1880s (he received £10,000 for his invention). The band is a continuous ring of copper which is slipped over the shell during manufacture and then pressed into a groove cut in the shell wall. The bottom of the groove is incised in various patterns — hatchures, wavy lines, axial lines — so that the copper flows into these incisions and is locked securely to the shell; this ensures that when the rifling grips the band the rotary motion is transferred to the body of the shell.

The external surface of the band is usually shaped to suit various requirements; it is desirable that the band should be sloped at its forward end so as to grip the interior of the gun when loaded, preventing the shell from slipping back when the gun is elevated. It is also desirable that the rear should be capable of being expanded outwards into the rifling, but it should not be so thin that it can be thrown out by centrifugal force in flight, so presenting an obstruction to the airflow.

Other materials can be used; gilding metal (an alloy of copper and zinc) is favoured by the USA, while the German Army in World War II used sintered iron in order to economize in copper. Nylon has been used on some modern designs, particularly for the bands on APDS shot, but gave problems with dimensional stability in its early days.

When the band engraves into the rifling it places considerable crushing stress on the shell; it also places a heavy shear stress on itself as it attempts to transmit the rotation from the rifling to the shell. In cases where either of these stresses may be too great, double or even triple driving bands can be employed. This distributes both types of stress more evenly.

Shells are occasionally seen which appear to have an extra driving band at the shoulder; these, on examination, will always be found to be

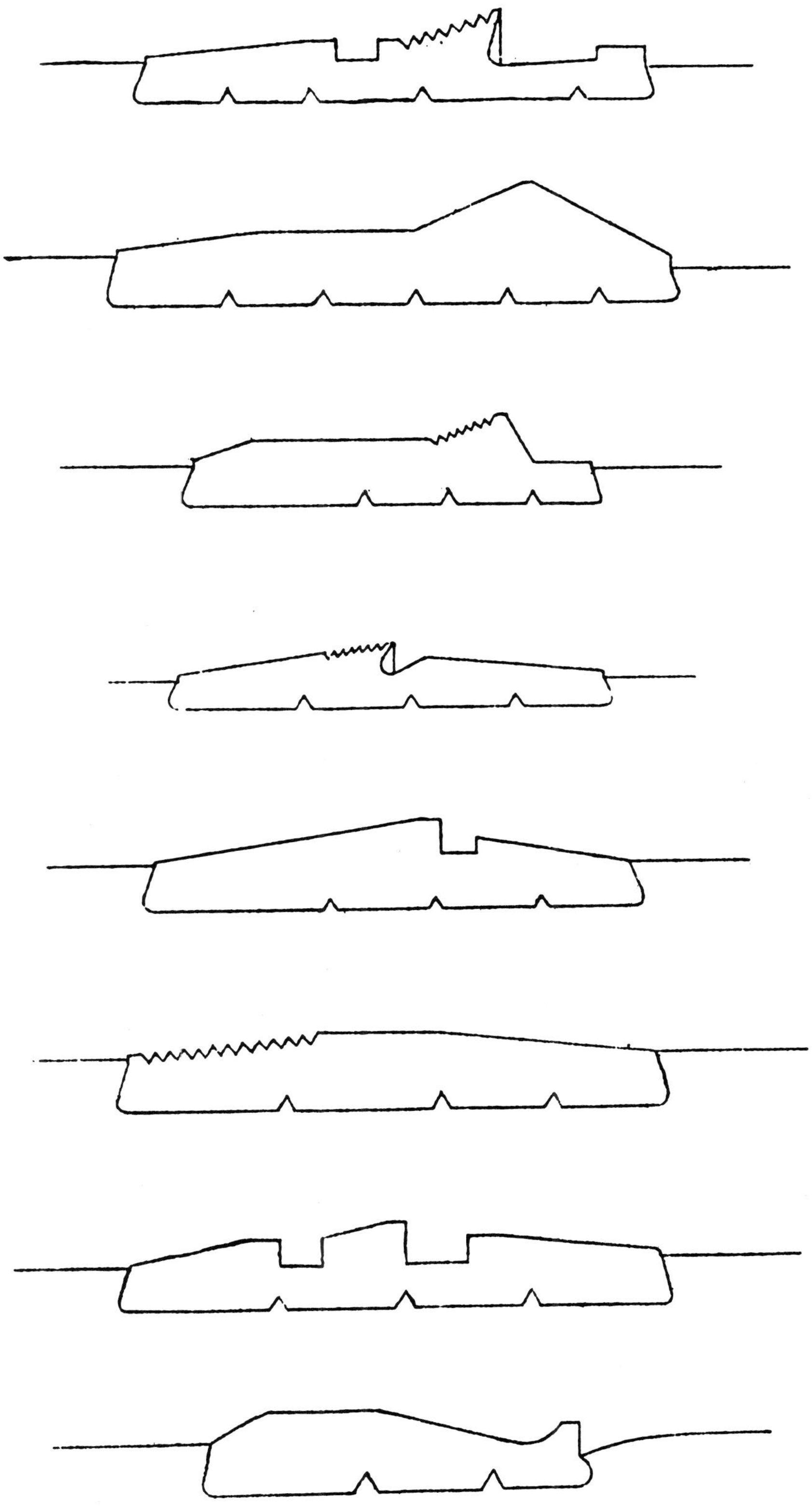

A selection of driving band designs showing various systems of sealing in the propellant gas. The head of the shell is to the left.

below rifling diameter and are there solely as steadying bands on shells with long noses. They do not engrave and play no part in spinning the shell.

DYNAMITE Dynamite is a blasting explosive invented by Nobel; it is simply nitroglycerine absorbed into diatomaceous earth in order to make the substance more stable and easier to handle. It gains a mention here because it was once proposed as a projectile filling for use with a gun, albeit a somewhat unusual gun.

The 'Dynamite Cannon' was the invention of Mr Mefford of Chicago in 1883, though his idea was usurped and became the 'Zalinski Dynamite Gun'. The object in view was to be able to use dynamite as the bursting charge of the shell, and to do this required a very gentle acceleration. The Dynamite Gun was therefore a gigantic air-

gun, normally of 8in (203mm) calibre, which pitched a fin-stabilized dynamite-filled projectile to a range of about 5 kilometres (3 miles). A number were installed as coast defence weapons around New York and San Francisco, and one was tested in Britain. One was also installed in a warship, the USS *Vesuvius*, and used, to little effect, in the Spanish-American War. The principal defect was that the gun required a massive air-compressing installation to provide the 'propellant', and just as the weapon was about to be adopted in quantity, advances in conventional gun design and the mastering of the use of other types of high explosive as shell fillings meant that far cheaper and less bulky conventional guns could outperform the Dynamite Cannon with ease. By 1900 they had all been scrapped.

The Zalinski Dynamite Gun showing the compressed air cylinders on the mounting. One was installed on the USS *Vesuvius* and used during the Spanish-American war.

E

ELECTROSTATIC FUZE Shell fuze for use with explosive projectiles which relies upon electrical impulses generated within the fuze. The principle was widely used by the Germans in aircraft bombs during World War II, charging the fuze immediately before releasing it so that it formed one plate of a capacitor, the surface of the earth acting as the other plate. When the bomb reached a point close to the earth where the electrical charge was balanced by the earth's electrical field, the bomb detonated.

The Germans also developed an electrostatic fuze for small-calibre (37–50mm) projectiles, sometimes known as the 'Dust Fuze'. It was hollow and contained a quantity of talcum powder; slots in the fuze nose allowed air to rush in during the shell's flight and agitate the dust. When the dust particles collided with one another and with the casing of the fuze they generated static electricity and this was extracted and used to charge a capacitor. On impact, a simple contact switch in the nose of the fuze was closed, allowing the capacitor to discharge into an electric detonator and thus detonate the shell.

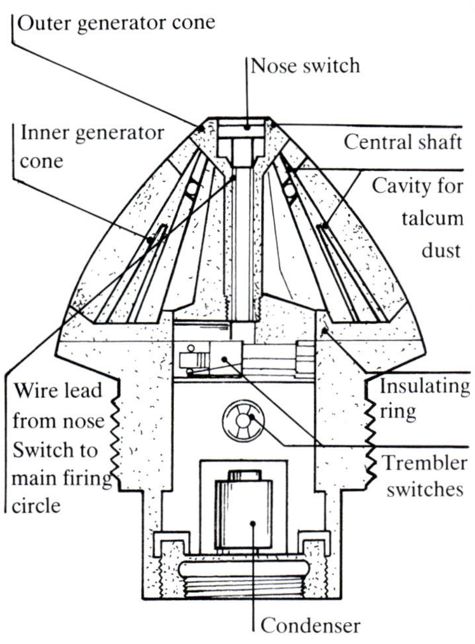

An early electrostatic fuze developed for small-calibre projectiles, the Dust Fuze used talcum powder to generate its electrical charge.

EXPLODER General term used for the intermediary or amplifying charge in a high-explosive shell which accepts the detonation from the fuze and in turn detonates so as to pass on, to the main bursting charge, a much more powerful detonation.

The main bursting charge of a shell needs to be relatively insensitive so that it will withstand the knocks and bumps of transportation and violent acceleration as the shell is fired from the gun. However, the most sensitive part of the system, the detonator in the fuze, is exceptionally small — perhaps weighing as little as 7–10 grains (0.5–0.7grm) — and its detonation is insufficient to make much impression on several kilograms of TNT; it *might* set it off, but it is not to be relied upon. The fuze detonator, in fact, sets off a small charge of explosive in the magazine of the fuze but this, too, is not thought to be sufficient for reliable initiation of the main charge. Beneath the fuze, therefore, is the exploder, a substance more sensitive than the main bursting charge so that it will unfailingly be detonated by the fuze, and with a high rate of detonation so that it will administer the necessary impulse to the whole of the main charge and detonate it efficiently. It was the mastering of this sequence of graded detonations which was the greatest problem in perfecting reliable high-explosive shells during the early years of the 20th century.

The explosive chosen for exploders is usually 'Tetryl', known in British service as 'Composition, Exploding' or 'CE' (it will be noted that

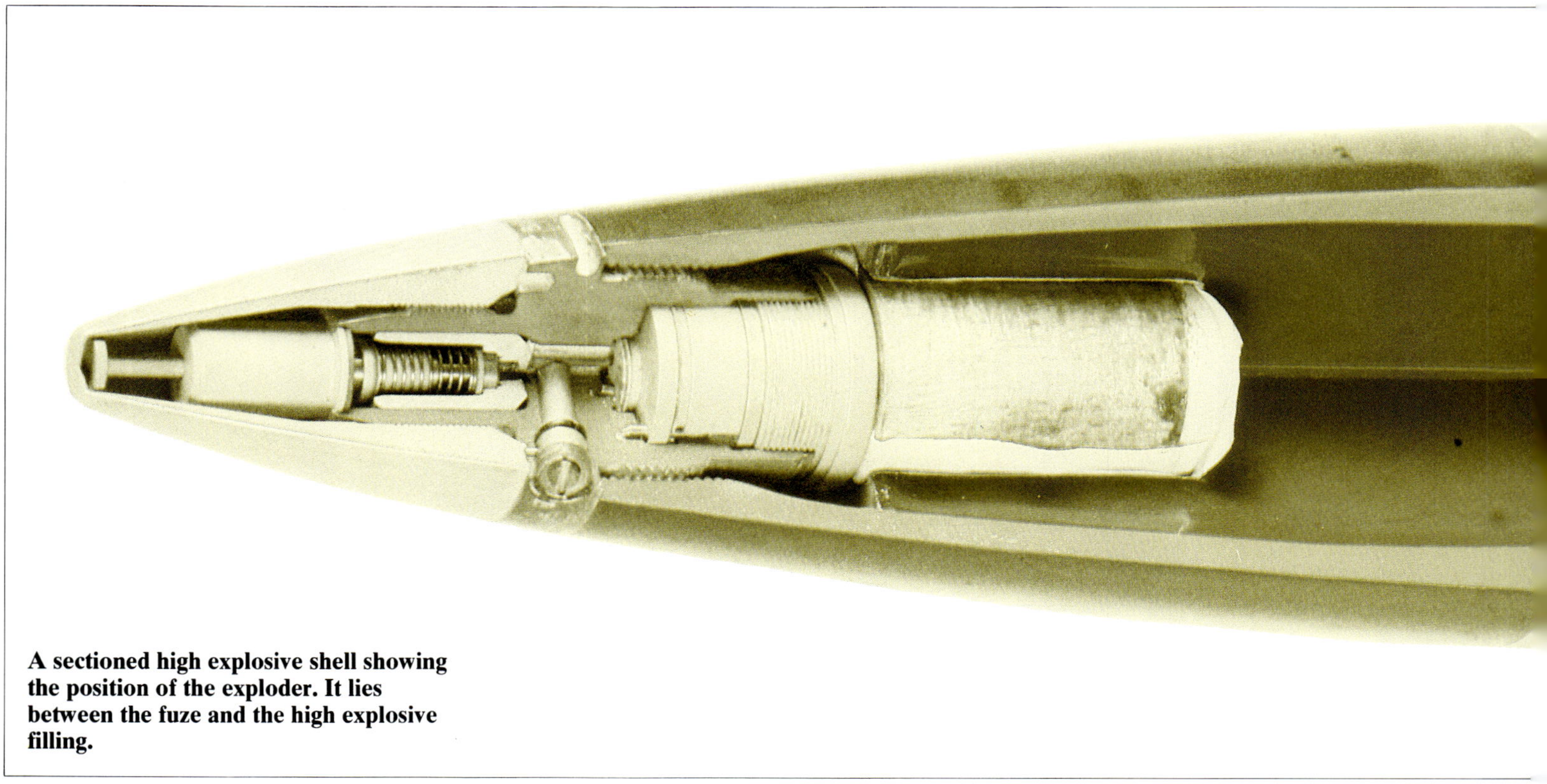

A sectioned high explosive shell showing the position of the exploder. It lies between the fuze and the high explosive filling.

the operative word throughout is 'explode' although the purpose is actually to detonate). Tetryl is correctly Tri-nitro-phenyl methyl nitramine and is one of the most powerful of military explosives. It is too sensitive to be used as a bursting charge, but when securely confined in a small container and buried in the head of the main charge, so that there is no chance of movement, it is safe to use in this role.

An earlier system of explodering used cloth bags of picric acid crystals; these could be initiated by a flash, so that detonating fuzes were not required, and once ignited the rate of burning accelerated until it became a detonation.

EXPLOSIVE D American name for Ammonium Picrate; also called 'Dunnite' after its inventor. Slightly less powerful than TNT, it was particularly prized for its insensitivity to shock, which made it a good filling for armour-piercing shells used in coast defence, and naval guns for shooting at armoured ships. Explosive D would withstand not only the acceleration out of the gun, but also the much more sudden deceleration as the shell struck the target and began penetrating. The defect was that the substance, when wet, forms dangerously sensitive compounds with copper and lead; it was necessary to varnish the interior of the shell to avoid it corroding the steel, and the fuze had to be made of copper- and lead-free metal. The fuze-to-shell joint also had to be carefully sealed against the ingress of moisture, particularly in naval shells which, naturally, need more protection.

EXPLOSIVES Military explosives fall into two broad categories: low explosives, and high explosives. Low explosives are defined as those explosive substances which have rates of detonation below 3,000mps (9,840fps); in fact they do not, as a rule, detonate but 'explode', ie, burn at an extremely fast rate. In ammunition their use is entirely confined to *propellants.*

High explosives detonate — that is, they undergo rapid molecular disintegration, and at a rate exceeding 3,000mps (9,840fps); in fact, the usual rate for military HEs is 7,000–10,000mps (22,960–32,800fps). They are used for main bursting charges in high-explosive shells and also as *exploders* and in fuze magazine.

The principal military high explosives are: TNT (2,4,6-Trinitrotoluene, also known as Trilite, Tritol, Trotyl, etc.) TNT is a yellowish crystalline substance with a velocity of detonation of 6,900mps (22,645fps). Its principal attraction for military use is its melting-point of 80.8°C, which allows it to be safely melted by water-baths and poured into shells in order to fill them. Since there is a 12 per cent loss of volume when molten TNT solidifies, precautions have to be taken to ensure that no cavities form in the filling; such cavities can collapse under acceleration and cause premature detonation.

Amatol is a mixture of TNT and ammonium nitrate, the salt being mixed with molten TNT. The addition of the salt lowers the rate of detonation to about 5,000mps (16,400fps) but provides a valuable economy measure. Amatol was widely used during both Great Wars but is rarely met today.

Picric Acid (2,4,6-Trinitrophenol; Lyddite, Melinite, Schimose, Ecrasite, Pertite etc,). Picric acid was originally a dyestuff, but it has a velocity of detonation of 7,350mps (24,100fps) and melts at 122.5°C. In its normal form it is crystalline or in sheets, but this form is rather sensitive and is highly susceptible to water, on contact with which it forms undesirably sensitive explosive compounds with most metals. For artillery use it was therefore melted by steam and poured into shells, and in this form was known as Lyddite by the British Army and by other countries under the various names quoted above. It has not been used since World War I, having been entirely replaced by other types of explosives.

PETN (Pentaerythritol tetranitrate: Nitropenta, Pentaryth, Pentryle, Pentolite, etc.) colourless crystals with a rate of detonation of 8,400mps (27,552fps) and a melting-point of 141°C. Too sensitive to be used alone in shells, it is usually desensitized by the addition of about 10 per cent wax. In this form it can be press-filled into shells, the melting-point being too high for convenience.

Hexogen (Cyclo-trimethyline-trinitramine; RDX, Cyclonite, T4, etc.) A colourless crystalline substance, Hexogen has a rate of detonation of 8,700mps (28,536fps) and a melting-point of 204°C. Since it will explode at 240°C, there is insufficient safety factor to permit its being melted safely and it can only be processed in the solid state. It is most usually used in a mixture with TNT (see *Composition B*) or desensitized with wax.

Octogen (Tetramethyline tetranitramine; HMX, Homo-cyclonite.) This is very similar to Hexogen except that it has greater density and is much more powerful, with a rate of detonation of 9,300mps (30,504fps). It is, in fact, a by-product of Hexogen production. Like Hexogen it cannot be used alone, and so far has seen little application in shell fillings, largely because of its cost. It is used in some shaped-charge projectiles, suitably denatured, but is principally used in demolition devices and in missile warheads.

Composition B is a mixture of Hexogen and TNT; the TNT is melted and the Hexogen stirred in in varying proportions from 60/40 to 40/60. With the 60/40 mix a velocity of detonation of 8,000mps (26,240fps) is achieved. It is virtually the international standard for filling artillery shells and is used in many other military explosive applications.

Hexotol is similar to Composition B, but in the proportion 80 per cent Hexogen to 20 per cent TNT. So far this has not been used as a shell filler.

Hexal is a mixture of Hexogen and aluminium powder desensitized with wax. It is particularly used in small-calibre anti-aircraft projectiles in order to take advantage of the incendiary effect and additional blast created by the addition of aluminium.

Trinalite (Tritonal) is a mixture of TNT and aluminium powder; it has similar properties and is used in much the same way as Hexal.

Plastic explosives are usually made of about 80–85 per cent Hexogen or PETN in a suitably soft plastic binder. They are used with *HESH* (squash-head) shells.

EXUDATION In the early days of high-explosive shells filled with TNT, the TNT could be of various degrees of purity; this was particularly so during World War I when production was more important than perfection. Unfortunately the lower grades of TNT contain impurities known as 'isomers' which have a low melting-point and at tropical temperatures separate from the TNT in the form of an oily explosive liquid. This was known as 'exudation' and was dangerous, because of the likelihood of the oil seeping into screw-threads and being nipped during the firing of the shell, leading to detonation in the gun barrel. The adoption of higher standards of purity ended the exudation risk, though it was not unknown in old stocks of ammunition even after 1945.

F

FIN STABILIZATION Method of stabilizing a projectile in flight — ie, keeping it point-first and on its correct trajectory — by using fins at the rear end. Commonly used with mortar bombs, it was only considered applicable to artillery projectiles if they were fired from a smooth-bore gun, which was exceptionally unusual. However, during World War II the Germans designed various projectiles which were much longer than usual; spinning a shell will only stabilize it satisfactorily if its length is less than about seven times its calibre, and therefore the German designers began to look at fin stabilization. In order to use this system with a rifled gun it was found necessary to develop a type of driving band which was not firmly connected to the shell but which could slip. It thus dug into the rifling to provide the necessary gas seal, but did not communicate much of the spin to the shell. So when the shell left the muzzle the fins came into play and what little spin the shell had picked up was soon lost. In order to obtain good stability and a rapid loss of spin the fins were usually arranged to expand on leaving the gun.

Fin stabilization was abandoned after the war, but was revived by the Soviets who applied it to armour-piercing discarding sabot projectiles fired from the smooth-bored 115mm tank gun. The idea was not entirely successful and the gun was later given a slow rifling in order to add roll stabilization and make the projectile more accurate. Since that time fin-stabilized ammunition has been widely adopted for tank guns, some smooth-bore and some rifled.

A German fin-stabilized HE shell. The driving band is on a roller bearing.

FIXED ROUND A complete round of artillery ammunition (shell and fuze, cartridge case, propellant and primer) which is assembled into a single unit, allowing it to be loaded in one movement.

The advantage of fixed ammunition lies solely in the speed of loading, and thus it is to be found in weapons where this is desirable — anti-tank and anti-aircraft guns, for example. It is also well-suited to mechanical loading systems, though some fast systems need to have the shell extremely firmly fixed into the case so that it is not shaken out during the loading actions. The drawback to fixed ammunition is that the propelling charge cannot be adjusted, and thus if the weapon requires different types of charge it becomes necessary to provide a variety of different complete rounds.

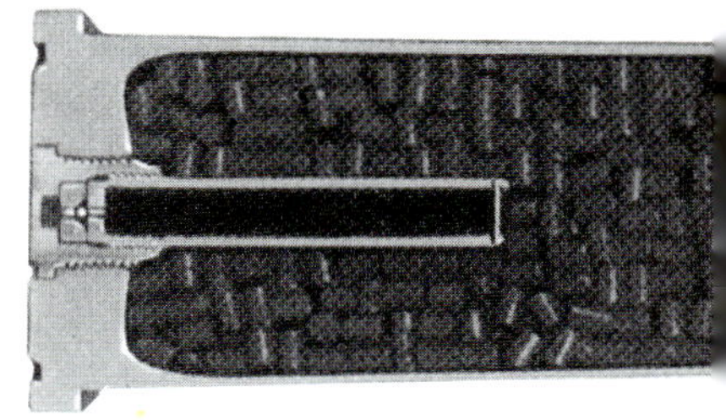

A complete fixed round of ammunition comprisi

Fixed rounds in large calibres can be awkward. Th

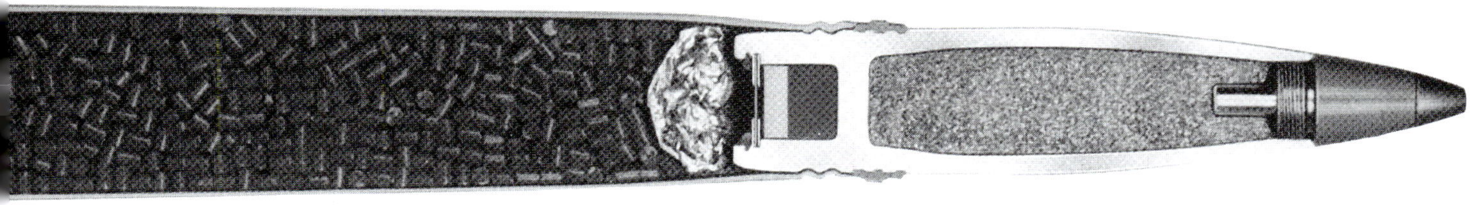

ıell, fuze, cartridge case, propellant and primer.

carry a single 15cm round in this French engraving of the 1880s.

FLARE SHELL Carrier shell containing canisters filled with a pyrotechnic mixture which burns with a bright coloured flame. The shell is a base-ejection type and ignites the canisters before ejecting them in mid-air to fall to the ground. Used for indicating targets to aircraft or for other marking purposes. Not to be confused with illuminating or star shells which provide general illumination of an area.

FLASH REDUCERS Substances added to propelling charges in order to reduce muzzle flash so that the location of the gun cannot be discovered by enemy observers.

Flash reducers may be of two types: either a substance is added to the propellant during its manufacture, or substances can be contained in bags and attached to the cartridge before

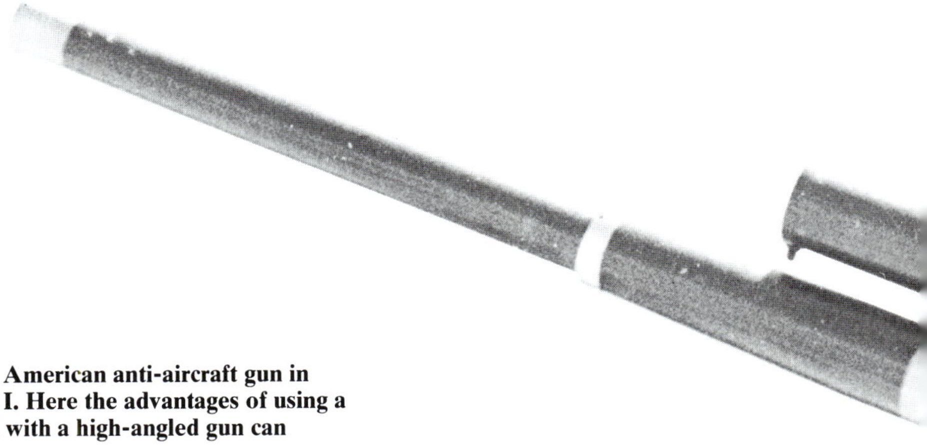

Loading an American anti-aircraft gun in World War I. Here the advantages of using a fixed round with a high-angled gun can clearly be seen.

The rate of loading — and consequently the rate of firing — can be increased when a fixed round can be carried by a single man, as with this 37in anti-aircraft gun.

loading. The substance is generally a metallic salt such as potassium nitrate which, when burned with the propellant, generates nitrogen. This, when ejected from the muzzle with the other combustion products, has the tendency to shroud the ejected gas with an inert envelope, so preventing the oxygen in the air from meeting the hot combustible gases and give rise to flash.

When not incorporated in the propellant powder, the salts are enclosed in a cloth bag which is attached to a bag charge by a string or placed inside a separate-loading cased charge.

Flash reducers are rarely encountered today, since modern propellants almost all incorporate flash inhibitors or are inherently flashless.

FLECHETTE An arrow-like inert projectile, small in size, used as a filling for *Beehive* shell. Weighing between 0.5 and 1.0grm, flechettes are lethal anti-personnel devices when launched at high velocity and in great numbers, and they have also been found useful in scything undergrowth to reveal attacking troops and ambushes.

FLYING DUSTBIN Nickname for an anti-fortification bomb launched from a spigot mortar carried on a special tank, the 'AVRE' or 'Armoured Vehicle, Royal Engineers'. The bomb was fin-stabilized and weighed about 18kg (40lb); the warhead was flat-fronted and cylindrical and the bomb had a low velocity, all of which led to the nickname. It was developed for use in the invasion of Normandy in 1944, with the specific purpose of demolishing beach defences and similar strongpoints.

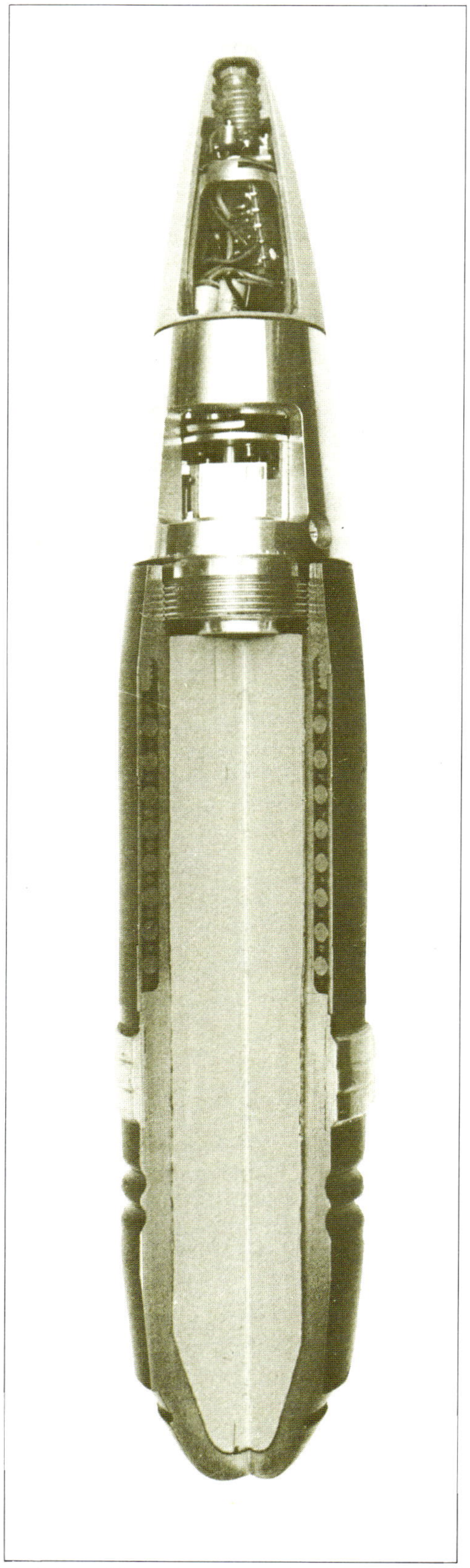

A Bofors 40mm fragmentation shell cut away to show the pre-formed fragments inside the double-skinned wall.

FRAGMENTATION Term used to describe the shattering of the steel body of a high-explosive shell. The fragments so produced become anti-personnel missiles, since they are driven off at the speed of detonation of the explosive in the shell. They also have a certain amount of anti-*matériel* effect against unarmoured targets.

The size of fragment is important; too large a fragment will soon lose velocity and become relatively harmless, and there will be too few of them to give adequate coverage. Too small a fragment means that although there will be more than adequate coverage the individual fragments will not cause sufficient damage. The ideal size of fragment is much smaller than might be imagined — about 2–3gm is generally considered the optimum, and a 105mm shell will provide several thousand of this size.

The size of fragment can be controlled, to some extent, by careful matching of the explosive and the tensile strength of the steel shell body. Modern practice is to rely less upon this than upon making fragments of the right size and lining a thin shell wall with them. A typical modern shell will have a thin wall and an inner lining of several thousand pre-formed fragments or steel balls, held in a plastic matrix. The explosive is then poured inside the matrix. On detonation the shell body is shattered — the fragmentation of this is now not important — and the pre-formed fragments or balls are blown outwards. This type of construction, however, is only economical in small calibres.

FUZES The mechanism or device in any explosive munition which causes it to function at the desired time and place.

Fuzes for artillery shells can be classified in various ways: by position (nose or base); by function (detonating or exploding); by operation (impact, graze, time, percussion, proximity) or any combination of these. The nomenclature adopted by different countries differs, but is generally sufficiently descriptive to allow the general operation to be understood. Thus, British terminology refers to 'Percussion, Direct Action' while the same fuze in American terminology would be 'Point Detonating' and in German 'Aufschlagzunder' (impact fuze). For details of specific types of fuze, reference should be made to the appropriate entries in this section.

The fuze has to meet certain basic requirements. It must be absolutely safe, and not be liable to detonate the shell during transportation, handling, loading, firing, and during its passage through the gun barrel. It must then be able to function as required, either on striking the target, or at some pre-set time after firing, or when it is within lethal distance of the target. These requirements have led to some extremely complicated designs in order to provide 100 per cent assurance of both safety and effectiveness.

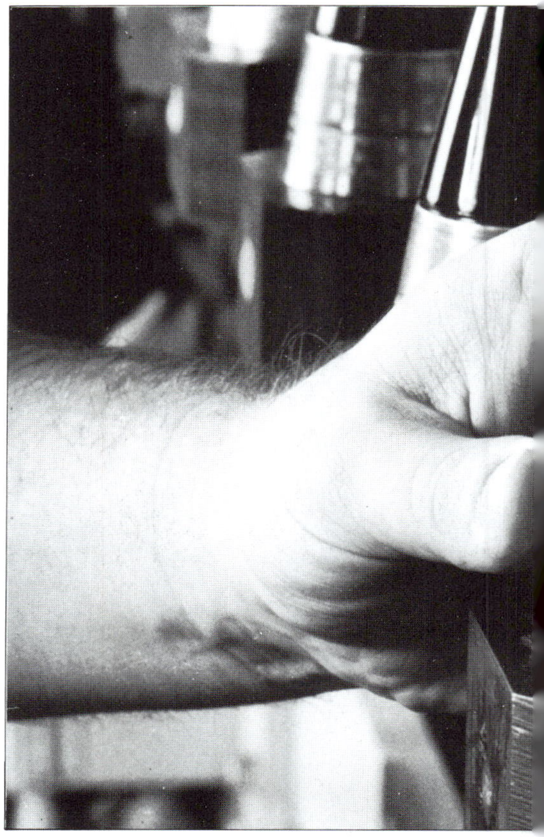

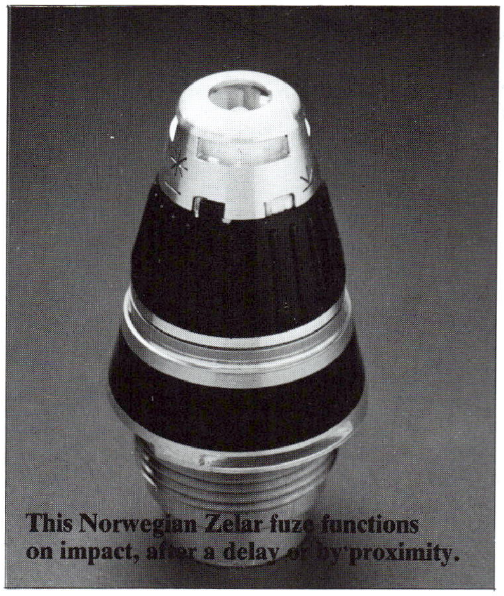

This Norwegian Zelar fuze functions on impact, after a delay or by proximity.

In general, the designer of the fuze makes use of the peculiar forces acting upon the fuze during firing to operate mechanical devices which will convert the fuze from the safe state to the functioning state, a process known as 'arming' the fuze.

In sequence these forces are: the acceleration of the projectile when fired; the rotation of the shell as it is generated by the gun's rifling; the slight check to the shell's flight as it leaves the

Above: A production line of modern proximity fuzes. The first design was patented in Sweden in 1938 and first made in Britain in 1940 in rockets used against daylight raiders.

Right: A German AZ23 impact fuze showing the centrifugally operated safety slide which governs whether the fuze operates instantly or after a delay.

gun barrel and acceleration ceases; and the sudden deceleration as the shell strikes the target. By making use of these forces in the correct sequence, safety devices can be unlocked and the fuze brought to the armed condition a safe distance from the gun muzzle. In effect, the fuze is a combination lock, the key to which is firing it from a rifled gun. Smooth-bore weapons, obviously, require a slightly different set of keys, since there is no spin, which both complicates design and simplifies manufacture.

A typical gaine. The detonator is carried in a centrifugal shutter which is locked by the vertical pin. On firing, this drops, allowing the shutter to turn. Its movement is controlled by the escapement to delay the alignment of the detonator until the shell is well away from the gun.

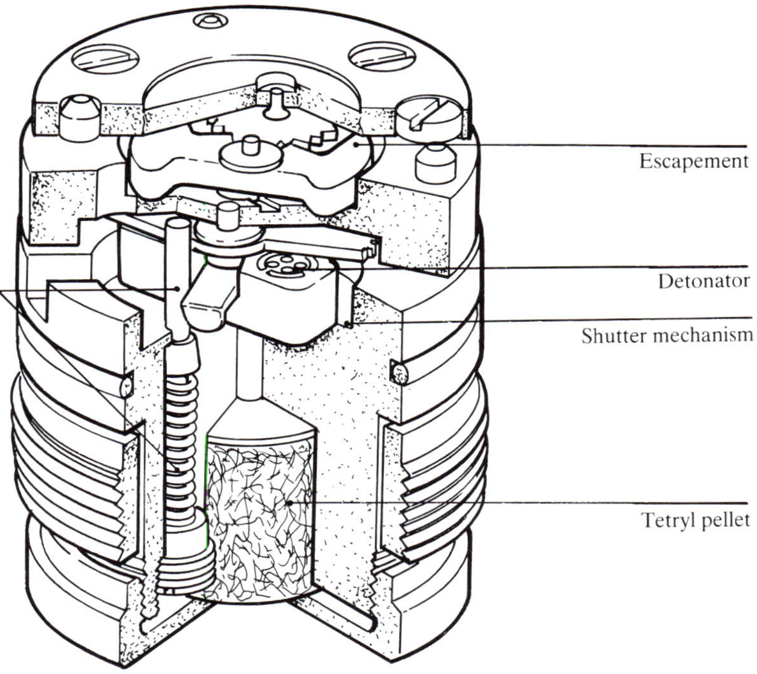

G

GAINE A gaine is a device which, fitted underneath a fuze, converts the flash from the fuze detonator to a detonation suitable for initiating a high-explosive shell. It may also incorporate a *bore-safety* device; the American term 'booster' is synonymous with 'gaine' and a booster exhibits both these primary features.

The need of a gaine arose in the early days of high-explosive shells when the existing fuzes were all designed to deliver a flash, from a gunpowder-filled magazine, in order to initiate either powder-filled explosive shells or shrapnel shells. At first the flash-to-detonation requirement was the only problem requiring solution, but when faulty fuzes detonated shells in the bore, blew up guns and killed gunners, the bore safety feature was added. For impact fuzes, gaines were soon outmoded, the fuzes being designed to produce the required detonation and also to incorporate bore safety. But the gaine stayed in service throughout World War II with some anti-aircraft guns, since the powder-burning time fuzes were all designed to produce a flash and the conversion and bore-safety feature was required. With the adoption of mechanical and electronic fuzes, advantage was taken of the new designs to incorporate the desired features, and the gaine was finally discarded from British service in 1945-50. It has remained in US and other services, as the booster, though these are now generally permanently attached to the fuze and are no longer separate items.

GAS GUN Not, as might be thought, a gun for shooting gas shells, but a gun which uses inert gas as part of the propellant system. Construction of the gun is more or less conventional, with a chamber and rifled barrel, but the chamber is perfectly cylindrical and ends in an abrupt step when the rifled portion begins. There is a loose piston head in the chamber which makes a gas-tight seal around its edge. This is inserted after the projectile has been loaded and is placed fairly well back in the chamber. The space between the piston and the projectile is then filled with inert gas under pressure. A conventional bag charge is then loaded behind the piston, the breech closed, and the charge fired. It drives the piston down the chamber at high velocity until it meets the step, whereupon it stops.

In its movement it has compressed the gas by forcing it from the wide chamber into the narrow bore behind the projectile. As a result the chamber pressure originally developed by the charge is considerably magnified and the projectile is shot from the barrel at an extremely high velocity.

As might be imagined, the process of loading the projectile, then carefully fitting the piston, then pumping the space full of gas, and finally firing the weapon, is a time-consuming business, and the gas gun can hardly be considered a quick-firer. Indeed, it never even sought acceptance as a weapon, being developed in the early 1960s purely as an experimental tool for research into high velocities. Nevertheless, its ballistic performance was impressive. It is cited here as an example that there are ways of making guns work besides those in common use.

GAS SHELL Common name for shells which contain chemical warfare agents — ie, poison gas. More properly called *chemical shell.*

GESSNER PROJECTILE The principal drawback of the shaped charge in conventional artillery use is that spinning it causes the jet to disperse, as a result of centrifugal force, thus degrading the penetrative performance. The Gessner Projectile (or 'G' projectile) was designed to counter this defect. It consists of two projectiles, one carried inside the other. The outer one is no more than a hollow sheath of the requisite ballistic shape, carrying a driving band. The inner one is a cylinder carrying a shaped charge, and the two projectiles are separated by two or more sets of roller bearings. Thus when the shell is fired, the outer jacket takes up spin

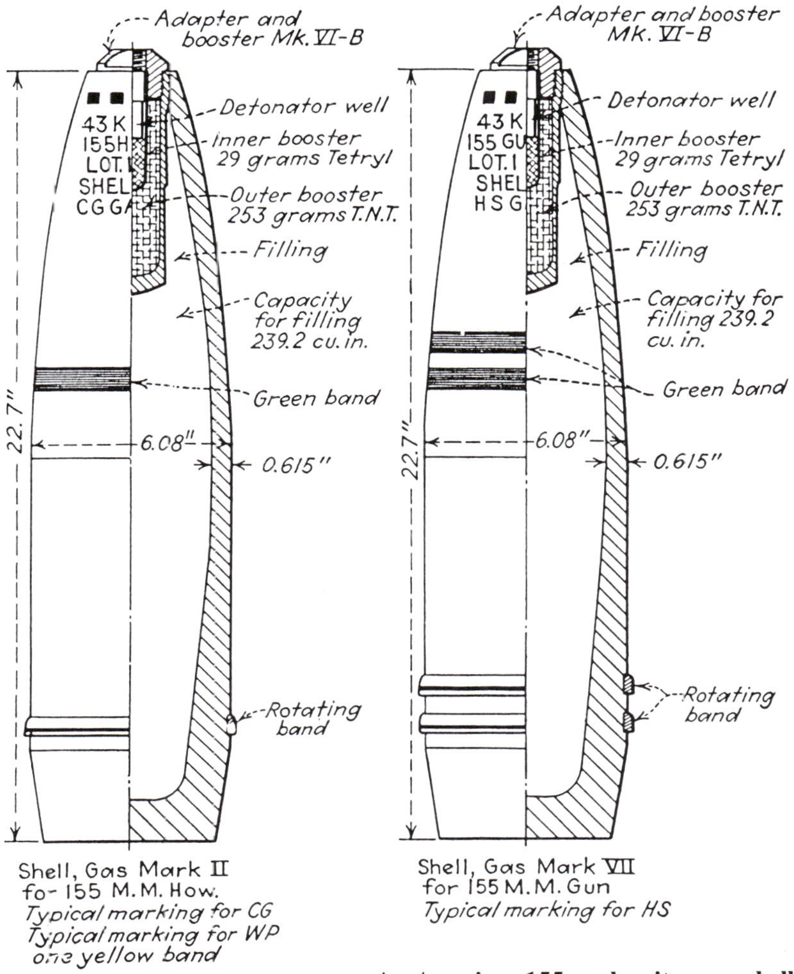

An American 155mm howitzer gas shell from the 1930s showing the typical method of filling used in that period.

and stabilizes the unit, while the inner jacket, because of the roller bearings, takes up very little of the spin. The inner unit is usually connected to the tip of the outer shell, which is loose and has vanes or is otherwise arranged to take advantage of the air flow to damp out any spin which the inner unit may have. When the projectile hits the target the outer shell does nothing, while the inner unit detonates its shaped charge and does the penetrating of the target. Since the shaped charge has little or no spin, the effect on the target is near-optimum.

The drawback to the Gessner shell is simply that it is expensive to manufacture and cannot be used with weapons having very high acceleration rates, since under very severe 'g' forces the roller bearings jam and the inner unit takes up the full amount of spin. It is used by the French Army for their 105mm tank gun and field howitzer.

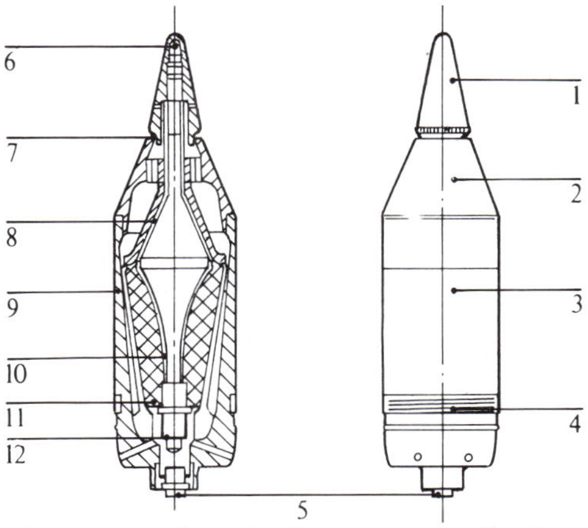

The construction of a Gessner projectile: 1. Light ballistic cap. 2. Steel nose. 3. Body. 4. Driving band. 5. Tracer. 6. Fuze. 7. Interior steel cap. 8. Guidance cone. 9. Interior steel envelope. 10. Shaped charge cone. 11. Explosive. 12. Fuze safety device.

One of the few Gessner projectiles in service is this French 105mm shaped charge shell used by the gun on the AMX-30 tank.

GRAZE FUZE A fuze intended to operate should the shell 'graze' the target, as for example when arriving at a low angle and striking the shoulder of the shell rather than the nose. The fuze mechanism reverses the normal impact operation by having a loose 'graze pellet' capable of moving longitudinally. This pellet is heavy and may carry either a firing pin or a detonator, at the designer's choice. In front of it is either a detonator or a firing pin, and the two are kept apart by a light spring during the shell's flight. Prior to firing, the pellet is normally locked away from its companion pin or detonator by a locking bolt which can only withdraw when a high rate of spin is applied. When the shell bearing the graze fuze is fired, centrifugal force withdraws the locking bolt, leaving the pellet free to move. On the shell striking the target, or even being checked in its flight, the inertia of the pellet causes it to fly forward so that firing pin and detonator meet. The subsequent flash is channelled past the pellet to strike a second detonator which initiates the filling of explosive in the fuze magazine and thus the shell.

It will be appreciated that there is a finite period of delay during which the pellet is moving forward and the flash moving back, and during this time the shell could be travelling farther, so that a graze fuze will function as a delay fuze and allow a shell to break through light obstacles and detonate inside. It also allows the shell to bounce, or ricochet, into the air so that the effect of detonation is a low air-burst. Base fuzes, since they cannot impact the target directly, are always graze fuzes. Many impact fuzes carry a graze element to act as a 'clean-up' feature should the shell not strike the target nose-on.

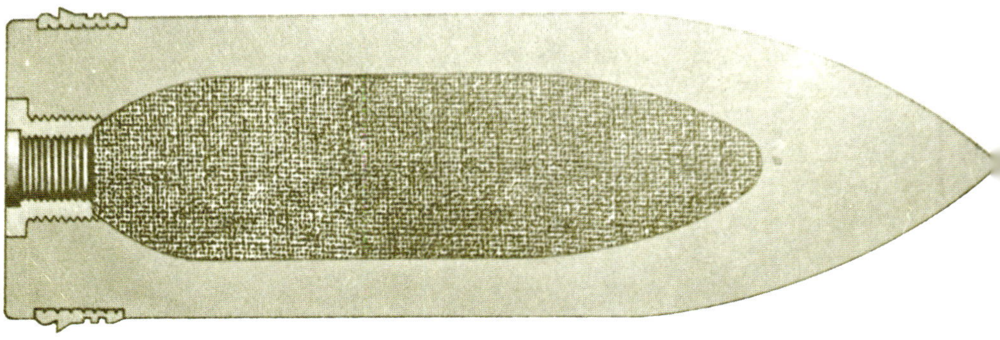

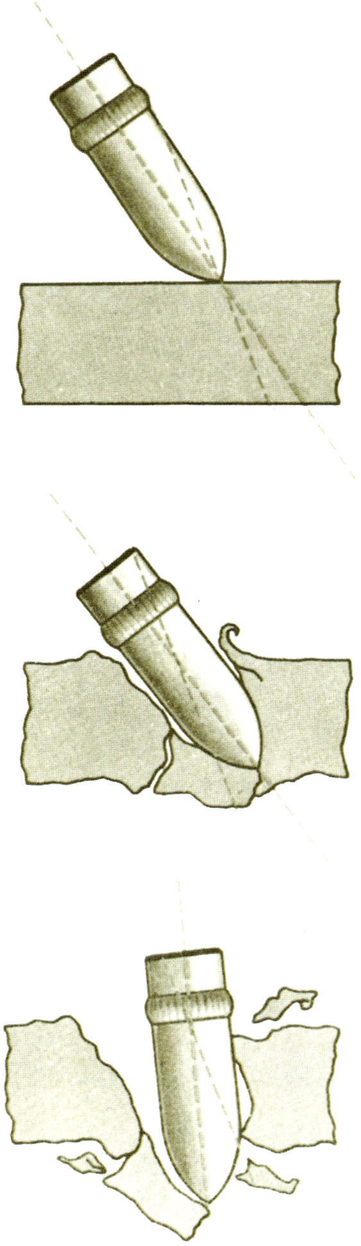

A typical high explosive armour-piercing shell with a sequence showing how it pierces armour.

H

HE/AP Abbreviation for 'High Explosive, Armour Piercing' and refers to a shell with a hard steel point and a small cavity in the rear which is filled with high explosive and fitted with a base fuze. The shell is intended to break through armour by the impact of its hard point, and then detonate inside the target, behind the armour, by the graze action of the base fuze.

This type of shell was first developed in the 19th century for attacking warships. It was then brought into use against tanks, but the improved armour of modern tanks has rendered this type of shell almost useless except against light vehicles. No major Western army currently uses HE/AP in anything other than cannon ammunition below 30mm calibre for attacking aircraft and lightly armoured APCs; the Soviet Army is believed to still use some HE/AP in artillery and tank gun calibres.

HEAT Abbreviation for 'High Explosive, Anti-Tank', and always used as a synonym for *shaped charge* ammunition.

HE/I Abbreviation for 'High Explosive, Incendiary', a type of projectile used with 20–30mm cannon against aircraft.

Early designs of HE/I used HE fillings with pellets of magnesium incendiary material carried in the explosive, so that the detonation of the shell threw the incendiary pellets in all directions. Others used a shell partly filled with HE and partly with incendiary composition. The current fashion is to use aluminized explosives as the sole filling, since these combine the blast effect desirable to disrupt the target and generate intense heat which provides the required incendiary effect.

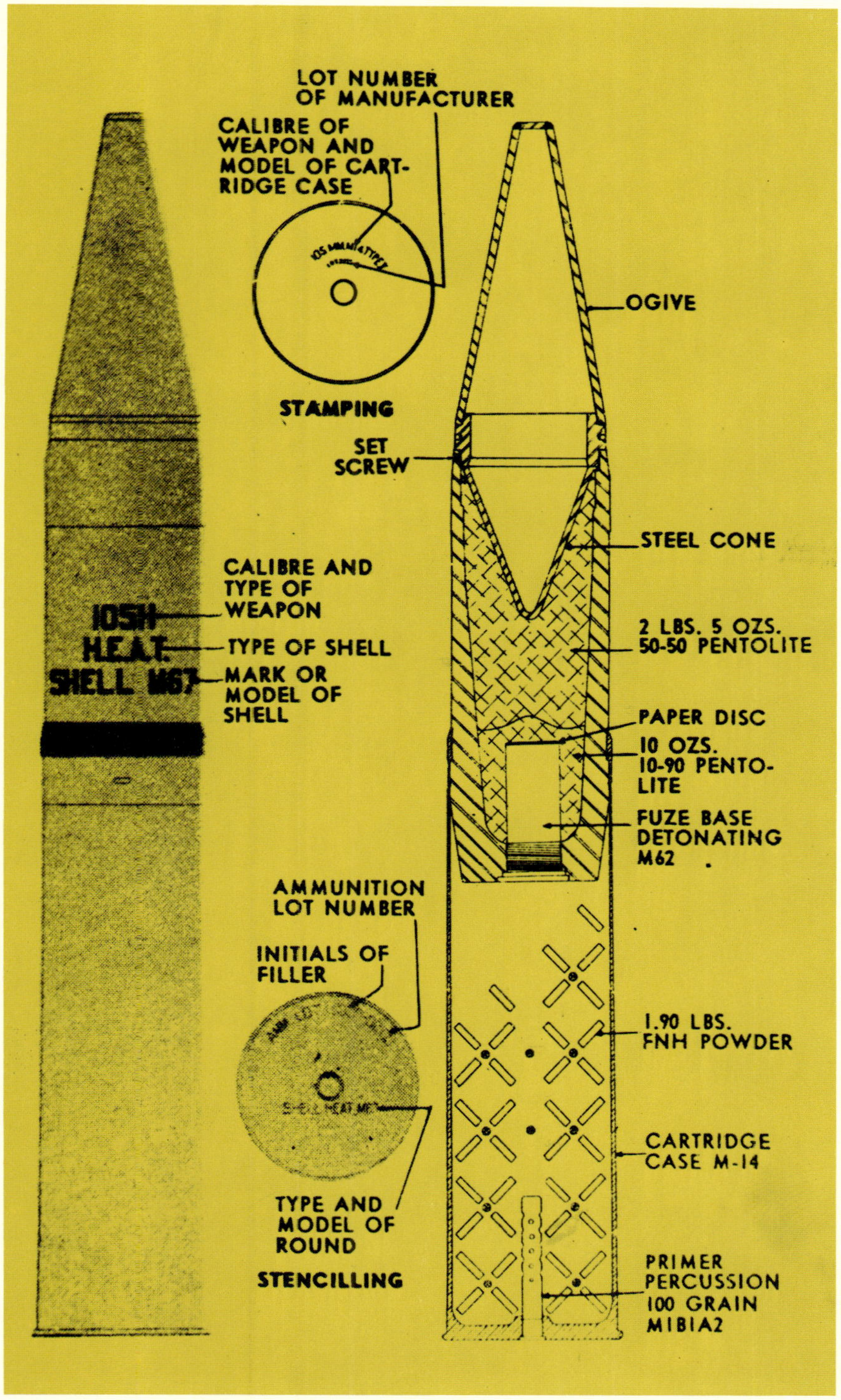

An American 105mm howitzer hollow high explosive anti-tank round of 1942 vintage showing its construction.

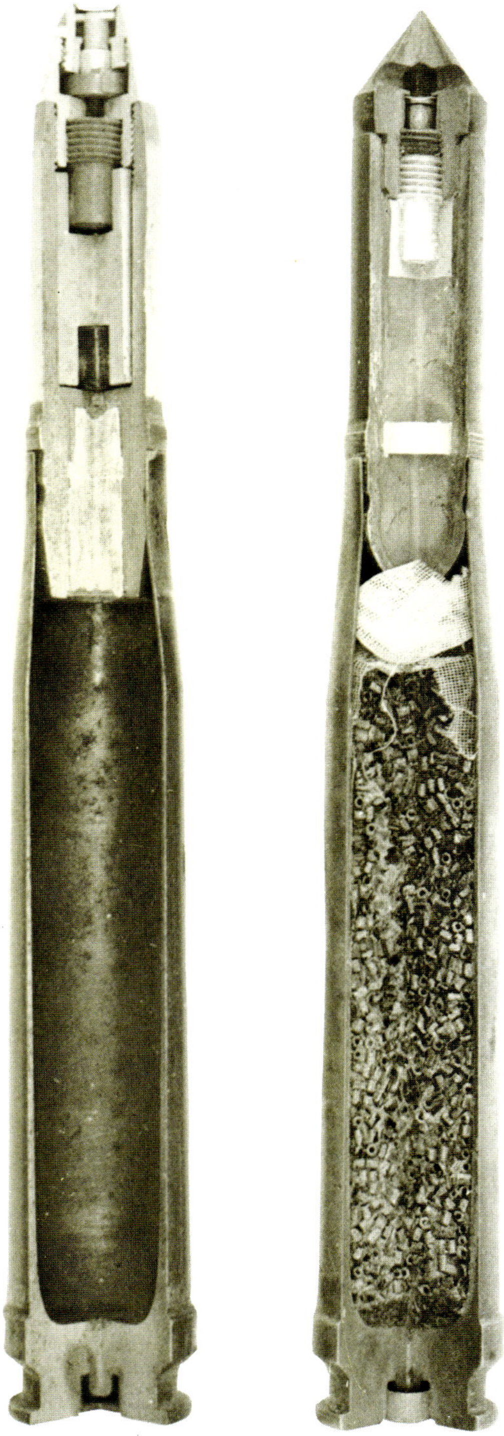

Two types of 20mm HE/I shells: Standard with self-destruction (left) and a fuel tank-piercing round.

A 120mm tank gun HESH shell, whole and sectioned, showing the plastic explosive filling and the base fuze.

HESH Abbreviation for 'High Explosive, Squash-Head', a type of shell used for attacking armour and other hard targets.

The HESH shell originated in 1943 with a design by Sir Denis Burney for his recoilless guns; at that time the shell was intended for the attack of reinforced concrete fortifications, and was known as the *wallbuster*. It consisted of a thin steel casing with a sharp point, a wire mesh inner lining, a filling of plastic high explosive, and a base fuze. On striking the target the thin outer casing collapsed and allowed the wire mesh bag and explosive to 'squash' on to the surface of the target. The base fuze then detonated the explosive and the detonation wave was transmitted into the fabric of the target as a shock wave. What happened then is rather involved, but can be best described by saying that the initial wave front passed through the material until it reached the other side, where it was reflected; this reflected wave was then met by the next shock wave, which amplified the effect and cause a large portion of the inner surface of the target to break free. This broken portion was blown off at high velocity, accompanied by a cloud of fragments.

The Wallbuster shell was successful against concrete and was tried against armour plate, and to everyone's surprise worked just as well, blowing off a large 'scab' from the inner surface. The design was not used during World War II and after the war the idea was researched more thoroughly and a fresh design, using a rounded nose and no wire mesh bag, was developed. This was put into British service in the early 1950s, and it was later adopted by the US Army under the name 'HE Plastic' or 'HEP'. It is still in service with these armies and with the German Army, principally as a tank gun projectile but also with some recoilless anti-tank guns.

HESH makes very little impression on the outside of the tank, but blows a massive scab from the inner surface; this then whirls around inside the tank, and, together with several hundred fragments, effectively wrecks the inside of the vehicle and kills or injures all the crew. The filling is selected for its rate of detonation and is usually Composition B or PETN/Wax. The only drawback is that the thin wall restricts the stresses the shell can take during firing acceleration, and velocities are usually kept to about 800 metres per second (2,625 feet per second). This means that the trajectory is distinctly curved, and errors in the estimation of range, particularly against such a small target as a tank, can easily lead to a miss. On the other hand, like shaped-charge munitions, the HESH shell is not dependent upon velocity for its effect, since the armour-defeating capability results from the explosive carried in the shell. The shell can also be used against other hard targets, such as concrete defences, and in emergency makes a passable high-explosive anti-personnel shell, since it gives ample blast but little fragmentation.

HE SHELL High-explosive shell, the standard projectile for most types of artillery and used for anti-personnel and anti-*matériel* fire against the majority of day-to-day targets on the battlefield.

The HE shell is descended from the 'common shell' of early breech-loading days, which was so-called because it was in 'common' use against all sorts of targets. The common shell was a hollow cast-iron shell filled with gunpowder and fitted with an impact fuze. With the adoption of high explosives such as Lyddite, the same shell was used, merely being filled with the new explosive, though it was found necessary to varnish the interior to prevent the explosive corroding the metal. When TNT was adopted, it was found that the violence of the explosion shattered the shell almost to dust, and that few usable fragments were produced. Cast steel therefore became the shell body material, a move that was hastened by the need of greater strength in the shell body in order to resist the greater accelerations as guns of higher velocity were developed.

Throughout World War II cast steel remained customary in many shell designs, since it was cheap and made no demands on the higher grades of steel. But the use of cheaper grades of steel meant making the shell body quite thick in order to support both its own weight and the weight of the nose fuze during acceleration, and this meant that the cavity for the explosive was

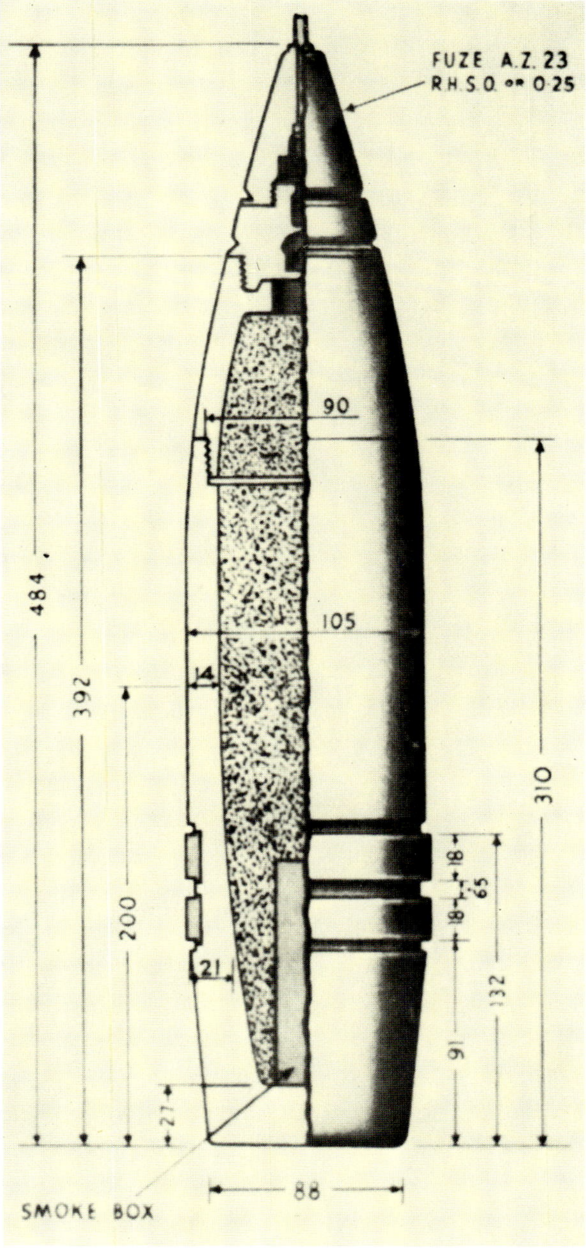

A German 105mm HE shell from World War II. It has double driving bands and a smoke box to aid observers.

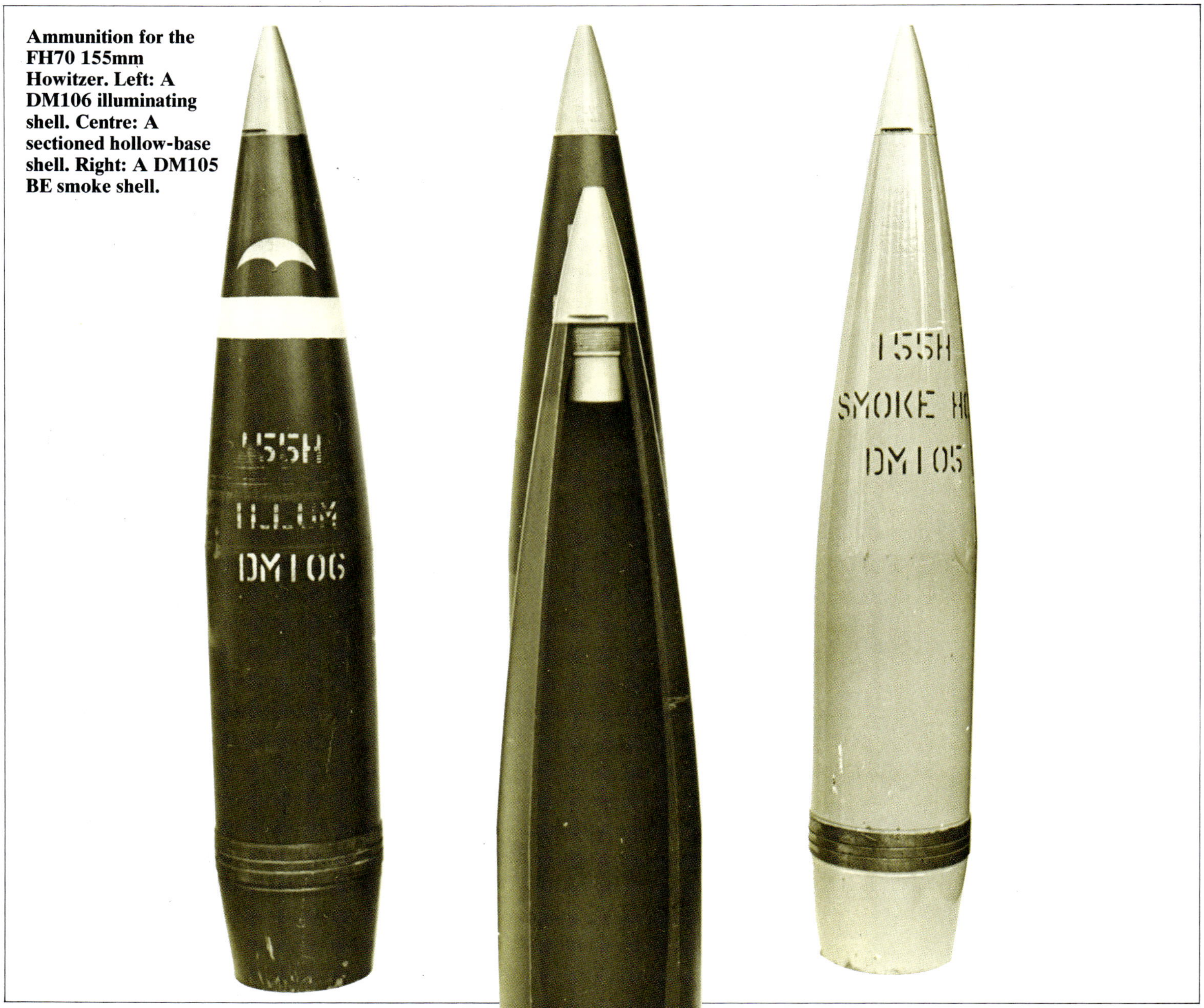

Ammunition for the FH70 155mm Howitzer. Left: A DM106 illuminating shell. Centre: A sectioned hollow-base shell. Right: A DM105 BE smoke shell.

relatively small; the average HE shell of 1939-45 had an explosive filling which was no more than 5 to 10 per cent of the total weight of the fuzed shell.

After the war, one demand made by all artillery forces was for greater effectiveness of the shell, and as a result it is now standard practice to manufacture HE shell bodies from forged steel, of greater tensile strength, to allow a larger cavity and a higher ratio of payload to weight.

The shell body is internally varnished, partly to discourage any interaction between the explosive and the steel (although this is unlikely with modern explosives) and principally to ensure good adhesion of the filling to the walls of the shell, since it is vital that there be no movement of the filling under acceleration. The explosive is poured into the shell in a molten state and a former is inserted into the nose in order to make a cavity in the forward end of the filling. When the filling has set solid the former is removed and the cavity carefully machine-cut to the correct dimensions. This cavity is for the exploder, and the cavity may be lined with an aluminium or waxed-paper tube. The exploder is then fitted and the shell fuze screwed into position, or the shell may be closed with a transit plug, the fuze not being fitted until just before firing.

HEXOGEN A high explosive, variously called RDX (for 'Research Department Explosive — a British term), Cyclonite, or T4. For further details see under *Explosives* above.

HOLLOW BASE SHELL Type of shell, developed in the 1970s, which has the base recessed to form a hemispherical cavity. Due to the extended 'skirt' effect at the sides of the shell, the low-pressure area which gives rise to base drag tends to be concealed within the cavity, and thus the drag is reduced. Hollow-base shells are used by the Anglo-German-Italian FH70 155mm howitzer and by the French GCT155 and TR155 155mm howitzers, and they give an increase in range of some 10–12 per cent over flat-based designs.

HOLLOW CHARGE A widely used alternative term for *shaped charge.*

I

IGNITER BAGS Cloth bags stitched to one or both ends of bag charges and containing gunpowder. Their purpose is to be ignited by the flash from the tube or primer and amplify that flash so as to efficiently ignite the entire propelling charge. The igniter bag consists of a double thickness of cloth stitched into pockets, each pocket holding a filling of gunpowder; this system ensures that no matter how the charge may be shaken while being handled, nor at what attitude it may be placed in the gun chamber, the powder will always be distributed fairly evenly across the entire bag and will thus intercept the flash from the tube or primer. This bag is then stitched to the base of a bag charge. In some cases it may actually form the base of the bag charge, and in some cartridges an igniter is stitched to both ends; this relieves the gunner from the need to remember which end to load next to the breech, and it also provides a supplementary igniter which will be lit by the flashover from the rear igniter and which will then ensure all-round ignition of the propellant.

While the bag-end igniter is by far the most common type, since it appears on all bag charges, there are other special types of igniter.

Supplementary igniters may be (as described above) second igniters attached to the front end of the bag, or they may be smaller igniters carried inside the charge so as to ensure adequate ignition. Thus a 'core igniter' is a thin tube of cloth containing gunpowder which runs up the centre of the propellant charge; its bottom end is in contact with the main igniter so that it will ensure rapid ignition of the charge from the centre. A 'bandolier igniter' is wrapped around the charge, under the charge bag, and acts similarly to amplify the flame from the main igniter. An 'auxiliary igniter' resembles a main igniter bag but is supplied separately and has a wide hem with a drawstring; its purpose is to replace the main igniter on a bag charge which has become damp. The auxiliary igniter is placed over the damp main igniter, the hem pulled around the bottom of the bag and the drawstring drawn tight. The cartridge is then loaded, and the auxiliary igniter makes sure that it fires; it generates more smoke, but that is preferable to having a misfire.

ILLUMINATING SHELL Projectile fired for the purpose of illuminating an area at night, either so that movement can be detected or so that targets can be silhouetted.

The illuminating shell (sometimes called a star shell) is a base-ejection carrier shell. It contains an open-ended metal canister filled with magnesium compound which is attached to a parachute by means of a swivel and wire support lines. The canister is packed into the shell with its open end towards the nose, and the folded parachute and its lines inserted behind it. Around the parachute are two or more curved steel plates which act as struts between the canister and the base-plate of the shell, which is usually secured to the body by shearable pins or rivets.

Above the canister is a small compartment filled with gunpowder and closed by a loose plate, resting on the canister, with a hole in the middle. A time fuze is screwed into the nose of the shell.

When fired, the time fuze commences to run; at the end of the set time it flashes down into the shell and ignites the gunpowder charge. This explodes; the flash passes through the small hole in the loose plate and ignites the magnesium flare mixture in the canister, and the expansion of gas from the explosion presses the loose plate down on to the canister. This pressure is passed via the canister walls to the support struts and to the base-plate of the shell, causing the fixing pins or rivets to break. The pressure from the charge thus pushes the base-plate off and pushes the canister and parachute out of the shell. The supports fall away, the parachute opens, and the canister is then suspended, open side down, beneath the opened parachute. The assembly

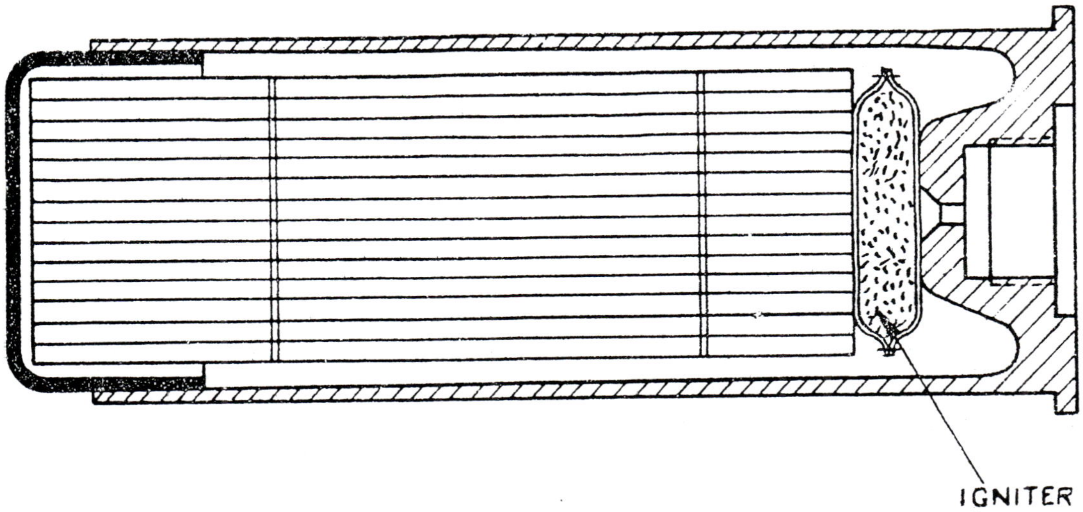

A gunpowder igniter used to boost the action of the primer and ensure rapid ignition of the propellant charge.

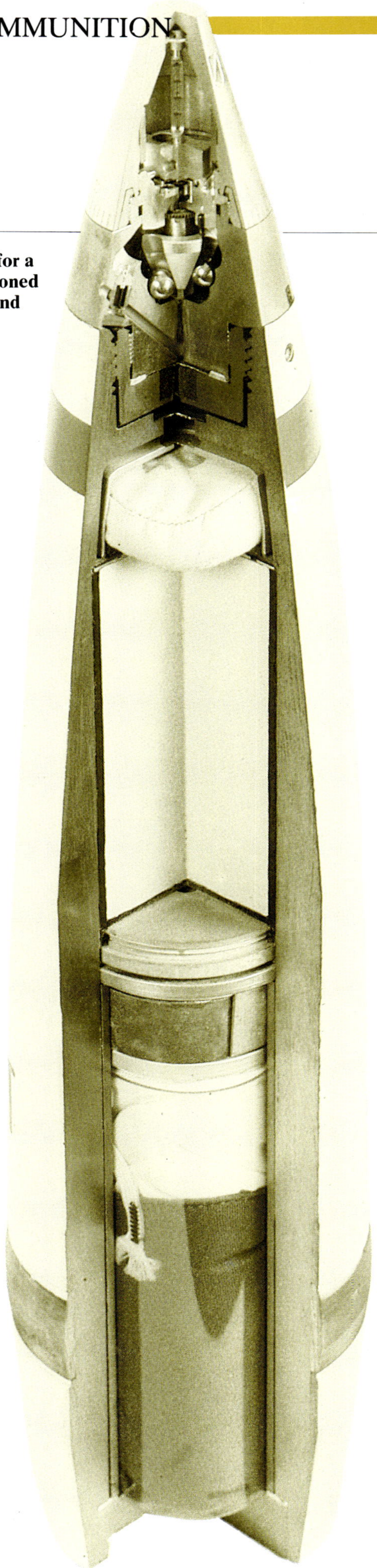

An illuminating shell for a 105mm light gun sectioned to show the star unit and parachute.

then descends to the ground, the magnesium mixture burning and illuminating the area beneath.

A typical 155mm howitzer illuminating shell unit will have a light intensity of 2 million candle-power, will fall at a rate of 4–5 metres per second, burn for about 60 seconds, and will give a level of illumination of 5 lux over an area some 800 metres in diameter during that time. (5 lux is somewhat brighter than a full moon, but not so bright as to permit a document to be read.)

There is also a parachute-less illuminating shell design, produced for use by tank guns. This simply ejects a number of illuminating units which fall to the ground before they begin to emit light. When fired from a tank gun with a flat trajectory the shell bursts at no more than 15 metres above the ground, and the light sources (four or five from one shell) serve to back-light or silhouette the target so that the tank crew can then open fire with normal ammunition.

IMPACT FUZE Any fuze designed to operate when it strikes the target. By definition it must be a nose fuze, and the British terminology 'direct action' is probably more descriptive.

Impact fuzes usually rely upon a striker or firing pin being driven back, by the impact, into a detonator. Safety before firing can be arranged by placing physical blocks between the two, withdrawing them by centrifugal action, or by placing a block in the initiating train below the detonator. Some impact fuzes may have a *graze* action assembly as an auxiliary, to ensure detonation should the shell fail to strike at an angle which will ensure impact on the nose of the fuze. Impact fuzes with graze action only have also been produced at various times, largely in order to have a graze fuze without going to the trouble of designing a completely new impact and graze type. Present-day practice is to have both types of action in the same fuze, with a selector mechanism which allows the direct action impact operation to be locked out, so allowing only graze action.

INCENDIARY SHELL Carrier shell which is loaded with a substance which will set fire to the target.

Designs of incendiary shell have included base-ejection shells with magnesium canisters filled with thermite and bursting shells filled with napalm, both types being intended for ground firing against houses and similar targets. There have also been nose-ejection carrier shells which ejected numbers of incendiary pellets, intended for the attack of aircraft by major-calibre guns.

Today the white phosphorus smoke shell is used as an incendiary projectile; it is perhaps not so effective as the specialized designs, particularly the base-ejection thermite type, but the applications for incendiary shells are so few that most armies have decided that carrying highly specialized shells on the inventory is not

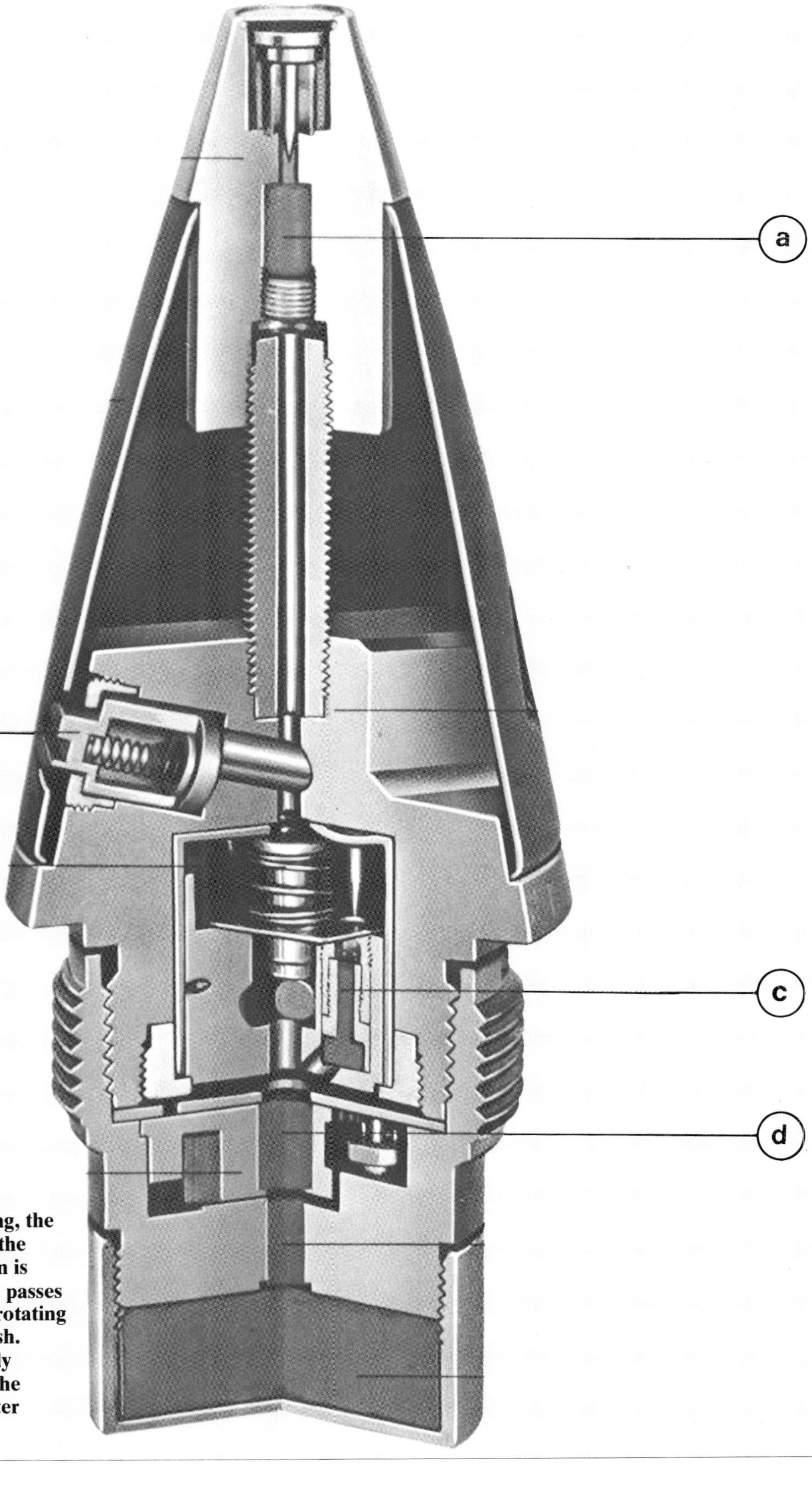

American M557 Impact Fuze. On firing, the centrifugal bolt B. slides out, opening the flash channel. On impact, the firing pin is driven into detonator A. and the flash passes down to the booster detonator D. By rotating B it is locked in place blocking the flash. The graze detonator unit C will then fly forward and hit the delay firing pin. The delay will burn and then fire the booster detonator.

worthwhile. In the cases where incendiary shells have been made, they have never been a general issue and have always had to be specially demanded for a particular operation, which is far from convenient. The only extensive use of specialized incendiary shells was by German anti-aircraft artillery during World War II.

INITIATOR General term for any substance which is used to start off an explosive reaction; eg, the sensitive explosive in a fuze detonator is classed as an initiator.

L

LETHAL AREA The area around a bursting shell in which a human stands a 50 per cent chance of being killed by fragments or blast. The area varies according to the angle at which the projectile is poised when it detonates; to take the two extremes, a shell balanced vertically on its nose will distribute its fragments in a circle around the point of burst, whereas a shell lying on its side on the ground will distribute fragments principally to the sides — those fragments which go straight up into the air and straight down into the ground will be non-effective. The problem is further complicated by the effect of the shell's velocity at the time of detonation, since this will have an effect on the direction taken by the fragments. A shell bursting in the air while travelling at several hundred metres per second will give a forward bias to all its fragments which will be reflected in the shape and direction of the lethal area. Generally, the steeper the angle at which the shell approaches the ground, the more evenly-distributed will be the lethal area; this is one of the reasons for the mortar being a highly dangerous weapon, since its high trajectory means that the bomb almost always comes in at this sort of angle.

LONG-DELAY FUZE Type of fuze fitted to high-explosive shells and used for sabotage, demolition or harassing bombardment. It resembles a standard impact fuze in appearance, but contains a chemical delay device which can function at any time from 2 hours to 6 weeks after impact; different timings are available and can be selected for different tasks. Long-delay fuzes can be used to bombard an area; the shells land but do not function, lying dormant for the selected time before beginning to detonate. thus a road junction or a potential airstrip can be 'seeded' so as to cause harassment at a future time.

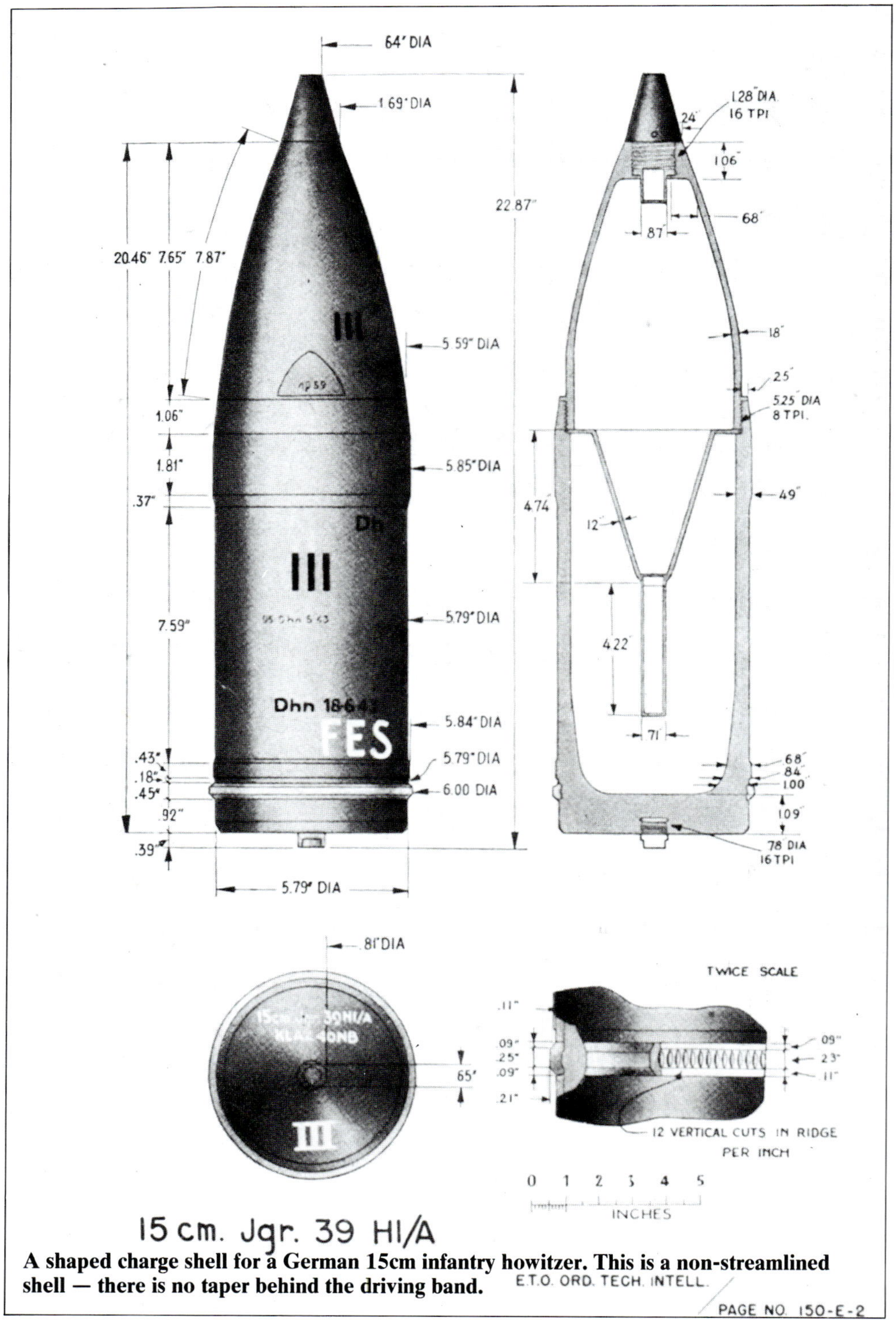

A shaped charge shell for a German 15cm infantry howitzer. This is a non-streamlined shell — there is no taper behind the driving band.

Another role is the destruction of ammunition dumps during a retreat; a few shells fitted with long-delay fuzes are placed in the ammunition stacks, and the fuzes are then hit with a hammer to begin the chemical reaction. With their safety caps replaced they are indistinguishable from other shells fitted with conventional fuzes. At the end of the selected time the fuzes detonate the shells and lead to the destruction of the dump. For immediate destruction the short-delay can be used; for maximum disruption of the enemy a longer delay can be used which will ensure that he has occupied the dump before it explodes.

Fuzes of this type were developed for British service in the 1930s; externally they resembled the standard Fuze No. 117. They were declared obsolete in 1943 and so far as is known no long-delay fuzes are held by Western armies.

LYDDITE British name for cast picric acid, used as a high explosive. It was the first high explosive to be adopted by Britain, in the 1890s, and was named 'Lyddite' because the trials which took place to perfect a design of shell were carried out on firing ranges at Lydd in Kent. It was also adopted by other countries at much the same time, being known as 'Melinite' in France, 'Shimose' in Japan and 'Granatfüllung 88' in Germany. For further details see under *High Explosive* above.

M

MINE SHELL High-explosive shell using a much thinner-walled body than standard so as to obtain the maximum capacity for explosive. This gives the shell a much greater blast effect and makes it more destructive of *matériel*, though the fragmentation suffers. The mine shell was developed in Germany during World War II for small-calibre aircraft cannon in order to develop the maximum destructive effect against aircraft targets. In order to withstand the acceleration stresses with a thin wall, high-grade steel was used and the material was drawn rather than being cast or forged. The improvement in payload was considerable; the standard 20mm high-explosive cannon shell carried 3.70gm of explosive, while a mine shell of the same calibre carried 17.0gm. In order to extract the maximum effect, aluminized explosives were used in these shells, and four 30mm mine shells were sufficient to completely destroy a four-engined bomber such as the B-17 Flying Fortress.

The mine shell was widely adopted after the war and is offered by almost all manufacturers of small-calibre cannon shells today, usually with an aluminized HE/Incendiary filling.

MONROE EFFECT Name given to the shaped charge in its early days, due to the fact that it had been first discovered by an American engineer named Monroe in the 1880s. Monroe was performing experiments with guncotton against steel plate and discovered that recessing the face of the guncotton reproduced the shape of the recess in the face of the steel plate. He also produced some artistic effects by placing leaves from trees between the explosive and the plate and so reproducing the vein pattern of the leaf in the plate. It remained a laboratory amusement until further experiments were performed by Neumann (see below).

N

NEUMANN EFFECT The second name applied to the *shaped-charge* effect. After Monroe (see above) the phenomenon remained a novelty until shortly before World War I when a German experimenter named Neumann began looking at the idea as a possible weapon. He determined that adding a metal liner to the cavity increased the depth of the impression in the steel, to the extent that it occasionally cut completely through the steel target, but he was unable to put this to any practical use before the war ended. However, Neumann's observations and published reports caused other researchers to begin investigations, which led to the development of the shaped-charge munition in the late 1930s and early 1940s.

NON-STREAMLINED A projectile in which the parallel sides of the body continue to the base; there is no tapered section behind the driving band. Nowadays only found in carrier shells where it is desirable to have the maximum size of aperture in the base to allow the payload to be discharged. Illuminating shells are almost the only non-streamlined shells commonly seen today, since base-ejection smoke shells can usually be made with some degree of streamlining. The principal objection to non-streamlined shells has always been that they range completely differently from the standard streamlined shell used with the weapon, and therefore considerable corrections have to be made to the calculated range; there is, perhaps, less justification for this argument now that computers are available to do the calculations.

A German 30mm mine shell (right) together with a 20mm shell of conventional type. Note the thinner walls — and thus the greater proportion of the cavity space.

O

OBTURATION The technical term for the sealing of the gun breech against the unwanted escape of propellant gas when the gun fires.

Obturation can be achieved by the ammunition, or by the gun. With cased charges, it is done by the ammunition, the metal cartridge case expanding outwards to form a tight seal against the chamber walls and thus prevent any leakage of gas. With bagged charges it is necessary to have the sealing performed by the gun, generally by using a resilient pad in the breech block. This is forced into the mouth of the chamber and when the gun is fired the pressure in the chamber causes the pad to be squeezed outwards to seal against the walls.

OGIVE The curve shape of the head of a projectile. It is defined by the *calibre-radius-head* measurement.

P

PAPER SHOT Early type of break-up shot, developed for use by coast artillery guns so that they could be fired in peacetime from sites where it was not possible to fire any type of projectile because of the danger to shipping in the vicinity.

The open breech of a 9.2in coastal defence gun. The screw block can be seen and at its front end the black line of the obturating pad.

The Paper Shot consisted of a thick paper or cardboard tube filled with fine sand. It was loaded into the gun in the same way as a normal projectile, and was then followed by a special propelling charge. On firing, the shock of discharge split the paper tube and ejected the paper fragments and sand from the gun muzzle. However, the initial check to the propellant force, which ruptured the tube, was sufficient to cause the gun to recoil as if a service round had been fired, and paper shot was used periodically to check that the gun's recoil system was in full working order. The debris ejected from the muzzle fell to the ground within a very short

Two specimens of paper shot for testing guns. The shorter one, for a 25-pounder gun, is filled with sand. The longer one, for a 105mm tank gun, is filled with water prior to firing.

An experimental Peenemunde Arrow Shell developed for the German 105mm anti-aircraft gun.

distance and there was thus no danger to any property in the line of fire. It was also, very rarely, used as training ammunition when it was not possible for the gun crews to go to some other location to fire.

PEENEMUNDE ARROW SHELL Projectile developed in Germany in 1939–40 by the Army Experimental Establishment at Peenemunde (where the V-1 and V-2 missiles were developed.) The projectile was in the form of a long sub-calibre dart, with full-calibre fins and a sabot around its centre and was intended to be fired from a smooth-bored 31cm barrel fitted to the K5 railway gun. The projectile body was 12cm (4.7in) diameter and 1.91m (6.2ft) long and had four or six fins. There was an impact fuze in the nose. The first design had a telescoping tail unit which was extended after the projectile left the gun, but after only one trial shot the idea was abandoned. Engineer Gessner then devised a shell which had a pusher piston loaded behind the fins which pushed the projectile up the bore and then fell to the ground; this also failed, since placing all the propulsive force at the tail of the shell set up enormous stresses in the long body. Gessner then proposed a three-piece discarding sabot around the centre, and this design proved successful, the sabot falling to the ground about 2 kilometres in front of the gun. Trial firings in the 31cm gun gave a maximum range of 151 kilometres (93.8miles), the longest ever achieved by a service gun and projectile.

Although the design was drawn up in 1940, production difficulties were such that it was not until 1944 that serviceable projectiles were made, and so far as is known the Peenemunde Arrow Shell was only used in action a few times. One of the only recorded instances was against the US Third Army of General Patton in late 1944 at a range estimated to be about 120 kilometres (75 miles). Smaller versions of the design were also developed for the 105mm anti-aircraft gun, the object here being to obtain high velocities and short times of flight.

The design was revived in the late 1960s and proposed as a long-range projectile for the Anglo-German-Italian FH70 155mm howitzer, but was later abandoned in favour of a rocket-boosted design. It was also used by Finnish designers for a long-range 160mm mortar bomb in the mid 1950s, also without success. There is nothing inherently wrong with the idea, as the Germans proved, but it is an expensive and difficult production engineering problem, and there are also difficulties in ensuring perfect detonation of a long and thin column of explosive.

PERCUSSION FUZE British term for an *impact fuze*, of either direct action or *graze* type.

PETN Penta-eryithritol-tetra-nitrate; a high

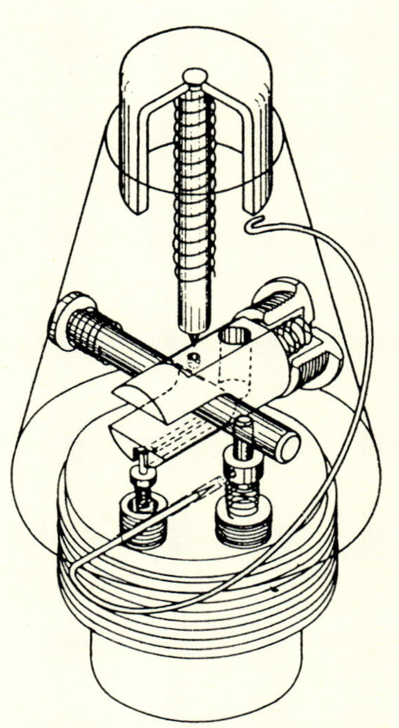

When the safety wire is removed, the safety pin is restrained only by the set-back pin.

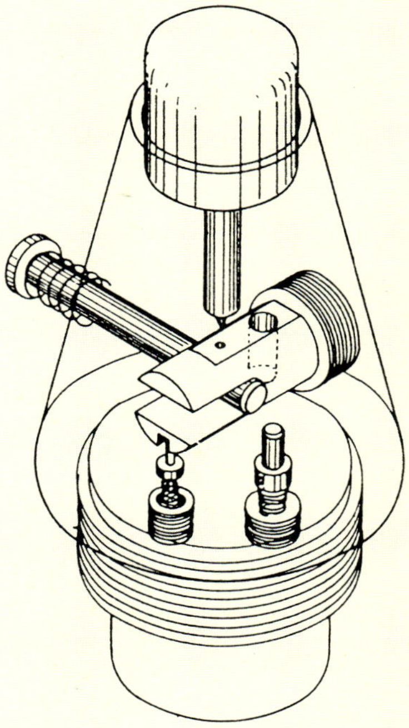

On firing, the set-back pin releases the safety pin which retracts under the action of its own spring.

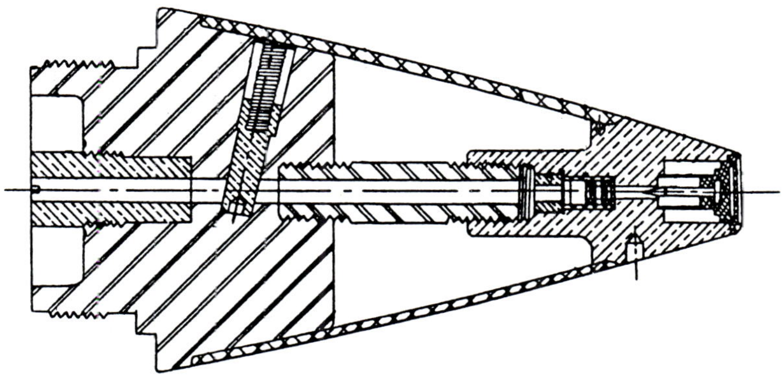

A simple percussion fuze for artillery shells. The interruptor (A) is spun outwards, clearing the central passage. On impact, the firing pin (C) is driven into the detonator (B) and the subsequent flash passes down the central passage to leave the fuze and initiate action in the booster.

THE FIRING SEQUENCE OF AN M52 FUZE

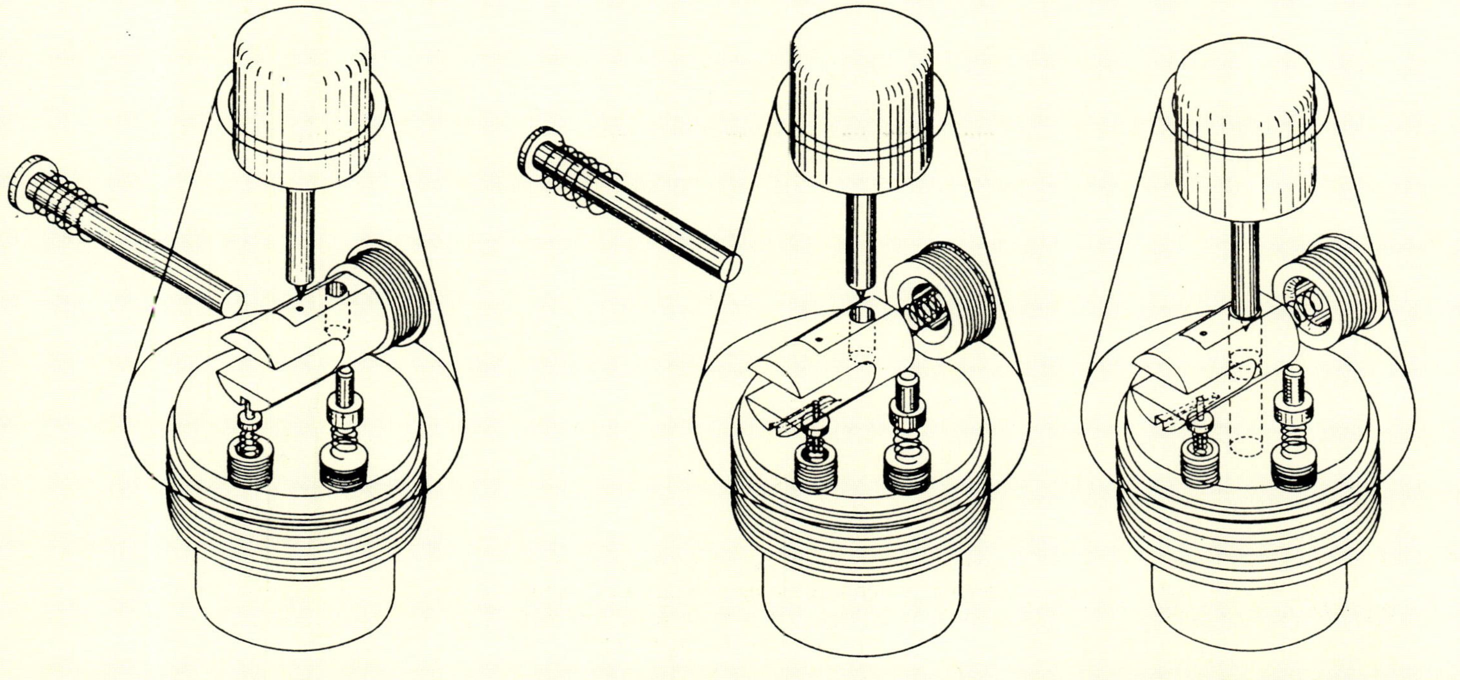

When the safety pin is clear of the fuze, the slider moves under the tension of its spring.

The locking pin then moves up in the action of its spring into a recess and locks the slider in position.

On impact, the striker and the firing pin are forced in and the firing pin crushes the upper detonator.

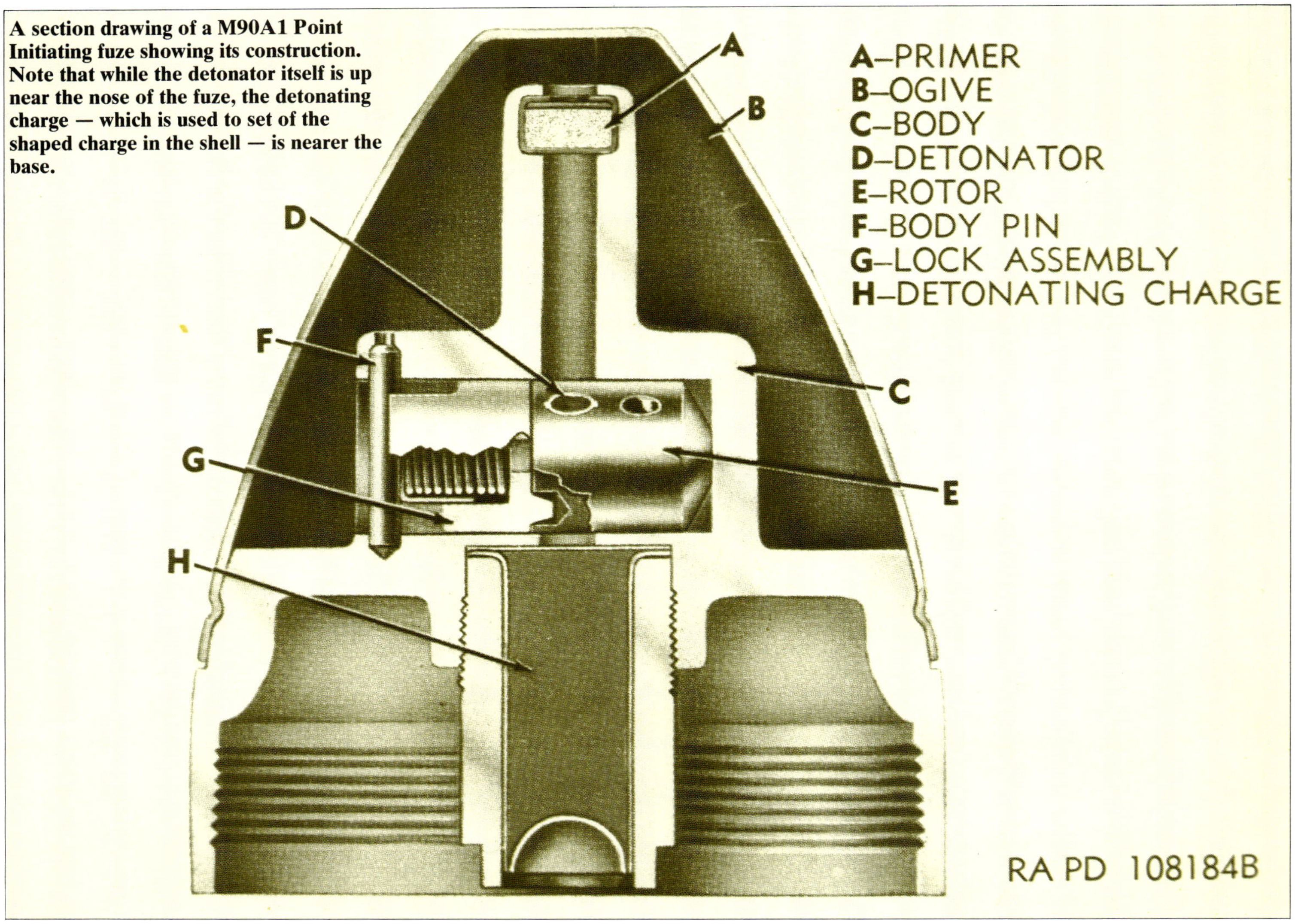

A section drawing of a M90A1 Point Initiating fuze showing its construction. Note that while the detonator itself is up near the nose of the fuze, the detonating charge — which is used to set of the shaped charge in the shell — is nearer the base.

explosive, also known as Penta, Pentolite, Penthrite, etc. Too sensitive to be used alone as a shell filling, it is usually mixed with 10 per cent wax to desensitize it. It is used by itself in some types of exploder and as a filling for fuze magazines. For further information see under *Explosives* above.

PIBD FUZE Abbreviation for 'Point Initiating, Base Detonating' an American term which is applied specifically to certain fuzes designed for use with shaped-charge projectiles.

The problem facing the designer of a shaped-charge projectile is that the detonation of the shaped filling must be initiated at the rear end of the charge so that the detonation wave moves forwards and so generates the special shaped-charge effect. But using a base fuze, which seems to be the obvious solution, means that there is a constant delay (see *Base Fuze*) due to the construction of the fuze, and this delay permits the shell to get too close to the target before it detonates. The ideal is to have an impact fuze in the tip of the projectile nose, but this introduces problems in shifting the detonation from the fuze itself to the base of the explosive charge.

The solution adopted by both American and British wartime designers was to give the base of the fuze a tiny shaped charge of its own; this, when detonated by the impact of the fuze with the target, fired an explosive jet into the shell and through a tube running down the centre of the main shaped-charge filling of the shell to a tetryl exploder placed at the rear. This then detonated and set off the operation of the shaped charge. Hence Point Initiating (by the tiny shaped charge in the fuze) Base Detonating (by the exploder behind the main charge).

The British designers did not give a specific title to this type of fuze; it was merely 'Fuze Percussion No 233' and was used only with a shaped-charge shell for the 3.7in mountain howitzer. The American design, the 'Fuze PIBD M90' was used with the HEAT shell for the 57mm recoilless rifle. This weapon is still used by one or two South American countries and it is probable that they still use PIBD fuzes copied from the American design.

PIERCING CAP A steel cap placed over the tip of an armour-piercing shot or shell in order to improve penetration, particularly against face-hardened armour plate. A plain steel projectile is liable to shatter at impact velocities above 853mps (2,800fps) but below that necessary to achieve penetration. By fitting a cap over the tip, designed so that it is anchored to the shoulders of the shell and does not press on the tip, the force of the impact is loaded on to the shoulders of the projectile rather than being concentrated on the point. The cap then splits apart, allowing the tip of the projectile to begin penetrating, and at the high velocities involved tends to melt and act as a lubricant to help in penetration. Furthermore the collapsed cap supports the projectile for a brief period as it begins to penetrate, so preventing any unsteadyness which would tend to degrade the penetration.

Rarely seen today, since armour-piercing ammunition relies on other systems for its operation, but some capped projectiles are still used in the Soviet Army. See also *APC*.

A 17-pounder anti-tank gun armour-piercing shot with its piercing cap removed and laid alongside it.

PIEZO-ELECTRIC FUZE Type of fuze used with modern shaped-charge ammunition. As explained above under 'PIBD Fuze', the designer is faced with the problem of having to detonate the shaped charge from the rear, while relying upon an impact fuze in the tip of the shell some distance in front of the charge. The piezo-electric fuze relies upon the electric phenomenon associated with certain types of crystal which, when violently crushed, generate an electrical current. The impact fuze carries such a crystal, and a wire runs from this along the wall of the shell to an electric detonator set into the exploder unit at the rear of the shaped charge. When the shell hits the target the fuze collapses and crushes the crystal; this sends an electrical

A typical shaped charge projectile showing a piezo-electric element fitted in the tip, connected to an electric detonator which initiates the charge from the rear end.

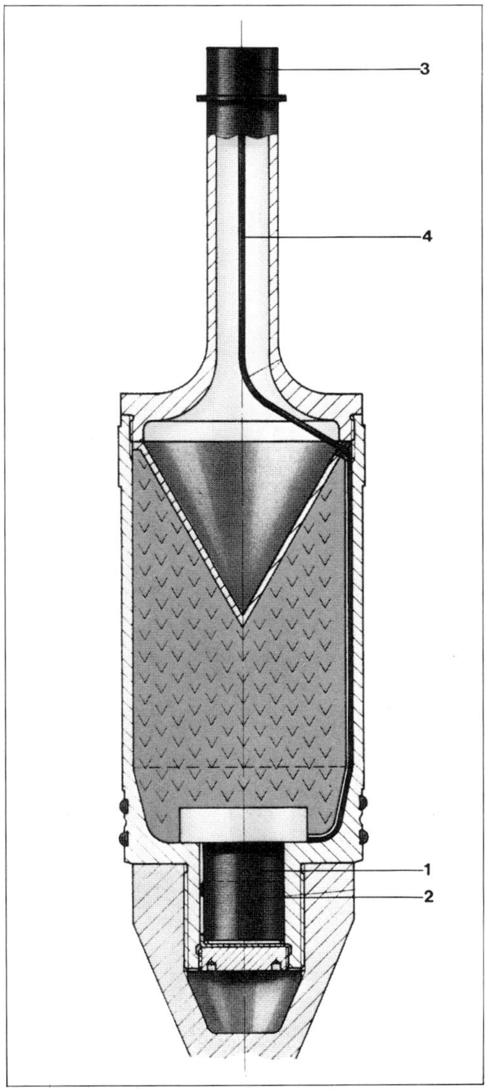

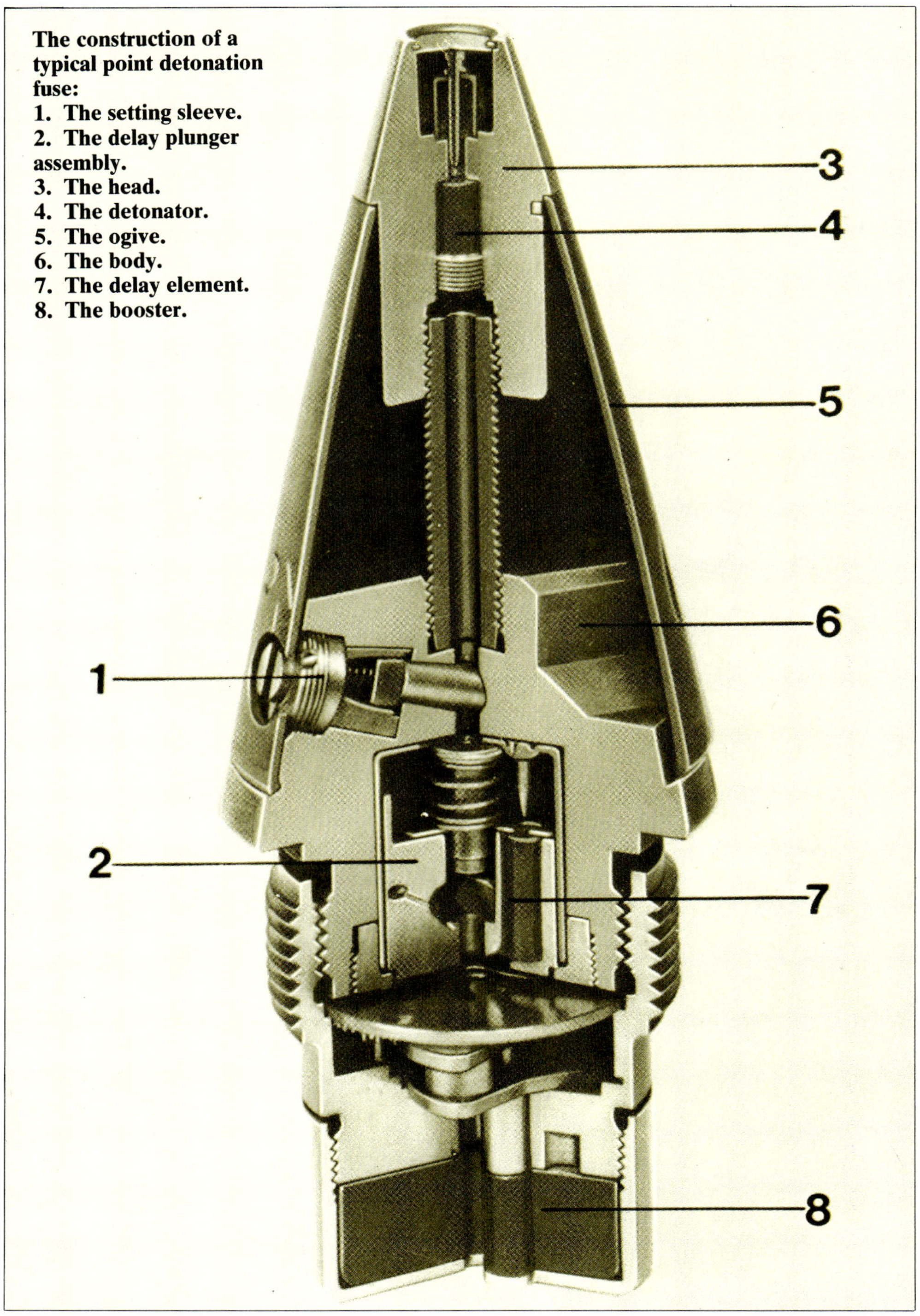

The construction of a typical point detonation fuse:
1. **The setting sleeve.**
2. **The delay plunger assembly.**
3. **The head.**
4. **The detonator.**
5. **The ogive.**
6. **The body.**
7. **The delay element.**
8. **The booster.**

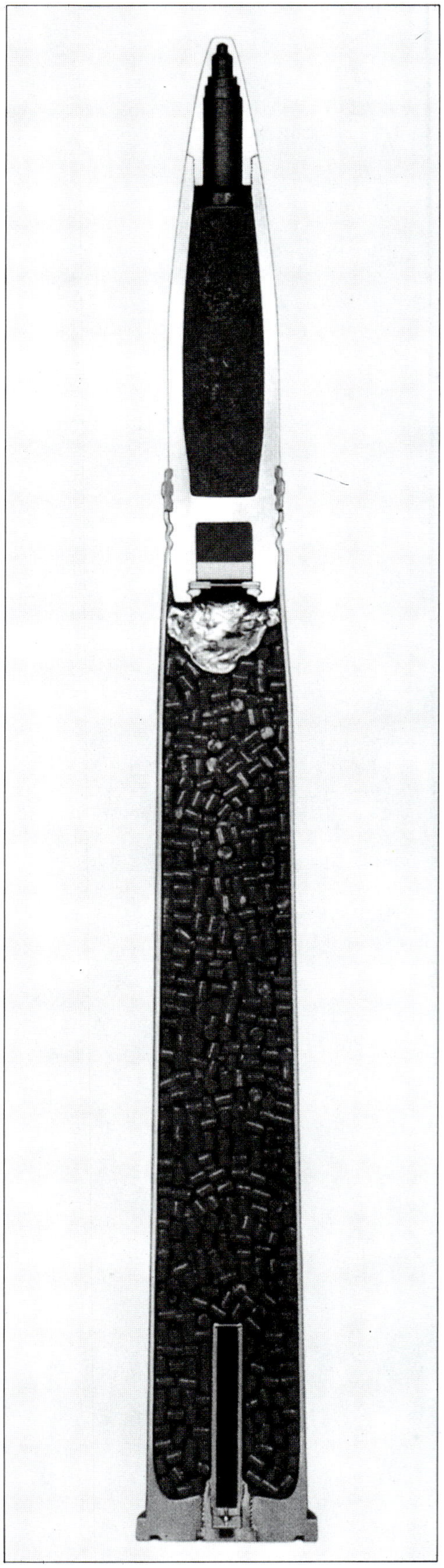

impulse down the wire to fire the electric detonator and thus initiate the exploder and the shaped charge. Since the impulse from the fuze moves at the speed of electricity, the delay between impact and charge detonation is negligible, and the shaped charge fires before the nose of the shell has had a chance to collapse to any significant degree. This type of fuze also has the advantage that it requires few mechanical safety devices, since the impact necessary to generate the electrical current is far beyond anything likely to be sustained by accidentally dropping or knocking the shell in the course of handling it. It does, however, require some careful circuitry; early designs used the body of the projectile as one 'leg' of the circuit, and it was found that screening the target with chicken wire was enough to short-circuit the fuze as the current leaked out to earth when the wire was in contact with the shell body.

POINT DETONATING FUZE American terminology for an *impact fuze* carried in the nose of the shell; to distinguish from 'base detonating fuze'.

PRACTICE SHELL Projectile intended for use in training exercises. It resembles the standard high-explosive shell for the particular gun, but does not contain a high-explosive filling. Instead it has a filling which is largely an inert material, but with a small 'blowing charge' which is sufficient to give a puff of smoke and a flash when the shell strikes. The ballistics of a practice shell must be precisely the same as the standard service projectile so that the same sights and fire-control calculations are used.

Practice shells, are not often seen, since it is usually cheaper and more effective to fire the standard high-explosive shell in training; these are mass produced, and it is more economic to use them than to set up a special production line to make practice shells.

PRACTICE SHOT As with practice shells, practice shot are used for training instead of armour-piercing shot. In this case, though, they are much more common since they economise on expensive material by not using the depleted uranium or tungsten carbide cores of service discarding sabot shot or, in the case of conventional shot designs, are not made of high-quality steel.

A major problem with practice shot, particularly with high-velocity APDS and APFSDS, is that if the sub-projectile misses the target or ricochets off it, it may well fly for a very long distance before coming to earth. Early APDS shot were prone to fly as far as 10 kilometres (6 miles); modern APFSDS are likely to ricochet as much as 60–80 kilometres (37–50 miles) before landing, and this poses immense problems in finding a range area big enough to contain ricochets so that they do not become a danger to the surrounding civil population. An alternative solution, currently being introduced, is to construct the practice shot so that it degrades ballistically during its flight, so ruining the shape and making it more affected by air resistance in order that it will then come to earth in a much shorter distance. A design by the German firm Diehl has the sub-projectile in two pieces, pinned together; an explosive charge is fired by a delay system 3,500–4,100 metres (3,800–4,500 yards) from the gun, and this breaks the joint, splitting the sub-projectile into two pieces. The forward section is unstable and tumbles to

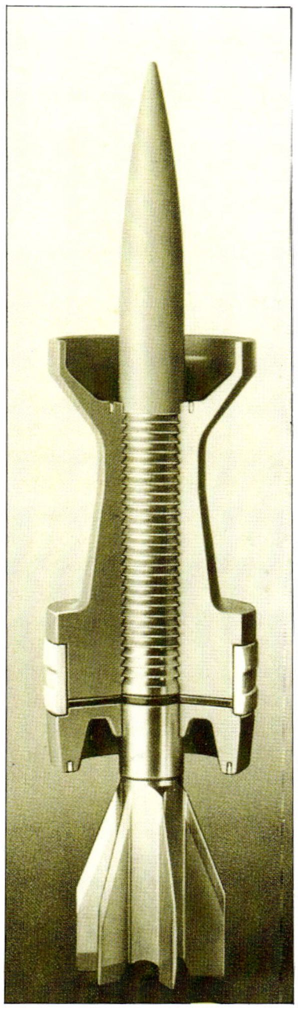

Left: A practice shell. Note that the fuze is empty and the filling is inert, but a live tracer is used so that its flight path can be observed.

The firing sequence of a practice APFSDS projectile.
Below: In its sabot.
Right: In initial flight.
Far right: A small explosive charge ignited by a delay breaks the penetrator and spoils the projectile's ballistic shape so that it falls to the ground within a safe distance.

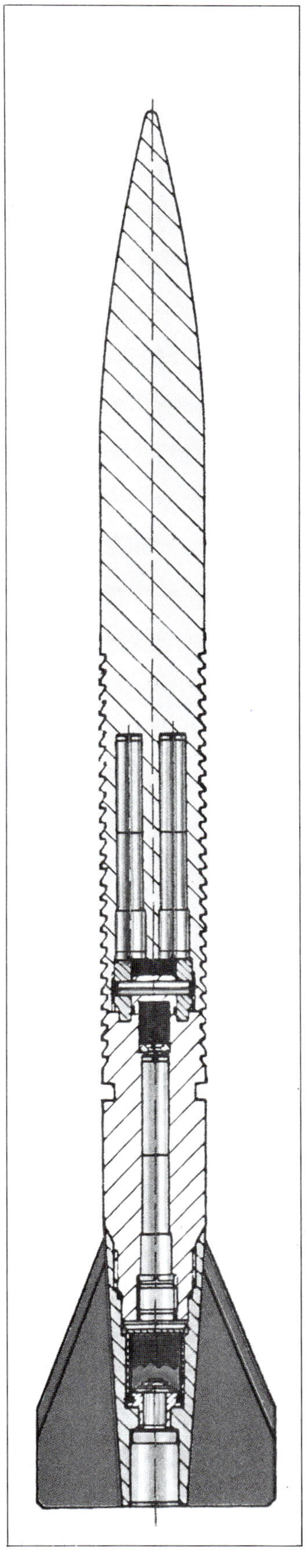

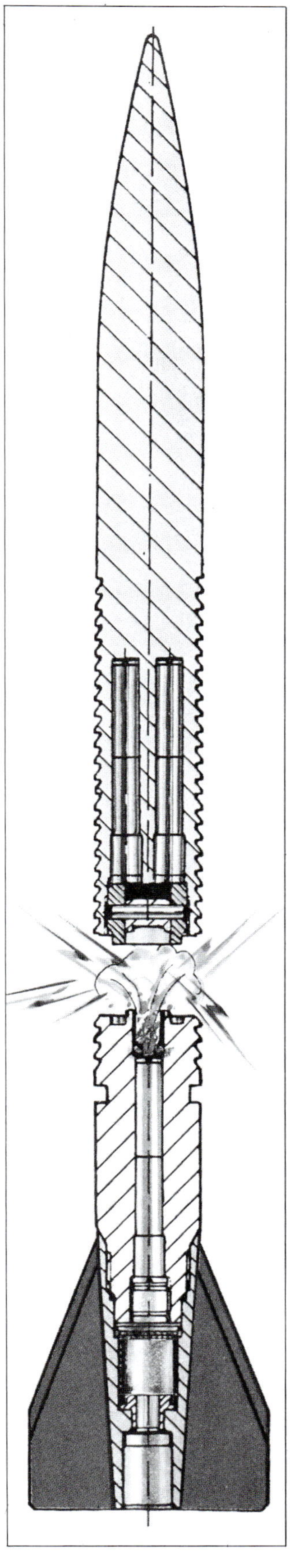

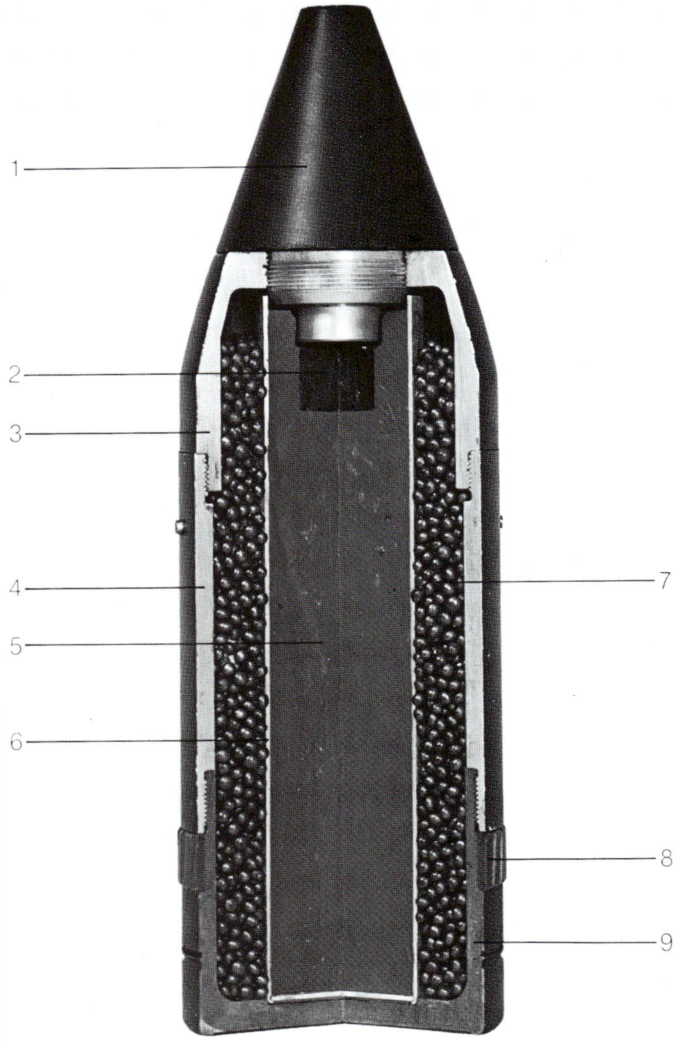

The construction of the projectile for a recoilless gun showing the use of pre-formed fragments:
1. Impact fuze.
2. Exploder.
3. Nose section.
4. Body section.
5. Bursting charge.
6. Bursting charge container.
7. Steel balls packed in plastic resin.
8. Driving band.
9. Base section.

the ground very quickly, while the rear section, carrying the fins, is rather more stable and will fly for a further 2,000 to 3,000 metres (2,200–3,300 yards) before falling to the ground. In any circumstances the entire shot will have landed with a range of 7.5 kilometres (5 miles) from the gun, which reduces the size of the firing area required.

PRE-FORMED FRAGMENTS Small pieces of steel, of a weight considered to be optimum for producing casualties, which are packed into an anti-personnel munition so as to be blown out and distributed around the point of burst when the explosive is detonated. They may be regularly- or irregularly-shaped pieces of metal, small steel balls, a coil of steel wire suitably notched to predispose it to break into the required size of fragment, an internal sleeve of metal also notched to the desired shape and size or any other method which takes the designer's fancy. The aim is to produce regular and precise fragments, rather than rely upon the haphazard breaking up of a shell body on detonation. Pre-formed fragments are principally used on small-calibre munitions where the shell body cannot be relied upon to fragment regularly. In such cases the shell body may be relatively light, but reinforced internally by the fragmenting system.

PREMATURE An artillery term which, strictly, means the bursting of a shell at any point on the trajectory prior to the intended point of burst. In practice, however, it is usually applied to a burst inside the gun barrel, bursts on the trajectory being often merely called 'short bursts'.

A bore premature can result from a variety of factors, but is most often caused by defective ammunition. Faulty assembly of a fuze or faulty filling of a shell are perhaps the most likely causes. Faulty performance of drills by the gunners — eg, wrongly assembling a fuze into a shell, using the wrong propelling charge — can also produce prematures. Finally, mechanical defects in the gun can, but rarely, cause them. Whatever causes them they are usually fatal for some of the gun crew and totally wreck the gun.

A 75mm gun after the premature detonation of the shell inside the bore. The ragged edge of the cut-off barrel can be seen at the front of the carriage and the entire breech end of the gun is almost vertical due to the loss of the balancing weight of the barrel.

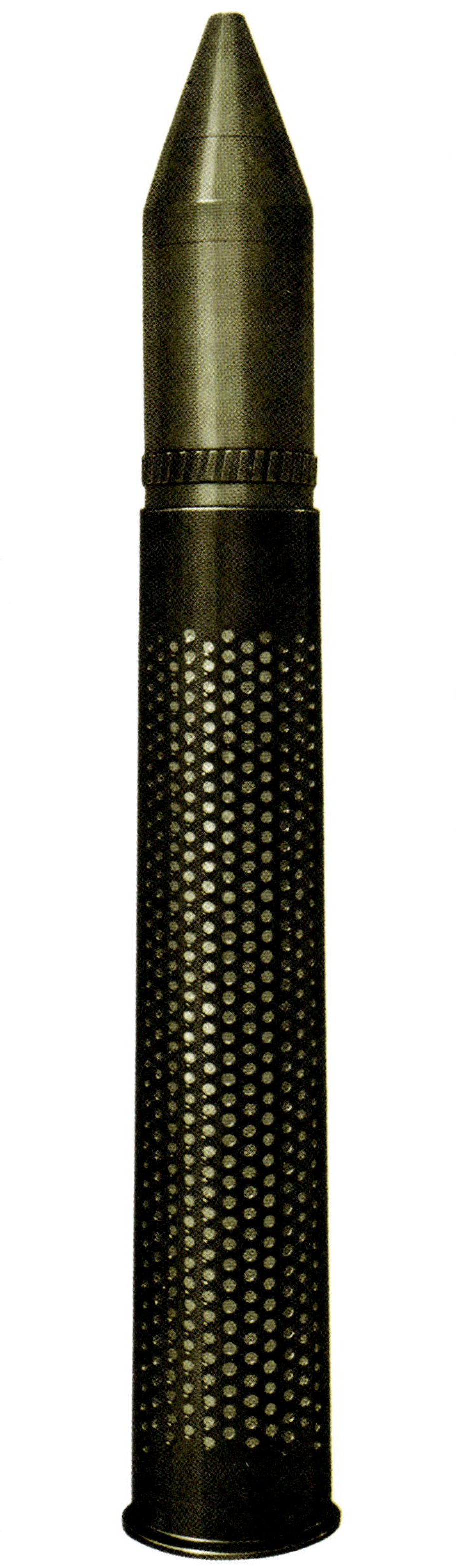

A pre-rifled shell for use with a recoilless gun.

PRE-RIFLED SHELL A projectile which has its driving band already cut with grooves to match the rifling of the gun.

In a normal projectile the driving band is uncut and is engraved — forced to conform to the grooves — by the pressure of the gas generated by the propelling charge. However, in some recoilless guns the pressure needed to engrave the driving band would interfere with the balance of the counter-recoil jet of gas. The projectiles are therefore pre-engraved so that the internal pressures are precisely predictable and the recoilless effect is always achieved. Pre-rifling the driving band means, of course, that the projectile has to be loaded so that the grooves in the band match the rifling, and registering devices have to be used, since the gunner cannot see the rifling at the forward end of the chamber as he loads the gun. Studs on the shell which locate in grooves in the chamber, or a raised portion of the cartridge case which fits into a recess in the chamber can be used.

Pre-rifled projectiles were first used in the American 57mm and 75mm recoilless rifles of World War II, and are currently used on the Carl Gustav 84mm recoilless gun.

PRIMER The initiating device which fires a cartridge upon receiving some impulse from the gun mechanism. In British terminology it refers only to the initiator in a cartridge case, but in other countries it also covers the initiator used with bag charges.

The primer for use in a cartridge case is screwed or press-fitted into the centre of the case base, except for one or two abnormal cartridges for recoilless guns. Initiation is done either by percussion, using a firing pin in the gun's breech block, or electrically, using a contact in the block which presses against the primer, or by using an induction coil.

In the case of a percussion primer, a thin copper cap, holding a sensitive composition, is fitted into a cap chamber at the bottom of the primer; this cap may be exposed or may be concealed within the primer, the base being thinned so as to permit the firing-pin blow to deform the metal and thus strike the cap. From the cap chamber a fire channel leads into the body of the primer which is usually filled with gunpowder. The size of the primer body depends upon the size of the cartridge which has to be ignited and on the type of propellant being used. Large cartridges, or those with propellant which is difficult to ignite, will need large primers that extend well into the centre of the charge. Smaller charges use small primers which may only protrude a few millimetres into the cartridge case. Ignition of the charge is often assured by the use of an *igniter* inside the case.

Once the ignition flame from the cap has passed through the fire channel and ignited the gunpowder, and thus the propelling charge, the pressure inside the gun is in the order of several

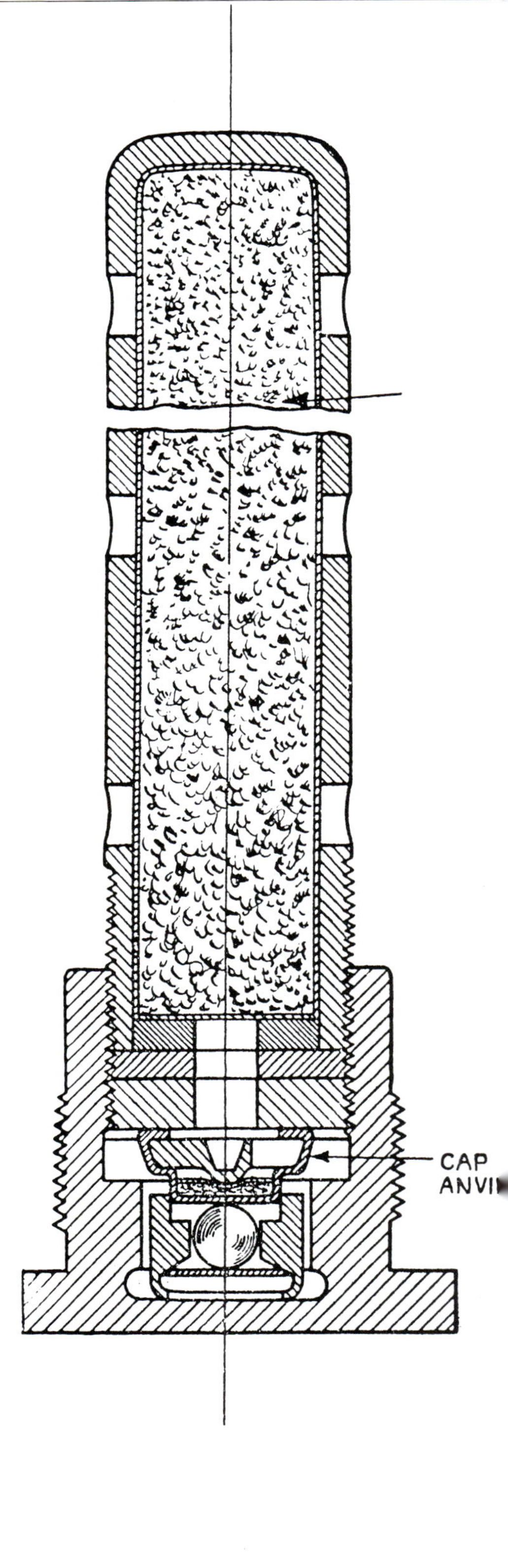

This German percussion primer has a solid base and uses the ball seal to transfer the firing pin blow to the cap.

A German C/23 electric primer. It is fired by the direct connection of a contact to the pole piece.

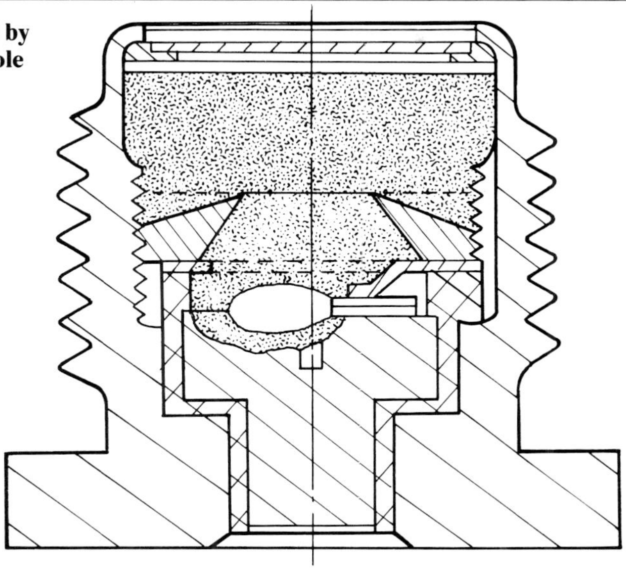

An induction-fired primer. A coil in the breech block induces a current in the secondary coil, firing the electric fuze head.

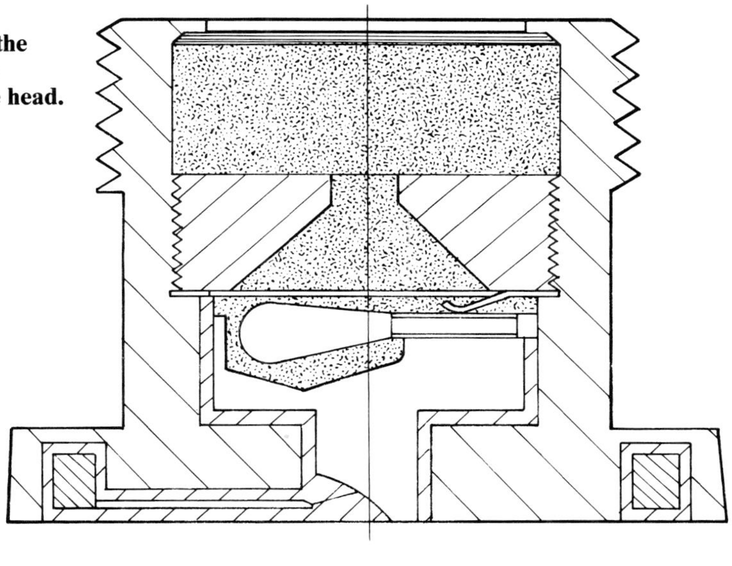

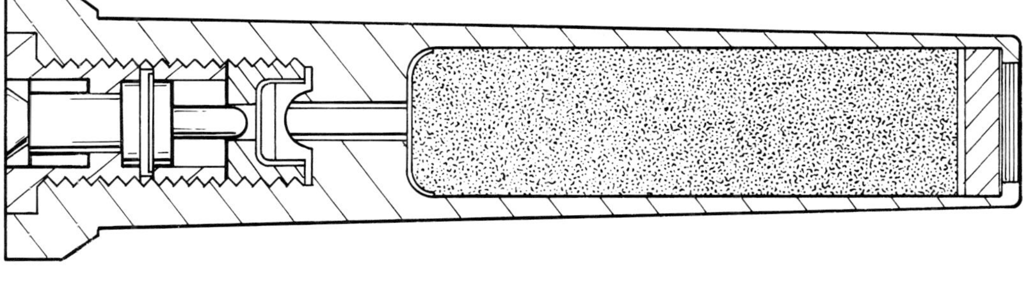

A primer for use with bag charges. The gun's firing pin hits the striker and breaks the shear wire, allowing the striker to hit the cap. When the powder explodes and ignites the cartridge in the chamber, the striker is blown back against the holder to make a gas-tight seal.

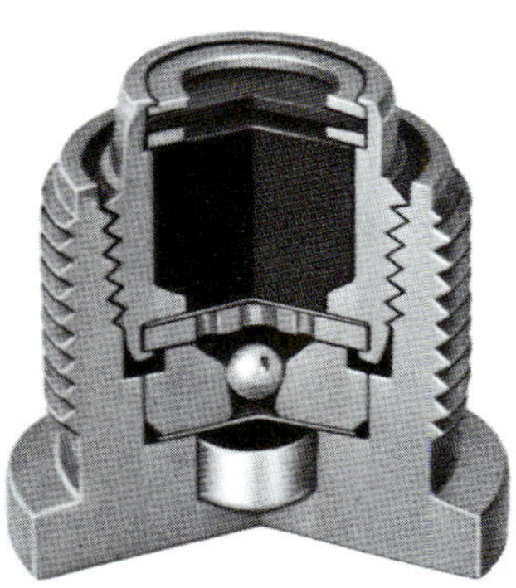

Above: A German DM 124 primer showing the cap, ball valve and fire holes.

Left: A German DM 201A2 primer showing the cap, ball valve and the extended primer magazine filled with black powder.

tons and there is a serious danger that this will pass back through the fire channel and blow out the soft copper cap, releasing flame and debris into the firing mechanism of the gun. To guard against this there is usually a simple flame and pressure trap; above the fire channel is a loose ball which is blown out of the way by the cap flame but, when pressure in the reverse direction develops, is driven firmly back to block the fire channel and prevent the pressure passing any further.

Electric primers are similar in general construction but do not have a cap. Instead the centre is filled by a 'pole-piece' insulated from the body of the primer; from this a thin wire passes to an even thinner filament which is surrounded by some heat-sensitive explosive — usually wisps of guncotton. From the other side of the filament the wire continues to the body of the primer. Around the 'squib' — filament and its surrounding guncotton — is the gunpowder filling. When the cartridge is loaded, a spring-actuated contact pin in the breech mechanism presses against the pole-piece, and electric current is passed through this to heat the filament. This ignites the guncotton and thus the gunpowder and the propelling charge.

Another electrical system is the 'conducting cap' in which there is no filament and squib, but simply a filling of some sensitive composition heavily loaded with graphite. The electric current is delivered to the bottom of this filling and passes through, due to the conductivity of the graphite; as it does so the high resistance generates heat and the igniting composition is fired.

The use of an induction coil is quite rare; this relies upon a filament and squib system, but instead of having a pole-piece and contact, there is a wire coil buried in the body of the primer and connected to the filament. The breech block of the gun carries a similar, though larger, coil connected to a source of electricity. When this coil is energized it sets up a force field which is picked up by the coil in the primer, thus generating current in the primer coil which is sufficient to fire the squib. The system is simply that of a transformer, and it has the advantage that unless the breech is closed and the two coils in alignment, there is no danger of firing the cartridge.

Primers for bag-charge guns are entirely separate components from the cartridge and in shape and size resemble blank cartridges for small arms. Internally they are similar to case-charge primers, having a cap, ball gas check and gunpowder filling. The filling, however, is generally a mixture of compressed gunpowder pellets and fine powder. The bag-charge primer is fitted into a firing lock on the rear of the breech screw, and discharges through a 'vent' or channel which passes axially through the screw and to the chamber. In large calibres of artillery this vent may be 25cm (9.8in) or more in length, and thus in order to ensure that the flash entering the chamber is of adequate strength, compressed powder pellets are used; these are lit by the fine powder and shot down the vent so that they arrive in the chamber and strike the rear of the bag charge while still burning fiercely. It will be recalled that there is more gunpowder, in the shape of the *igniter*, on the rear end of the bag charge so that ignition is made easier.

Bag-charge primers may be percussion or electric, using similar systems to those described above, and there are also combination primers which have both percussion and electric systems so that they can be used in a variety of weapons. Combination primers were originally for coast defence and naval guns which used electricity as the primary method of firing, since this could be controlled from some central point; if, however, the electrical system failed, percussion firing could be resorted to without having to remove the primer and replace it.

The induction firing system is currently used on the French GCT155 155mm howitzer; in this case there is no primer, the induction coil and squib being installed inside the base of the bag charge and responding to a transmitting coil in the breech block of the gun. This has the advantage that no primer need be loaded into the breech, and since this particular weapon uses a remote-controlled mechanical loading system the absence of a separate primer simplifies the mechanical design.

PROBE FUZE General term for projectile fuzes which have a long and sensitive element sticking out ahead of the main body of the fuze. The purpose is to ensure that the fuze begins to function before the shell actually makes contact with the target. Originally proposed for howitzer shells to ensure that they burst above the ground rather than bury themselves before detonation, they were rarely used since the danger of

A fin-stabilized shaped-charge projectile for a shoulder-fired recoilless weapon.

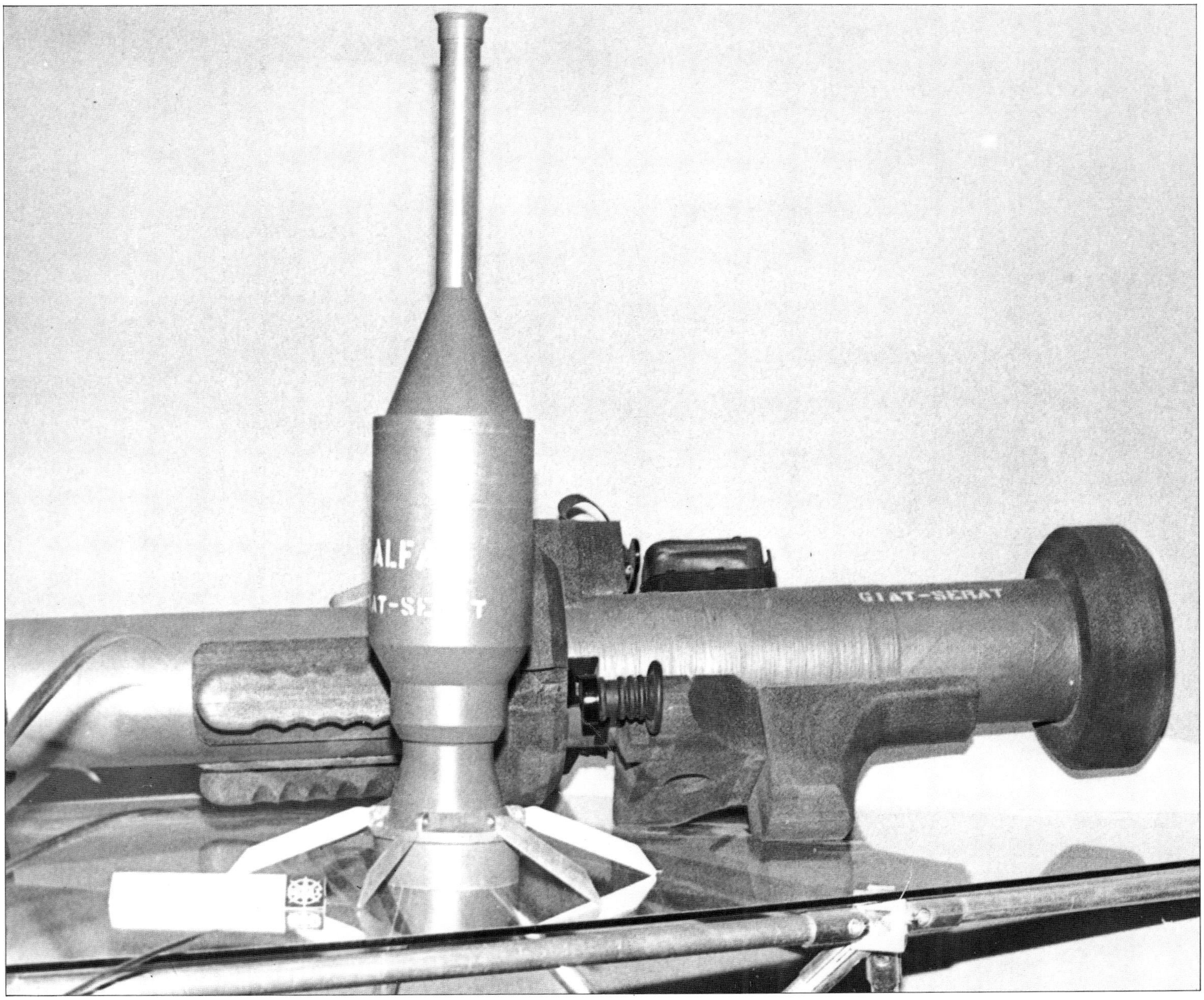

Proof shot need to resemble the real projectile inside the gun but don't need to be shaped ballistically for flight outside it.

Below: A proof shot representing a non-streamlined 25-pounder shell.

Right: A modern proof shot representing an APFSDS sabot shot.

damaging the probe during handling and loading of the shell was too great. In modern *HEAT* shells a form of probe fuze is often used in order to obtain the required stand-off distance without having to use a large ballistic cap which would upset the balance of the projectile. In these cases the shell nose is formed into the probe and at the tip is a *piezo-electric fuze* element.

PROOF SHOT An inert projectile used to test guns after manufacture and before they are taken into service. The proof shot need not have any ballistic shape since its sole function is to stress the gun above its normal working limits and expose any weaknesses. It is, therefore, a simple cylinder of iron or steel, with the correct type of driving band, and weighing some specified percentage more than the service projectile for the gun. This is usually 10 per cent, and this, together with a special over-strength propelling charge, ensures that the pressure inside the gun is above that to be expected in normal use. The gun is loaded with the proof round, and fired 'under precaution', ie electrically by remote control or by a very long lanyard pulled from behind a protective wall. After the gun has fired it is carefully examined to ensure that no damage has been done, that the barrel has not expanded, or the rifling become damaged. A number of proof shots are usually fired, with examination after each one, after which the gun is certified as having been 'proved' and can be issued for service. The proof shot is usually fired into a stop butt of earth at a short range, and the flat front prevents it going too deep; it can then be dug out, re-banded, and used again.

PROPELLANT Propellant is the low explosive which forms the cartridge and which propels the projectile out of the gun.

The original propellant was gunpowder, which had certain defects; it was highly susceptible to damp, easily ignited by accident, and because of its fragile grains was practically impossible to control once ignited. It generated a cloud of white smoke when fired and deposited a black fouling, the products of burning, on the bore of the gun.

It was replaced by 'smokeless powder' a name used as a general expression for propellants based on modern chemistry. The basic smokeless powder is nitro-cellulose (NC), obtained by the action of acids on cellulose materials such as cotton or wood. Nitrocellulose is relatively smokeless and, since it is a colloidal material, hard and impervious, can be shaped and sized so as to burn in particular ways, allowing some control over the process of explosion. For example, a square block of NC powder will burn on all surfaces when ignited and so the block will retain its shape but will gradually become smaller; as it does so, it will generate less gas, and therefore the rate of rise in pressure inside the gun will gradually decrease. The shell, therefore, will receive a sudden acceleration from the initial explosive force of the gas and then the pressure will drop and the rate of acceleration will drop with it. On the other hand, if the piece of NC is in the form of a hollow tube, it will burn on all surfaces, but as the outside surface decreases, so the inner surface area will increase, so that the rate of generation of gas will remain constant and thus the rate of acceleration of the shell will also be constant. If the pieces (or 'grains') of NC are small they will be rapidly consumed and convert all the charge into gas very quickly, so giving a very fast acceleration; if the grains are large they will take longer to burn, generating gas for a longer time and so giving a slower but sustained acceleration. (It should be appreciated that 'fast' and 'slower' in this context is a matter of micro-seconds; to the observer the gun goes off with a bang, irrespective of whether the powder charge is fast- or slow-burning.) By juggling with size and shape the designer can deliver precisely the performance required from any gun/shell combination.

Propellant powders made from nitro-cellulose alone are known as 'single-base powders'. For a greater amount of energy it is possible to mix nitro-glycerine with nitrocellulose, producing a 'double-base' powder. this is much more powerful, bulk for bulk, but it is also much hotter and has a tendency to melt away the steel inside the gun barrel, a process called 'erosion'.

'Triple-base powders' are those made with nitro-cellulose, nitro-glycerine and nitro-guanidine. This latter substance has explosive value, but also lowers the flame temperature so that a powder almost as powerful as a double-base is achieved, but with far less erosion of the gun.

'Glycol powders' are double-base powders which used diglycol dinitrate in place of nitro-glycerine. These are efficient, since the diglycol produces more gas than an equivalent amount of either NC or NG, but are also cool-burning and cause less erosion.

All types of powder contain various substances in addition to the basic explosives; these include such things as mineral jelly, Akardite, potassium salts, diphenylamine and camphor. The potassium salts are added in order to generate inert nitrogen gas which helps to prevent excessive flash at the gun muzzle. The other substances are added in order to keep the propellant chemically stable during storage or to assist in manufacture.

PROXIMITY FUZE A fuze which reacts to the proximity of the target and detonates the shell, instead of having either to strike the target or have some timing mechanism set before firing.

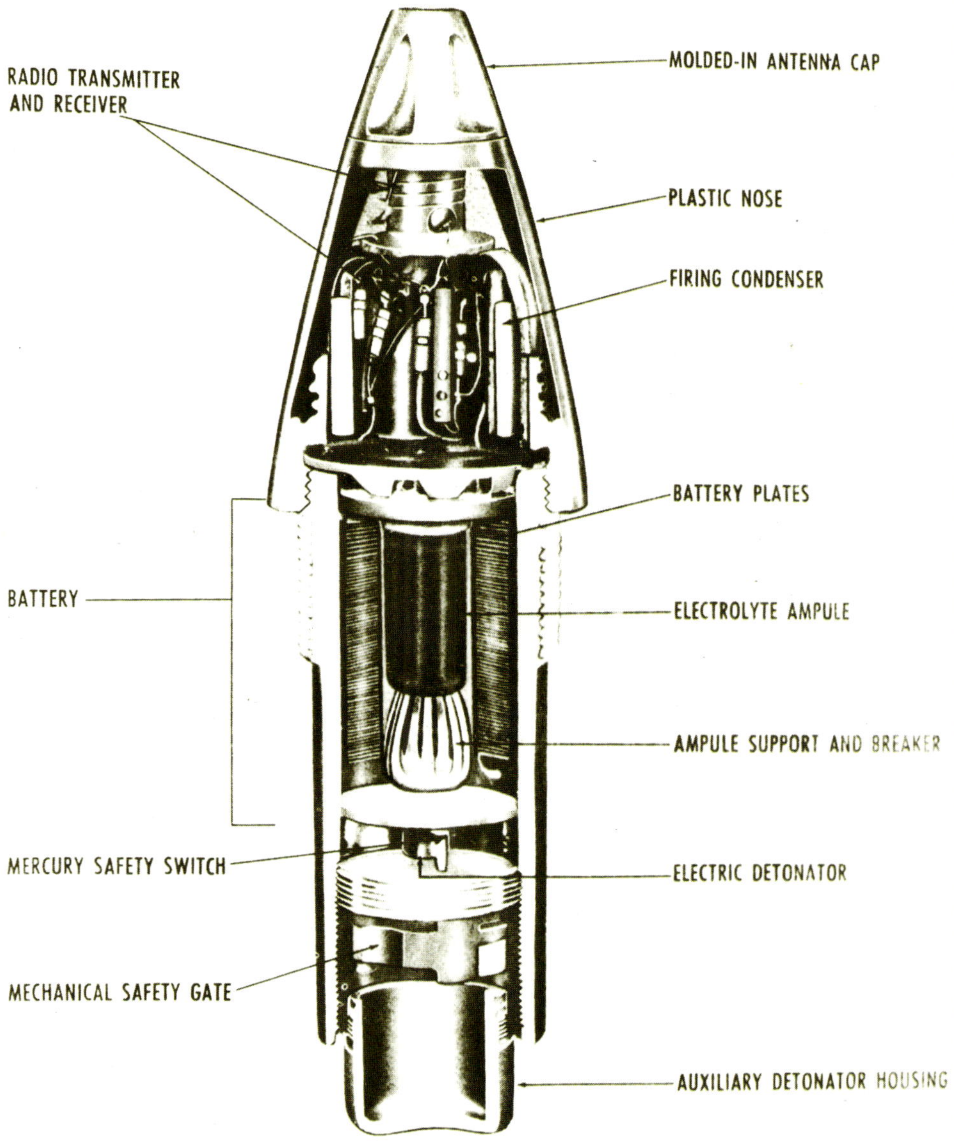

An American proximity fuze of World War II vintage. The radio transmitter sent out a signal which was reflected by the target. The receiver picked up the reflected signal and, when the signal strength indicated that the target was close, the detonator was fired and the shell exploded.

Left: A Thomson-CSF naval proximity fuze.

The first proximity fuze to be announced was a Swedish design in 1938 which proposed having a light source and several photo-electric cells arranged around the shell body. The light source would project beams of light, and when these were reflected from the target into the photo-electric cells this would cause the shell to detonate. The idea was patented but never built.

In 1940 the British developed a photo-electric fuze for rockets; this was 'tuned' to average daylight, and if anything cast a shadow over the cells the fuze would operate and detonate the rocket warhead. It was adopted for

Above: A selection of Thomson-CSF proximity fuzes, each designed for a specific application.

use in anti-aircraft rockets and was used successfully against daylight raiders, but when the German attacks switched to night-time the fuze was useless.

At the same time, scientists working on the development of radar suggested putting a radar receiver in the fuze and using this to detect the reflections of the fire-control radar set coming from enemy aircraft. It was soon shown that, at that stage of radar development, such an arrangement was physically impossible to fit into a shell fuze. From this, however, came the idea of making a simple radio which transmitted a signal; this reflected off the aircraft and back to the receiver in the fuze. The signal strength was measured and when it indicated that the shell was within lethal distance of the target, a detonator was fired and the shell itself detonated.

Experiments showed that the idea was feasible, but at that stage of the war there was no possibility of developing and manufacturing such a fuze in Britain, so the idea was passed to the USA. There it was taken up by the US Navy, who gave a contract to the Eastman Kodak Company. The fuze was developed and was first used in June 1943 by the USS *Helena* to shoot down a Japanese aircraft in the Pacific. Proximity fuzes were used extensively in Britain against the V1 flying bombs in 1944, contributing enormously to the success of the defences. They were first used against ground targets in December 1944 during the Battle of the Ardennes.

The proximity fuze of that time was a large device, reaching several centimetres into the shell; this displaced a certain amount of explosive and there was a minimum size of shell below which it was uneconomic to use the fuze.

The nose of the fuze was of plastic, with a small antenna inset; inside the nose were the transmitter and receiver circuits, which used tiny glass valves. The remainder of the fuze body contained a wet-cell battery; this consisted of a stack of ring-shaped plates with a glass ampoule of acid inside. On firing, the ampoule set back and shattered, and the spinning of the shell spread the acid across the plates to generate the necessary electricity. Below the battery were mechanical and electrical safety devices, an electric detonator and the fuze magazine filled with tetryl.

Modern proximity fuzes have benefitted from advances in electronic technology, using transistors, printed circuits, micro-chips and compact batteries. They are now comparable in size and cost with conventional fuzes.

Proximity fuzes using other technologies are used with guided missiles, where there is more space available. These include fuzes based on lasers, radars, electro-optics and acoustic technology.

R

RADAR ECHO SHELL A base-ejection carrier shell which is loaded with several packets containing short lengths of tin-foil or wire. These are cut to half the wave-length of a specific radar frequency band and act as highly efficient radar reflectors. The shell is fitted with a time fuze and when this operates it fires a small expelling charge which blows off the base-plate of the shell and ejects the packages. They break open and clouds of the reflectors are formed. Due to their small size and weight they take a long time to fall to the ground and during their fall will reflect radar signals. They therefore 'blind' any radar set operating in their frequency band and their direction. Their use is analogous to the use of smoke as a counter to visual observation.

RAMJET ASSISTANCE Method of assisting the flight of a shell by incorporating a simple form of ramjet motor. Developed experimentally in Germany during World War II by Professor Tromsdorff, two types were proposed. The 'ring' type had a series of holes which admitted air to an annular space around the shell body; here the incoming air was mixed with carbon disulphide fuel and ignited, to give an exhaust through an annular port. This added thrust to the shell in flight. The other design was the 'tube' type which had a central tube through the shell body through which air was admitted and through the rear of which the thrust was developed. Several experimental shells were made and fired, and theoretical calculations suggested that a 28cm ramjet shell should be capable of a range of 400 kilometres (248 miles). Most of the German work was intended to increase the effective height to which anti-aircraft shells could be fired, but the war ended without a satisfactory shell being developed. The idea has been toyed with since then but nothing practical has come of it. Also called 'athodyd' shells in some reports.

A German experimental athodyd (ramjet) shell. Air enters the holes in the ogive and the jet exhausts through the mid-point shirt.

RANGING SHELL A projectile used to determine the range and elevation necessary to hit the target and used instead of the projectile

Austrian troops load a 106mm recoilless gun. Note the characteristic perforated cartridge case which allows the 'countershot' gas to escape.

which will be used for effective fire. The only application of ranging shell is in the case of nuclear projectiles, since it would be inadvisable to use a succession of such shells to determine the range. The ranging shell (called a 'spotting shell' in US terminology) has the same ballistic characteristics, but is filled with high explosive and flash composition so that it makes a very distinctive impact. Fire is then adjusted by observers until the ranging shell falls precisely where the nuclear shell is required, after which the nuclear shell is loaded and fired on the same data to hit the same place.

RDX British terminology for the explosive more commonly known as Hexogen. RDX was the code designation adopted during the 1930s when it was being developed in Britain, and stood for 'Research Department Explosive'. For further details see under *Explosives* above.

RECOILLESS Gun which, when fired, remains perfectly still and does not recoil.

The first recoilless gun was developed by Commander Davis of the US Navy during World War I and consisted of a single chamber with two gun barrels which pointed in opposite directions. One barrel was loaded with the service projectile, the other with an equal weight of lead shot and grease. The cartridge was loaded into the central chamber and fired, and the two projectiles were driven up their barrels at the same velocity, so that the recoil of each one cancelled out that of the other and the entire gun remained still. The service projectile went to its target, while the lead shot 'countershot' dispersed in the air just behind the gun. The idea was adopted by the British Royal Naval Air Service who fitted Davis guns of various calibres to aircraft in order to shoot at submarines, but so far as is known none were ever used in combat since they arrived too late in the war.

Above: A British 120mm BAT recoilless gun firing at night showing the highly visible rearward flash.
Below: An American hollow charge shell and cartridge for the US Army's 75mm recoilless anti-tank gun.

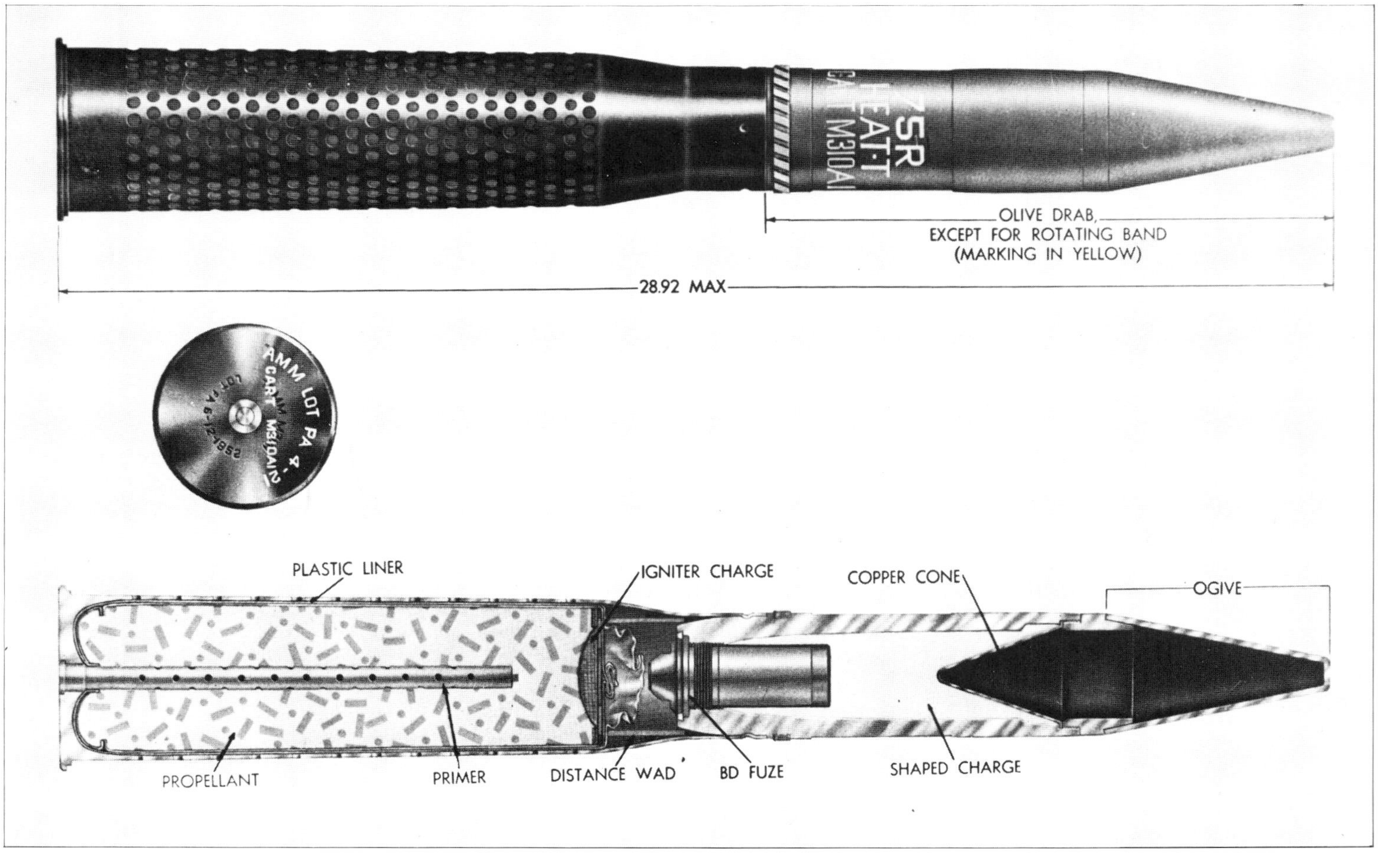

It will be appreciated that if two equal weights can be fired at equal velocity, it is also possible to use a countershot of half the weight and fire it at twice the velocity; provided the product of weight × velocity remains the same, the weapon will be recoilless. Taking this to the extreme, it is possible to use a stream of very light gas as the countershot, provided that it has sufficient velocity, and it is this principle which controls all current recoilless guns.

Two designs are currently used; the first, originated by the Germans in their airborne field guns of 1940, uses a cartridge case in which the base is cut out and replaced by a thick plastic disc. The primer may be in this disc, or may be inserted into the side of the case. The gun breech block, behind the case, is pierced with a hole which terminates in a venturi and nozzle. On firing, the disc remains whole for just enough time to allow the driving band to engrave in the rifling and the shell to begin moving; the disc then breaks and a stream of gas is ejected through the base of the case, through the hole in the breech and through the venturi which accelerates it and discharges it rearward through the nozzle. The velocity-accelerated gas, multiplied by its mass, equals the mass × velocity of the shell and the gun does not recoil. This system is used by the current Carl Gustav 84mm and British BAT 120mm guns.

The other system, used by American 90mm and 106mm recoilless rifles, has a cartridge case with a solid base but with holes perforated in its sides. It is internally lined with a thin metal sheath. The chamber of the gun is surrounded by an annular space which leads rearwards to a venturi and nozzle. On firing, the inner sheath ruptures and allows gas to flow into the annular space and then out of the venturi. In this design a pre-rifled shell is used, so that there is no need to make the liner very thick in order to allow a pressure build-up.

The object of all this is to save weight; the apparatus which absorbs the recoil in a normal gun is a complicated arrangement of hydraulic cylinders and is extremely heavy. Moreover the gun mounting must also be heavy and strong to withstand the force of recoil. By doing away with recoil, one does away with the need for a recoil system and for a heavy and strong carriage; the only stress the carriage now needs to be designed for is the stress of moving it across country. Recoilless guns can thus be mounted on very light vehicles or even fired from a man's shoulder.

The drawbacks are twofold: first, a recoilless gun demands much more propellant than a conventional gun, since four-fifths of the propelling charge is exhausted from the jet. Secondly the blast of flame and gas from the rear

The British 3.45in (87mm) shoulder-fired recoilless gun made in 1945 showing the venturi jets at the rear.

Ribbed shells for the German 28cm railway gun. The ribs are at the same pitch as the gun's rifling.

of the gun gives its position away as soon as it fires, and the gun must be carefully sited so that the flame is not reflected back on to the gun's crew.

RIBBED SHELL Projectiles which have inclined ribs on the body which engage in special rifling grooves in the gun. Used only on very long-range guns which have extremely large propelling charges, the ribbed construction was adopted because a conventional driving band would shear as it met the rifling due to the very high pressure behind it. The use of ribs which are pre-engaged in special deep grooves, ensures that there will be no resistance while a driving band engraves, and the shell will take up spin irrespective of the amount of pressure. It is necessary to place a soft sealing band at the rear of the ribs to make sure that the propelling gas is held behind the shell, but this band, although it engraves into the rifling to form a seal, does not transmit any spin to the shell and is not likely to shear.

The first practical use of ribbed shell was in the German 'Paris Gun' of World War I, a 21cm calibre weapon which had a maximum range of about 120 kilometres (75 miles). The same principle was then adopted for the 21cm K12(E) railway gun used by the Germans to bombard Britain from the French coast in 1940-41. This fired a 107kg (237lb) shell with 8 ribs to a range of 115 kilometres (72 miles). At much the same time an experimental British gun using a 203mm (8in) liner inside a 343mm (13.5in) barrel was developed; this fired a 116kg (256lb) shell with 16 ribs to a maximum range of 110 kilometres (68 miles) but it was never used in action, being developed solely as an experimental weapon.

RICOCHET FUZE A fuze which permits the shell to ricochet (bounce) into the air after striking the ground and then detonate. This gives a low air-burst effect which is effective against entrenched enemy troops or can be used to bounce a shell across a protective barrier such as a wall and there detonate to injure the men behind the wall. It saves having to use time fuzes, but the technique is only effective when the shell arrives at a very low angle so that it strikes the ground on its shoulder and then bounces. If it strikes at a steep angle it will detonate on hitting since it will have dug into the ground and will not bounce. To make the ricochet action work it is necessary to have a *graze* action fuze in the shell, so that the fuze begins to operate on impact, but the graze delay permits the shell to bounce before the detonation occurs. The technique was widely used by US artillery during World War II.

ROCHLING SHELL Special type of anti-concrete shell developed by the Rochling Eisen- und Stahl-werke of Düsseldorf, Germany, during World War II.

Conventional anti-concrete shells of the period were full-calibre shells with blunt points; the Rochling designers reasoned that better penetration would be achieved by a shell of small calibre, but delivering a heavier blow by virtue of being much longer than normal and thus concentrating its weight into a smaller area of contact. They accordingly designed a long sub-calibre shell with a discarding sabot at the shoulder and a sabot at the rear which enclosed a set of four flexible fins. When fired, the shell left the muzzle and discarded the two sabots, allowing the fins to spring out and stabilize it. It was fired from a 21cm rifled gun, but the fins soon damped out the spin.

The Rochling shell was not used against the Maginot Line during the 1940 invasion of France, since it was not necessary. After the occupation of France and Belgium a number of trials were made against fortifications and one record of these tests reports of a shell which passed through 3m of earth, 36m of concrete, a layer of broken stone, the roof of a subterranean chamber, then into the floor beneath and 5m into the earth beneath the floor. This was a test of

The Röchling anti-concrete shell. The front sabot has been partly disconnected and there is a drive piston behind the tail fins. Both were thrown clear as the shell left the gun.

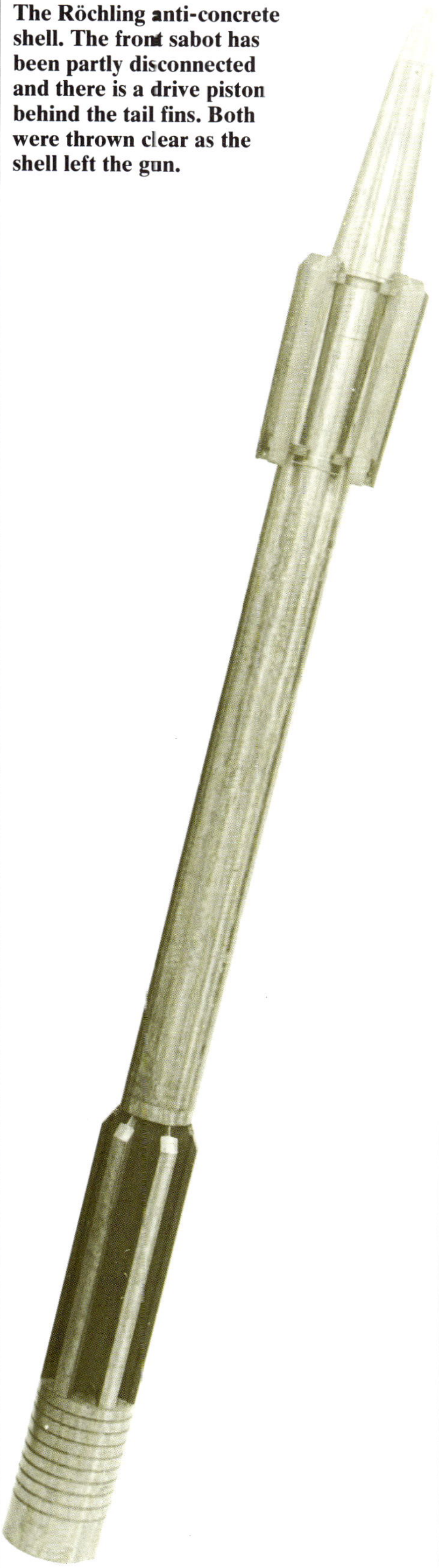

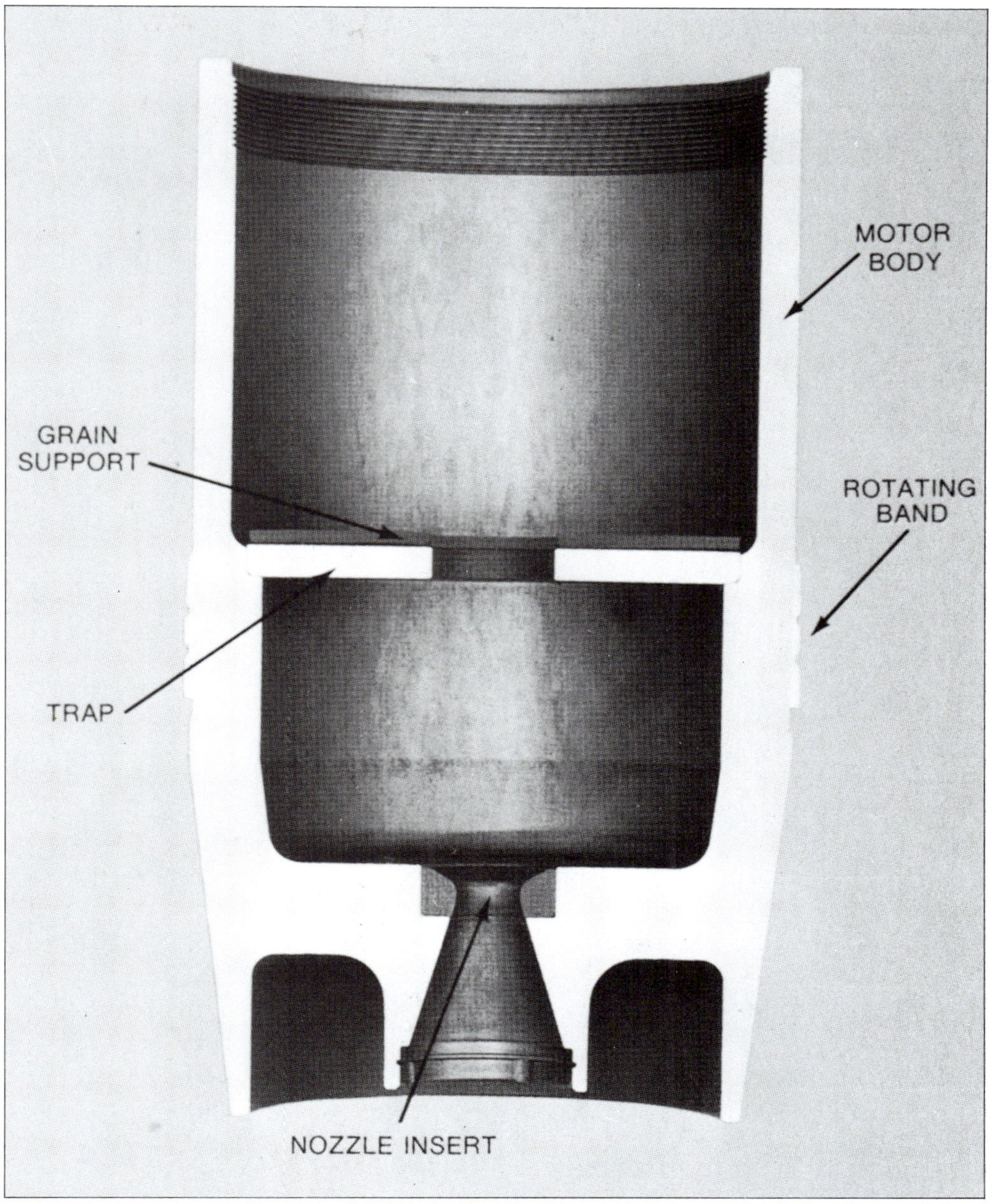

The motor sections of a rocket-assisted 155mm shell showing the propellant chamber and the rocket venturi.

an inert shell; a live shell would have detonated in the subterranean chamber if correctly fuzed.

Achievement of this kind of performance demanded high-grade chrome-vanadium steel for the shell, and precise manufacture. 8,000 shells were made and stockpiled, but after that it was hardly used. A few were fired against the fortress of Brest-Litovsk during the invasion of Russia in 1941, but use of the Rochling shell was then forbidden by Hitler on the grounds that a specimen might fall into enemy hands and be copied for use against Germany. From then on the shells could only be used with his permission, and since this was rarely requested and even more rarely granted, the existence of the shell was gradually forgotten.

ROCKET ASSISTANCE Method of increasing the range of a shell by incorporating a rocket motor in the body and igniting it during flight. The arrangements vary according to the ideas of the designer, but a typical system would be to have the forward half of the shell hold a solid-fuel rocket motor and then run a blast pipe down the centre of the rest of the body to exhaust at the base. The exhaust vent would be securely plugged, so that the flash from the normal propellant cartridge could not ignite the rocket motor, and a time fuze would then ignite the motor at some point on the upward leg of the trajectory. The blast would blow out the plug and the rocket exhaust would accelerate the shell so that it continued upwards for some

considerable distance, thus increasing the range. An impact fuze inside the shell would ensure detonation at the target.

A rocket shell designed on these lines was developed for the German 28cm K5(E) long-range railway gun during World War II. The standard shell for this gun ranged to 62 kilometres (38 miles), while the rocket-assisted shell ranged to 86.5 kilometres (53 miles). Rocket-assisted shells are currently provided for 155mm howitzers in US and German service.

The principal defect of the rocket-assisted shell is that if the shell happens to be yawing from its planned trajectory at the moment the rocket ignites, the thrust will send it off at that yaw angle, so that it might veer right or left, up or down, and thus not adhere correctly to the trajectory. In consequence the accuracy of a rocket-assisted shell is not comparable with that of a conventional shell.

RODDED BOMB Type of shaped-charge projectile which consists of a very large warhead, of much greater diameter than the gun's calibre, supported on a rod which is loaded into the muzzle of the gun. A propellant charge is then

Left: A 28cm rocket-assisted shell showing the motor and its blast pipe.

fired, and this pushes the rod and the warhead out of the muzzle and to the target. Stabilization is by fins, which are usually arranged on a sleeve which slides over the outside of the barrel when the bomb is loaded. Developed by Germany during World War II and used with a 37mm anti-tank gun in order to give it an effective projectile when its standard projectiles could no longer make any impression on Allied tanks. A similar design was also developed for a 15cm howitzer but saw very little use. This type of projectile suffers from a short range.

Below: Anti-tank gun with a ribbed hollow charge bomb loaded in the muzzle.

S

SALUTING CHARGE A type of blank charge used for firing salutes and on other ceremonial occasions. In many cases the standard blank charge is used, since this gives a satisfactory noise, but it is sometimes necessary to produce a special saluting charge. One of the best examples was the British 25pdr gun which had three blank cartridges designed for it; the standard blank was used for training and for most saluting purposes. But two others were manufactured, one 'special to Dover' had a larger charge of gunpowder than usual, since the saluting battery at Dover is at the Castle, a considerable distance away from the landing-stage used by visiting dignitaries, so that a very loud report was needed to make sure it was heard. Conversely, the second special was 'special to Gibraltar' and was rather weaker than standard since the saluting battery was immediately adjacent to the landing-stage and it was thought undesirable to deafen the visitor immediately on arrival.

SCHARDIN EFFECT If an explosive charge has its face formed into a concave hollow, and if this face is then lined with a heavy steel plate of the same curvature, when the explosive is detonated the plate is flattened out and propelled forward at high velocity. This was discovered by a German ordnance engineer named Schardin during World War II and some work was done by him on developing the effect into a munition. His ideas were principally concerned with anti-tank mines which blew a large circular slab of steel for considerable distances to penetrate armour, and some very impressive results were achieved. The war ended, however, before the ideas could be put into workable form. The principle was studied in postwar years and made its first appearance in the 'Claymore mine', which used a slab of explosive to propel several hundred steel balls in a fan-shaped swathe. The application of this effect to projectiles has not yet become general, but is being actively pursued, particularly in the USA, in various remotely-delivered munitions. For further details see the 'New Developments' section and *Self-forging Fragments.*

SELF-DESTROYING Projectile which detonates, destroying itself, at the end of some specified time of flight. Used with air-to-air and ground-to-air projectiles to ensure that they detonate in the air and do not fall to the ground in a live condition and thus become a danger to the populace beneath, which might very well be friendly.

Self-destruction is usually achieved by one of two methods: either by using the projectile tracer as a timing device or by using the fuze. Using the tracer involves dividing the projectile into two compartments seperated by a steel wall, the front holding the explosive bursting charge and the rear holding the tracer. A hole is drilled through the wall and into this goes a small steel capsule filled with an initiator which burns to detonation. When the shell is fired, the tracer is lit and burns in the usual way; when it reaches the

A 20mm cannon shell showing the relay capsule between the tracer filling and the high explosive.

end of its burning, the final layers ignite the initiator in the capsule which burns and then detonates, passing the detonation to the main charge which detonates and destroys the projectile. The object in using the capsule is to ensure that should the tracer be faulty and permit the propellant flash to pass around it, the capsule contents will take a short time to complete their burning and detonation, by which time the shell will be out of the gun barrel and a safe distance away.

Self-destruction by using the projectile fuze can be accomplished in one of two ways. In the first method there is a short delay train of powder which is ignited when the projectile is fired; this burns at a pre-determined rate and, at the end of its burning, ignites a detonator which then detonates the main filling to destroy the shell. The second method is to use 'spin decay', by having a mechanism in the fuze which, so long as the spin of the projectile is above a certain rate, keeps a firing pin away from a detonator. After a few seconds of flight, the spin rate decreases and centrifugal force is no longer strong enough to keep the mechanism operating, so that it allows the firing pin to strike the

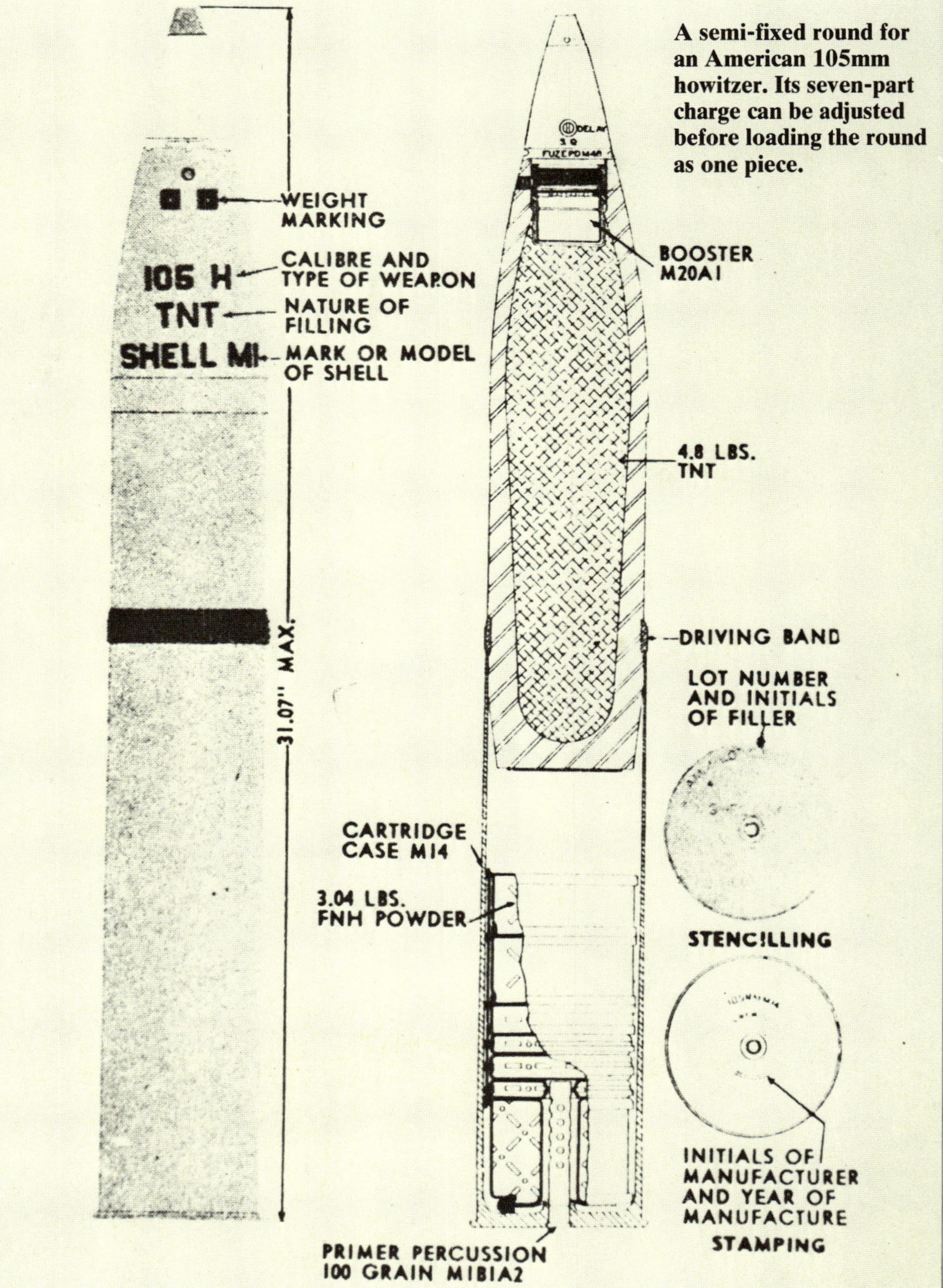

A semi-fixed round for an American 105mm howitzer. Its seven-part charge can be adjusted before loading the round as one piece.

detonator and detonate the shell filling. In both cases the self-destroying element in the fuze is independent of the normal impact action, so that if the shell strikes the target before the self-destruction time, it will detonate in the normal way.

The systems outlined above are those used on minor-calibre cannon shells. Major-calibre anti-aircraft artillery used time fuzes set to a specific time, so that self-destruction automatically occurred when the fuze reached the set time, irrespective of whether it had reached a target or not. In the case of proximity fuzes there was a form of spin decay incorporated in one of the electrical safety switches which detonated the fuze and shell if the proximity action had not taken place.

SEMI-FIXED ROUND A complete round of artillery ammunition which is supplied in two units — shell and fuze; cartridge, propellant and primer — but which is assembled into one unit, as a fixed round, before loading. The object is to permit the adjustment of the propelling charge, by removing propellant bags, before firing, but to have the advantage of being able to load the round as one piece, in one movement, and by one man. Probably the best-known example of semi-fixed ammunition is the round for the American 105mm howitzer.

SEPARATE LOADING AMMUNITION A complete round of artillery ammunition which is supplied in two units and loaded in two units. Cased-charge ammunition may be separate loading; bag-charge ammunition is always separate loading. Separate loading may be adopted in order that the propelling charge can be altered prior to loading (in the case of cased

Separate loading is necessary when the components are big and heavy. In this case, four men are carrying a 164kg (360lb) shell for an American 240mm howitzer. The bag charge weighs a further 35kg (77lb).

Loading a German 24cm railway gun. The heavy shell has to be hoisted up onto the gun's loading platform by crane.

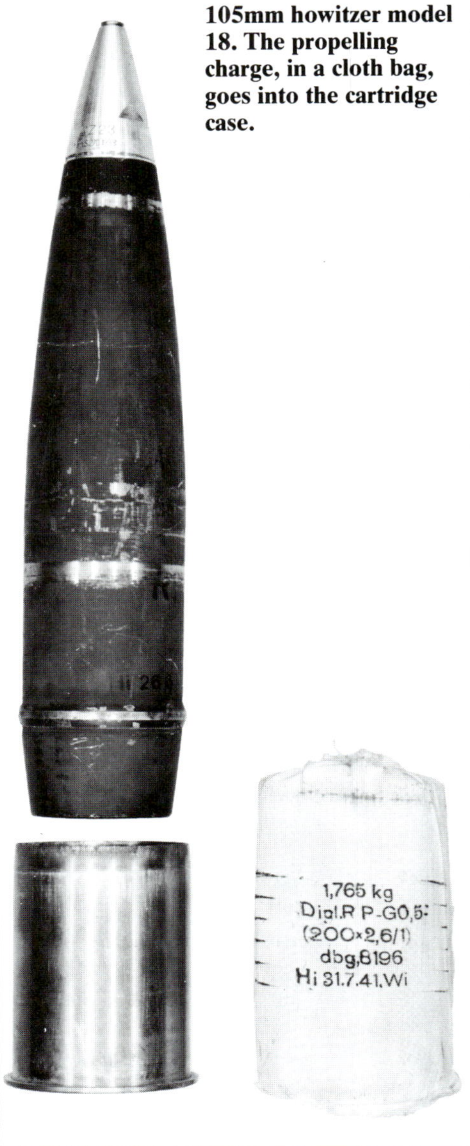

The separate-loading round for the German 105mm howitzer model 18. The propelling charge, in a cloth bag, goes into the cartridge case.

charge) or simply because a fixed or semi-fixed round would be too big to be conveniently handled (also in cased charges) or because it would be impossible to make a fixed round (ie, in the bag-charge round).

The limit of size for handling is not entirely based on weight, though this, of course, is important. Generally speaking, about 40kg is taken as the maximum weight which can be conveniently handled by gun-loaders for any length of time; as an example, the British 114mm (4.5in) anti-aircraft gun used a fixed round weighing 39kg, while the next larger weapon, the 133mm (5.25in) used a separate-loading round in which the shell weighed 36kg and the cartridge 21kg. However, bulk has to be considered as well; it would be possible to manufacture a fixed round within the weight limit, but which was so long that the loaders found it awkward to handle; moreover a long round, with a heavy projectile at the end of a long thin case, would place a great deal of strain on the case-shell joint, so that the shell might well droop and fall out of alignment during handling, leading to jamming during loading.

Separate loading cased charges may be designed so that the shell has to be rammed into the gun by hand before the case is loaded behind it, or the case mouth may be reinforced with a rigid cap and the entire case used to ram the shell into place. This latter system has been used on anti-aircraft and anti-tank guns where speed of loading is of paramount importance. Separate loading bag-charge rounds which use cloth bags require the shell to be separately rammed, after which the bag is loaded into the chamber, but those using combustible-cased cartridges of rigid plastic material can use the cartridge as a shell rammer.

Italian troops load a 152mm gun. One man heaves the shell into the chamber while another waits with a rammer. A third man carries the cased cartridge to the gun.

A British 5.5in gun with its complete round of ammunition — a shell and two bag charges — ready for loading.

The entrance hole made by a 75mm shaped-charge shell. The slab of armour plate is 250mm thick and about 300mm wide.

SHAPED CHARGE Also known as *hollow charge, Monroe effect Neumann effect, HEAT shell* etc.; in German 'Hohlladung', in French 'Obus à Charge Creuse'. A method of defeating armour or other hard targets by shaping an explosive charge so as to 'focus' the detonation.

The basic shaped-charge shell consists of a main explosive charge inside the cylindrical body of the shell; the front end of this charge is hollowed out in the form of a cone, and this cone is lined with a cone of copper or other metal. In front of the cone the shell's shape is continued by a hollow ogive, at the tip of which is a fuze, or the piezo-electric initiator for a fuzing device. To ensure the correct operation of the shaped charge the explosive must be initiated at the end away from the hollowed-out cone, and methods must be adopted to ensure that the impact of the shell on the target is translated into this detonation; see entries under *Piezo-electric* and *PIBD* for further information on this aspect of initiation. It is possible to use a conventional base fuze, but there are some problems associated with them.

The operation of the shaped-charge is as follows: when the projectile strikes the target, the detonation of the charge is initiated, from the base. The detonating wave passes through the explosive and as it begins to pass around the hollow cone, so the cone is collapsed towards the axis of the hollow. As a result, the molten metal of the liner, plus the detonating wave and the explosive gases, collects on the axis and moves forward, in the form of a jet, at very high speed — in the order of 10,000mps (32,800fps) — to strike the armour and pierce it simply by its momentum. (The actual mechanics of jet formation and performance are extremely involved and this is necessarily a somewhat simplified explanation.)

The jet requires a certain amount of space in which to form and accelerate; this is roughly twice to three times the diameter of the cone, and is known as the 'stand-off distance'. Firing the charge at less than this distance will lead to

The exit hole made by the 75mm shaped-charge shell after it had penetrated the 250mm of armour plate.

less than optimum results, as will firing the charge at excessive stand-off distances. This is one reason for not using base fuzes; the inherent delay in the fuze is likely to permit the shell's empty ogive to crush up and reduce the stand-off distance to below the critical figure before detonation of the charge takes place. The only answer to this is to use a false nose of considerable length so that it crushes down to the correct stand-off just as the detonation takes place, but this, obviously, has ballistic drawbacks. It is for this reason that PIBD and piezo-electric fuzes have been adopted.

The hollowed-out face of the charge can be of other shapes — hemispherical, conical-hemispherical — but experience has shown that a cone of about 100° included angle gives the best results. One defect of this system is that at high rates of spin the jet is diffused, due to centrifugal force acting on it, and thus does not penetrate so efficiently; for this reason the shaped charge is favoured for use in rocket-propelled projectiles which are fin-stabilized. However recent research indicates that an efficient, spun shaped-charge can be achieved when the cone angle is between 130° and 140°, that is, the so-called 'flat cone'.

Shaped charges can be used in combination with other effects; one early experiment was to mount a small shaped charge inside the ballistic cap of a heavy armour-piercing shell, so that the shaped charge would make an initial hole which the heavy pointed projectile could then exploit. Another variation was the 'follow-through shell' in which a small projectile was carried inside the shaped-charge shell, behind the shaped-charge unit. On impact the shaped-charge blew a hole in the target and the momentum of the rest of the projectile launched the small projectile through the hole made by the jet and thus into the target, there to detonate. They can also be used in other munitions; there are shaped-charge grenades, mines and demolition devices which use the same basic principles.

SHELL A hollow projectile, carrying some payload, fired from a gun or howitzer. They are broadly divided into two groups, high explosive and carrier shell; the former contain high explosive as the payload and detonate to give anti-personnel and anti-*matériel* effects from their blast and fragmentation. The latter are inert, but carry some payload which has tactical effect when released from the shell, eg, smoke, incendiary fillings, gas, leaflets, illuminating stars, etc.

SHOT A solid projectile, containing no explosive or other filling, fired from a gun and intended to overcome hard targets by kinetic energy.

The basic shot is a simple, hardened steel cylinder with a point; on impact the mass and velocity combine to deliver a blow to the target which defeats the protective hardness; it was developed for the attack of armoured ships and concrete fortifications. As a result the targets became harder, by advances in armour technology and constructional technique, and various sub-divisions of shot appeared. Capped shot (APC) was invented in the late 19th century to overcome face-hardened plate. Composite shot (APCR, APCNR) appeared during World War II to defeat greater thickness of homogenous plate, relying on the greater hardness of tungsten carbide to pierce metals which steel could not defeat. Discarding sabot shot (APDS, APFSDS) appeared later, in order to extract the maximum velocity possible from the gun, while concentrating the kinetic energy into a very small impact area. At the present time the recent introduction of compound armour (Chobham armour) appears to have put the protection on top, but further advances in technology will no doubt produce a shot capable of overcoming this.

SHRAPNEL SHELL A carrier shell in whch the payload is a quantity of sub-projectiles, usually in the form of lead balls or short lengths of steel (bar shrapnel) which are ejected towards the target.

The shrapnel shell was invented by Lieutenant (later Major-General) Henry Shrapnel (1761-1842) in 1784 under the name 'spherical case shot' since, in the inventor's words, it was intended to carry the effect of case shot to the enemy. In its original form it consisted of a hollow spherical shell packed with gunpowder and musket balls and with a time fuze; when the fuze fired the gunpowder the shell was burst open and the balls were liberated, but at low velocity so that instead of flying in all directions they followed the trajectory of the shell. When burst slightly above and in front of troops in the open it was a highly effective anti-personnel weapon. It was first used in the Siege of Surinam in 1804 and achieved fame during the Peninsular War.

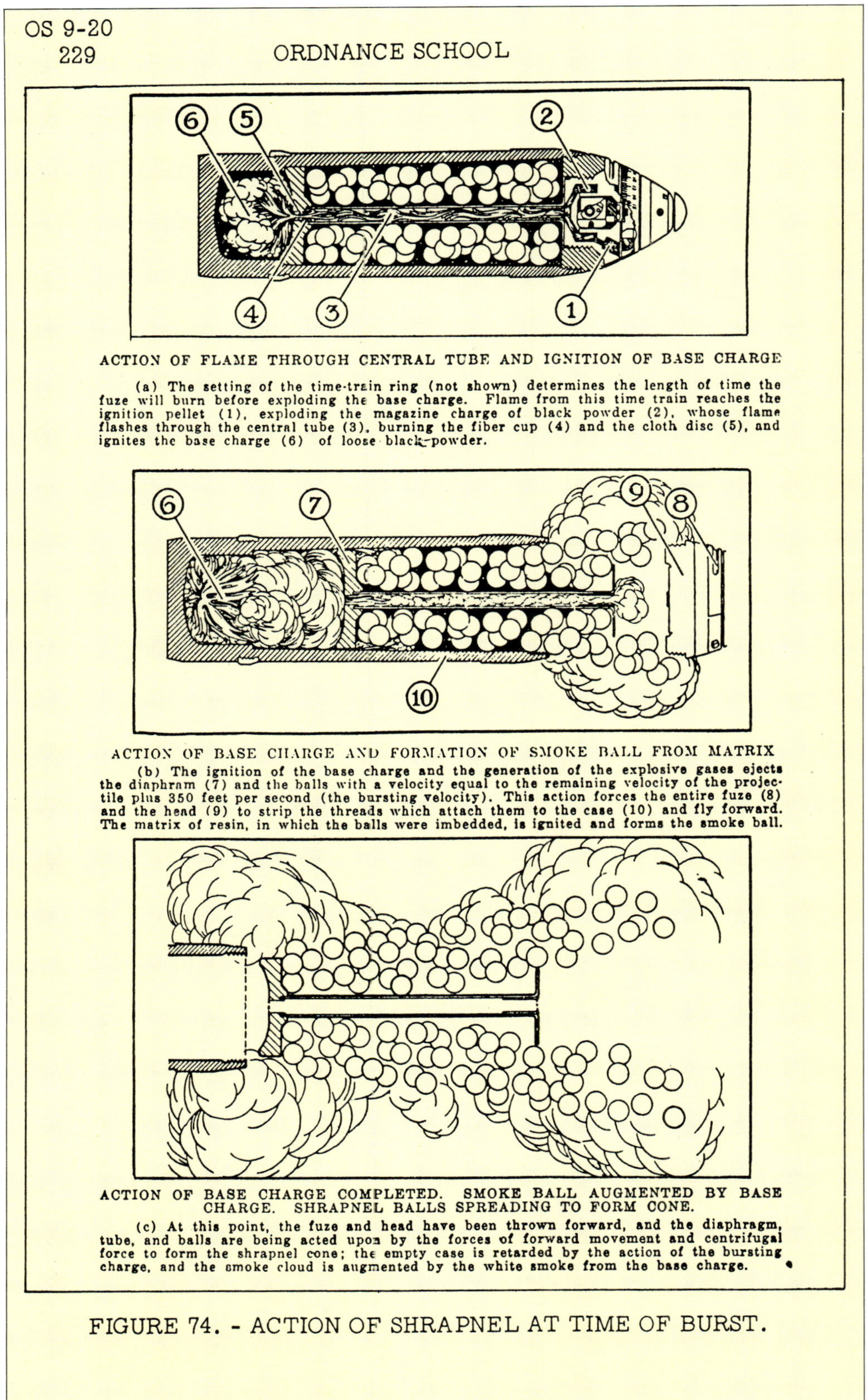
OS 9-20
229
ORDNANCE SCHOOL

ACTION OF FLAME THROUGH CENTRAL TUBE AND IGNITION OF BASE CHARGE

(a) The setting of the time-train ring (not shown) determines the length of time the fuze will burn before exploding the base charge. Flame from this time train reaches the ignition pellet (1), exploding the magazine charge of black powder (2), whose flame flashes through the central tube (3), burning the fiber cup (4) and the cloth disc (5), and ignites the base charge (6) of loose black-powder.

ACTION OF BASE CHARGE AND FORMATION OF SMOKE BALL FROM MATRIX

(b) The ignition of the base charge and the generation of the explosive gases ejects the diaphram (7) and the balls with a velocity equal to the remaining velocity of the projectile plus 350 feet per second (the bursting velocity). This action forces the entire fuze (8) and the head (9) to strip the threads which attach them to the case (10) and fly forward. The matrix of resin, in which the balls were imbedded, is ignited and forms the smoke ball.

ACTION OF BASE CHARGE COMPLETED. SMOKE BALL AUGMENTED BY BASE CHARGE. SHRAPNEL BALLS SPREADING TO FORM CONE.

(c) At this point, the fuze and head have been thrown forward, and the diaphragm, tube, and balls are being acted upon by the forces of forward movement and centrifugal force to form the shrapnel cone; the empty case is retarded by the action of the bursting charge, and the smoke cloud is augmented by the white smoke from the base charge.

FIGURE 74. - ACTION OF SHRAPNEL AT TIME OF BURST.

A page from an American Ordnance School textbook for 1942, giving a graphic description of how a shrapnel shell worked.

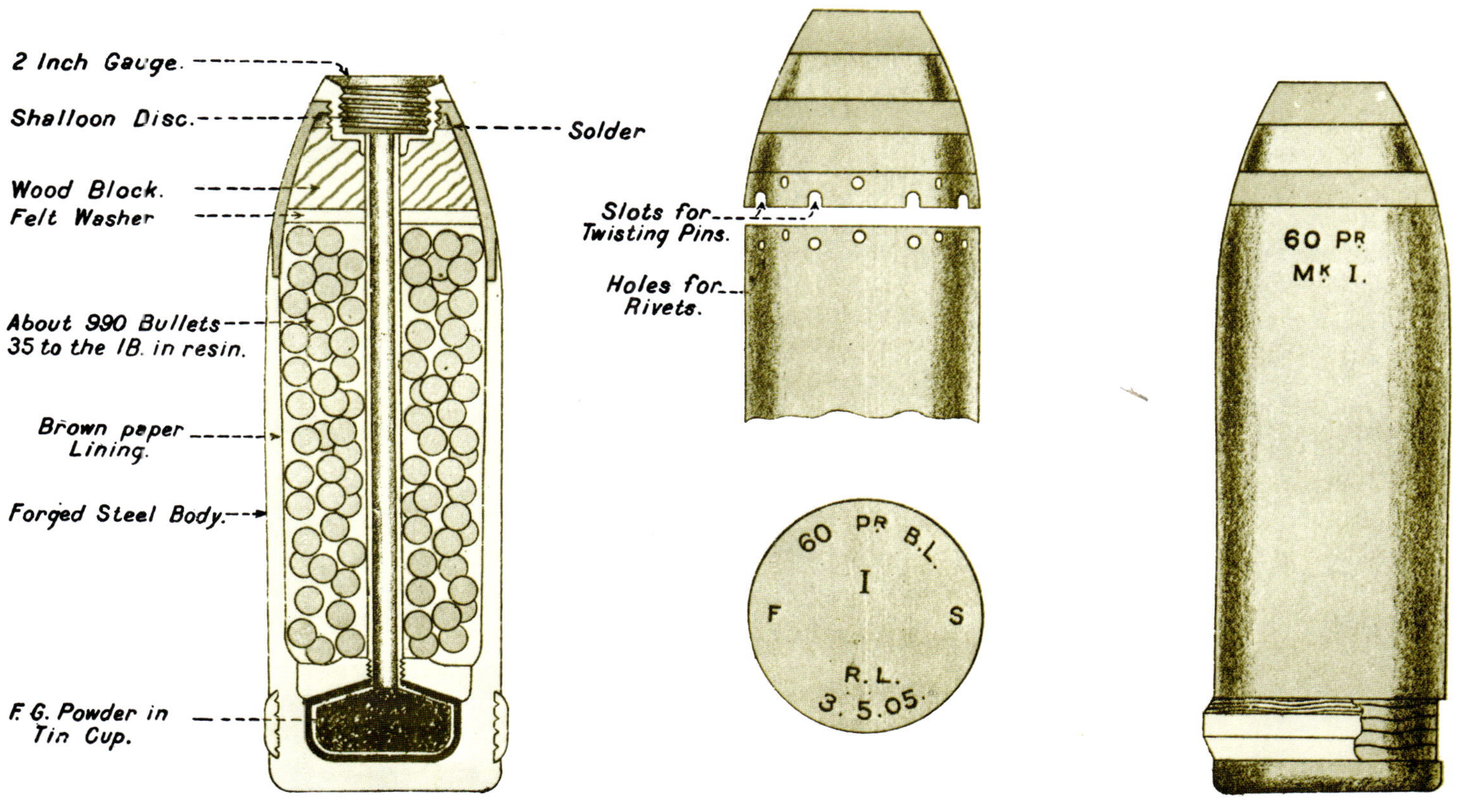

A shrapnel shell for the British 60-pounder (127mm) gun of World War I vintage, giving details of its construction.

With the coming of breech-loading guns, the design was altered by Colonel Boxer, Superintendent of the Royal Laboratory at Woolwich Arsenal, into the form which, except for minor modifications it retained thereafter. The cylindrical shell has a cavity at the base which contains a small charge of gunpowder; above this is a loose 'pusher plate' with a central hole to which is fixed a thin tube running through the shell to the mouth. Around the tube is packed the payload of musket balls, usually held in a resin matrix. This matrix holds the balls firmly so that they do not shift and upset the shell's stability in flight, and it is ignited by the gunpowder charge to give a visible puff of smoke which indicates the point of burst, so permitting corrections to be made. The ogival head of the shell is a separate component, fixed to the cylindrical body by either a weak screw-thread or by rivets.

The shell is closed by a time fuze which connects directly to the central tube. When the fuze functions at the set time it fires a flash down the tube which ignites the gunpowder charge at the bottom of the shell. This explodes and the pressure pushes up on the loose plate, on the filling of balls, and on the head of the shell which is broken free of the body. The balls are thus thrust out of the shell with minimal force and continue along the shell's trajectory to the target; the shell body, now unstable, is slowed by air drag and falls to the ground.

Shrapnel shell became the standard projectile for field artillery in the late 19th century and remained so until World War I. That conflict led to trench warfare and troops protected by trenches were in little or no danger from shrapnel balls, so that the high-explosive shell became more important. In the post-war years the shrapnel shell was gradually superseded by the high-explosive shell, although it managed to survive until the stocks were used up. The last shrapnel shell fired by British field artillery is believed to have been from 60pdr guns in Burma in 1943; shrapnel was retained as a short-range defence against low-flying raiders by anti-aircraft guns until about the same time. The Soviet Army carried bar shrapnel shells until the late 1940s.

One point which can never be repeated too often is that when modern high-explosive shells detonate, they produce fragments, and not shrapnel. It should be clear from the foregoing that the word 'shrapnel' has very strict meaning, and shrapnel cannot come from HE shells, mines, grenades, terrorist bombs or, indeed, anything other than a proper Shrapnel shell.

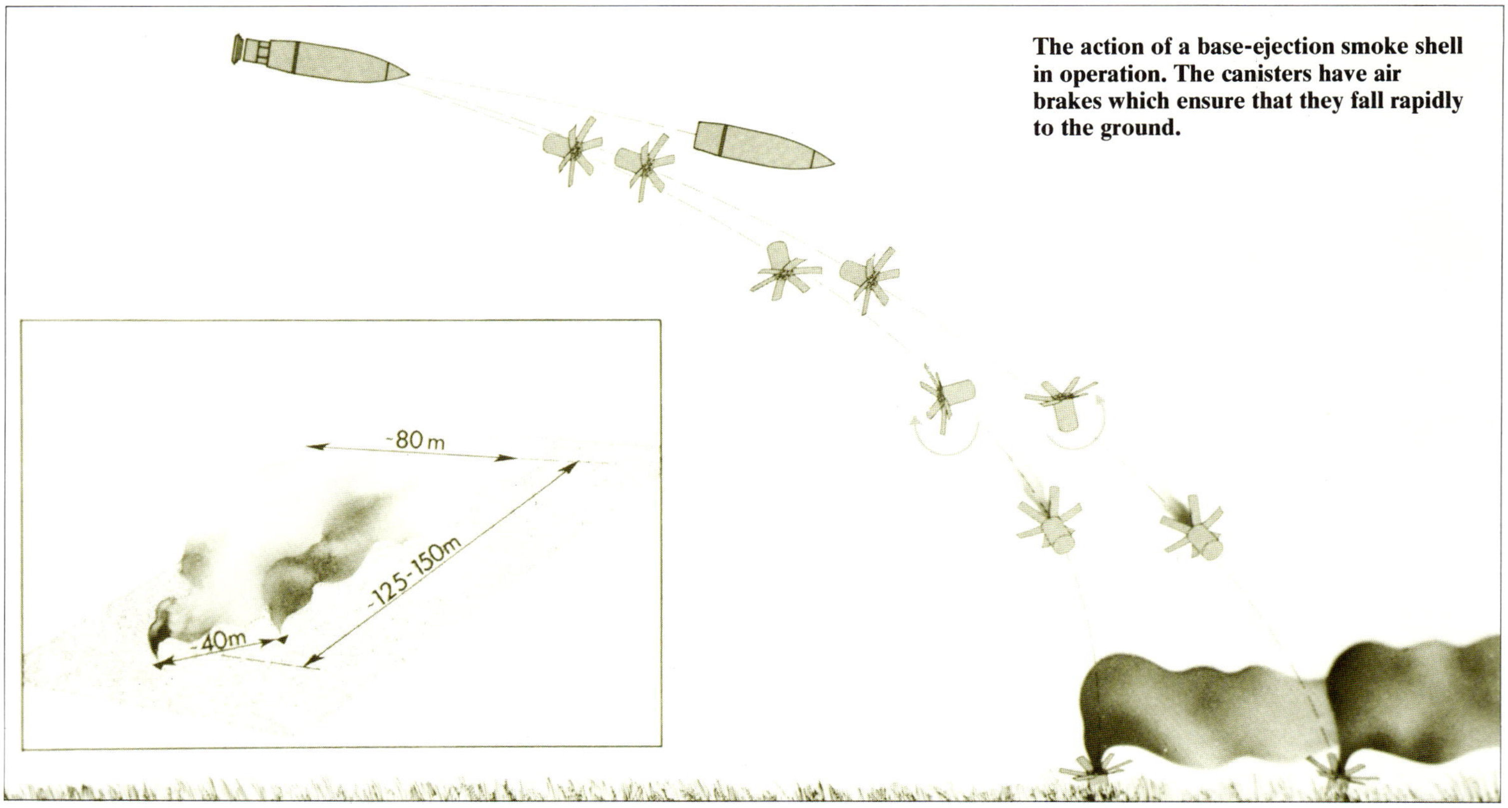

The action of a base-ejection smoke shell in operation. The canisters have air brakes which ensure that they fall rapidly to the ground.

SMOKE SHELL A projectile designed to deliver smoke; a carrier shell, the payload of which is a smoke-producing chemical. May be of the bursting or base-ejection type; the smoke produced may be for screening or for signalling. Screening smoke is white, since this has the greatest obscuring power, and is intended to mask movements from an enemy. Signalling smoke is coloured, for ample contrast with its background and easy visibility, and can be used to indicate targets to aircraft, to indicate zones of advance, to signal success or failure to observers elsewhere or for any other desirable purpose.

SPIN DECAY The gradual decrease in the rate of spin of a spin-stabilized projectile as it proceeds on its flight. As the velocity drops due to air resistance and gravity, so too does the rate of spin. This spin decay can be used to operate mechanisms inside fuzes for various purposes, but is most usually employed to operate *self-destroying* devices.

SPIN STABILIZATION Keeping a projectile pointing towards its target by spinning it at a high rate. The effect is due to the same principle as that which keeps a spinning gyroscope aligned and causes it to resist a change of direction. Once a shell is fired it literally becomes a gyroscope, and any tendency to change direction away from the ordained trajectory is resisted. The amount of spin which a projectile requires in order to be stable in flight can be calculated, the most significant factor being its length. A projectile in excess of about six calibres in length (eg, for a 15cm shell, six calibres length would be 90cm) is not easily spin stabilized, and projectiles longer than this are usually fin-stabilized. It is for this reason, as much as any other, that the first thing to be

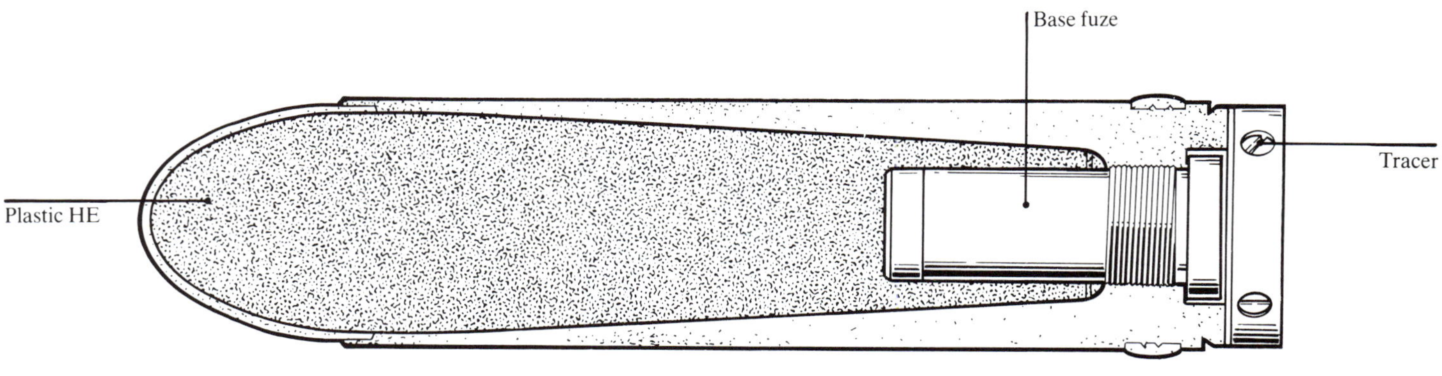

The construction of a typical squash-head anti-tank shell.

designed for a weapon system is the projectile; once this is settled the gun can be designed to throw it.

Spin causes the shell to drift; ie to move sideways through the air in the direction of the twist of the rifling. this is due to the effect of the rapidly-moving surface of the spinning projectile upon the surrounding air; to simplify, the air on one side of the projectile becomes rarified and on the other side becomes increased, so that the pressure pushes the shell towards the thinner air. Drift can be compensated for in the design of the weapon sight, but has to be adjustable in weapons which use variable propelling charges, since the different charges develop different velocities and thus different rates of spin.

SPITBACK FUZE Another name for the PIBD fuze which incorporates a tiny shaped charge; this small charge 'spits back' a jet of explosive vapour down the centre of the shell and through a tube in the centre of the main charge to initiate an exploder situated at the rear of the main charge. See under *PIBD Fuze*.

SPLINTER An alternative name for a fragment of steel thrown off a detonating high-explosive shell. Usually confined to descriptions of protection, such as a 'splinter-proof shelter'.

SQUASH HEAD SHELL Familiar name for *HESH* shell, a type of anti-tank shell which contains a plastic explosive filling. The shell squashes on to the target by the force of impact, depositing the plastic explosive in tight contact with the armour. The explosive is then detonated and sends a violent shock wave through the plate to shake free a large 'scab' of plate from the inner surface.

A 105mm base-ejection smoke shell with its various components.

A German experimental shell for use with a squeeze-bore gun. The starting calibre is 105mm, which is squeezed down to 88mm. The studs are hollow and both they and the skirt are soft enough to reduce in size during the squeeze.

Projectiles for squeeze-bore guns. Below: A shell for the German 28/20 anti-tank gun. Right: A German 75/50mm anti-tank shell.

SQUEEZE BORE Type of gun in which the diameter of the barrel is suddenly reduced at some point along the bore, so that the diameter of the projectile is similarly reduced. Used to fire *APCNR* shot, which consists of a tungsten or other hard metal core set in a light alloy body which is capable of being compressed to the smaller diameter as it passes through the squeeze. Used in some World War II designs of anti-tank weapon in order to achieve high velocities. Compare with *Taper Bore.*

STAR SHELL Alternative name for *illuminating* shell. A shell containing a brilliant flare whch lights up the target area at night. Early star shell deposited the illuminant on the ground; modern types use a parachute to suspend the illuminant above the ground so as to give a better spread of light.

STAR SHELL CHARGE Parachute illuminating shells function by opening the shell at the optimum height and ejecting the star unit and the parachute. If the shell's velocity at this point is too high, the opening of the parachute can be so violent as to wrench loose the connection between it and the star unit, so that the parachute floats away and the star falls straight to the ground. In some weapons with adjustable charges the maximum charge was too high for safe ejection, while the next lower charge was too low to achieve the optimum break-open height. The 'Star Shell Charge' was a small bag of propellant which was normally never touched but, when removed from the maximum charge make-up, lowered the velocity just sufficiently to make sure the ejection velocity was safe but gave the maximum possible height for operation.

STREAMLINED Description of a shell in which the ogive is tapered and the shell body behind the driving band is also tapered, so allowing the airstream around the shell to merge and reduce the amount of base drag. 'Boat-tailed' is the equivalent American term.

SWEDISH ADDITIVE A sleeve of fibre impregnated with metallic salts and other substances which is used to line a cartridge case or the inside of a cartridge bag. It is consumed in the explosion of the charge, but the chemicals reduce the temperature of the explosion and so reduce the erosive wear on the gun. Adopted in the 1960s, with some high-velocity tank guns.

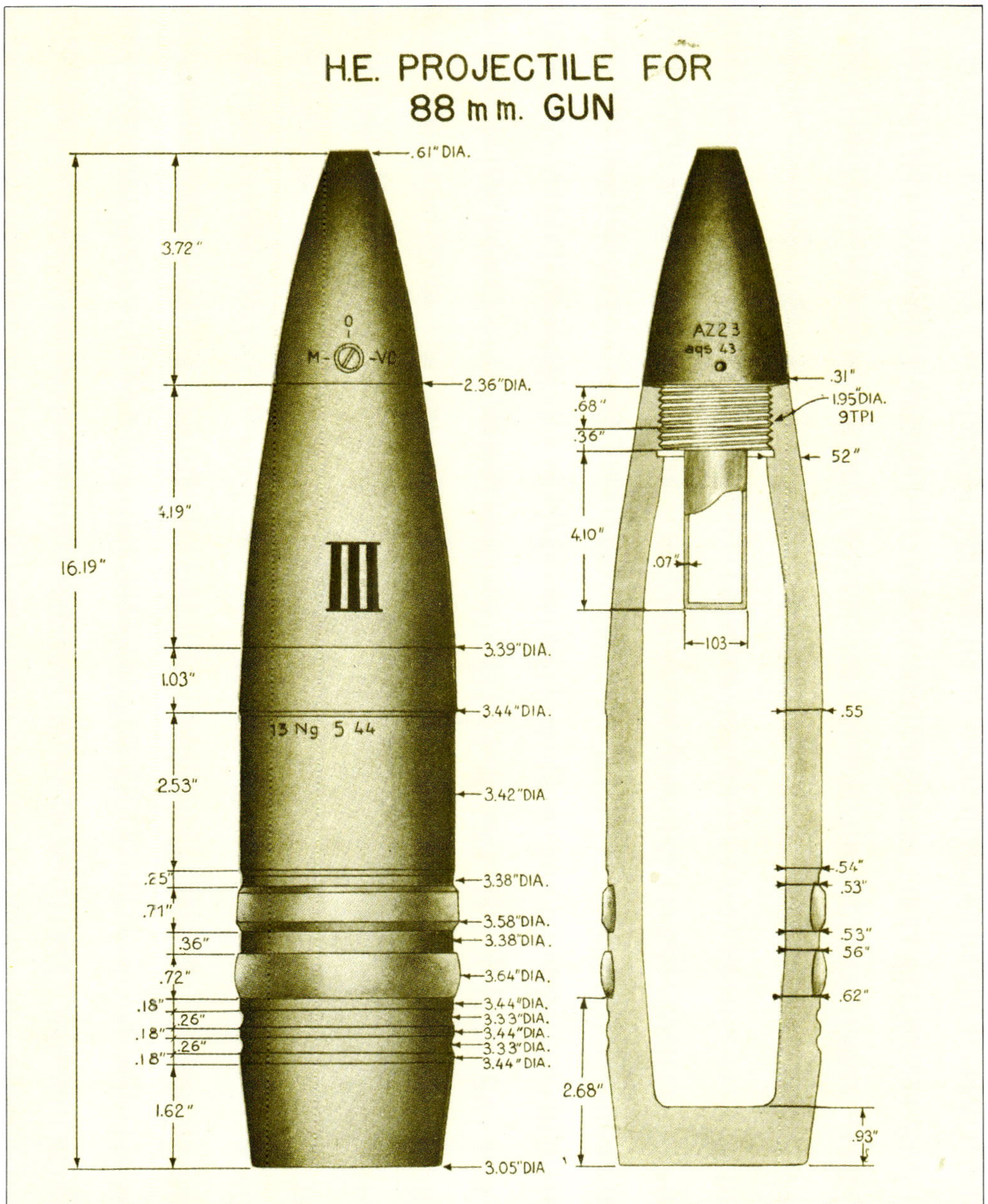

A streamlined shell for the celebrated 88mm German anti-aircraft gun. Note both the ogive and the base of the shell behind the drive band are tapered.

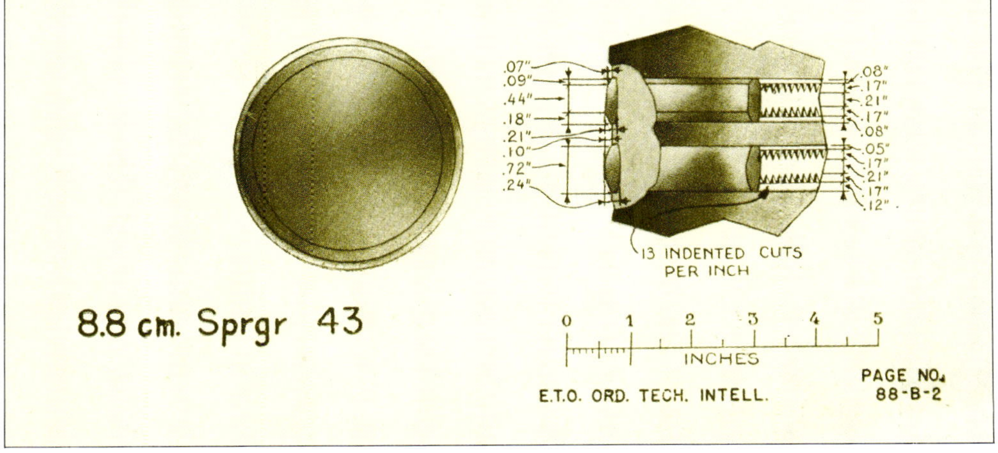

Rear view of a taper-bore projectile showing the skirted driving bands which are squeezed down when the shell is fired.

T

TAPER BORE Type of gun in which the calibre gradually decreases from chamber to muzzle. Used with *APCNR* shot, which decreases as it progresses through the tapered bore and so increases its velocity due to the surface area at the base decreasing while the propellant gas pressure remains the same. Differs from *squeeze bore* in being a constant decrease throughout the length of the gun, whereas squeeze bore is a short, sharp decrease in an otherwise parallel-bored weapon.

TARGET SHELL High-explosive shell which matches the standard HE shell, for the gun, but which has coloured dye mixed with the HE so that on detonation there is a vividly-coloured smoke cloud which is easily recognizable. Used to assist observation in difficult terrain, where an ordinary shell-burst could easily be missed, or to distinguish the firing of a particular gun when the target area is under bombardment by several other guns.

THINKING FUZE Common name for a *base fuze* which is designed to have a self-actuating variable delay.

Base fuzes to be fired against hard targets have an inherent delay, and often have additional, selectable, delay timings built in. The difficulty is that unless the thickness of the target and the actual striking velocity of the shell are both precisely known, it is impossible to assess the amount of delay required, and setting has to

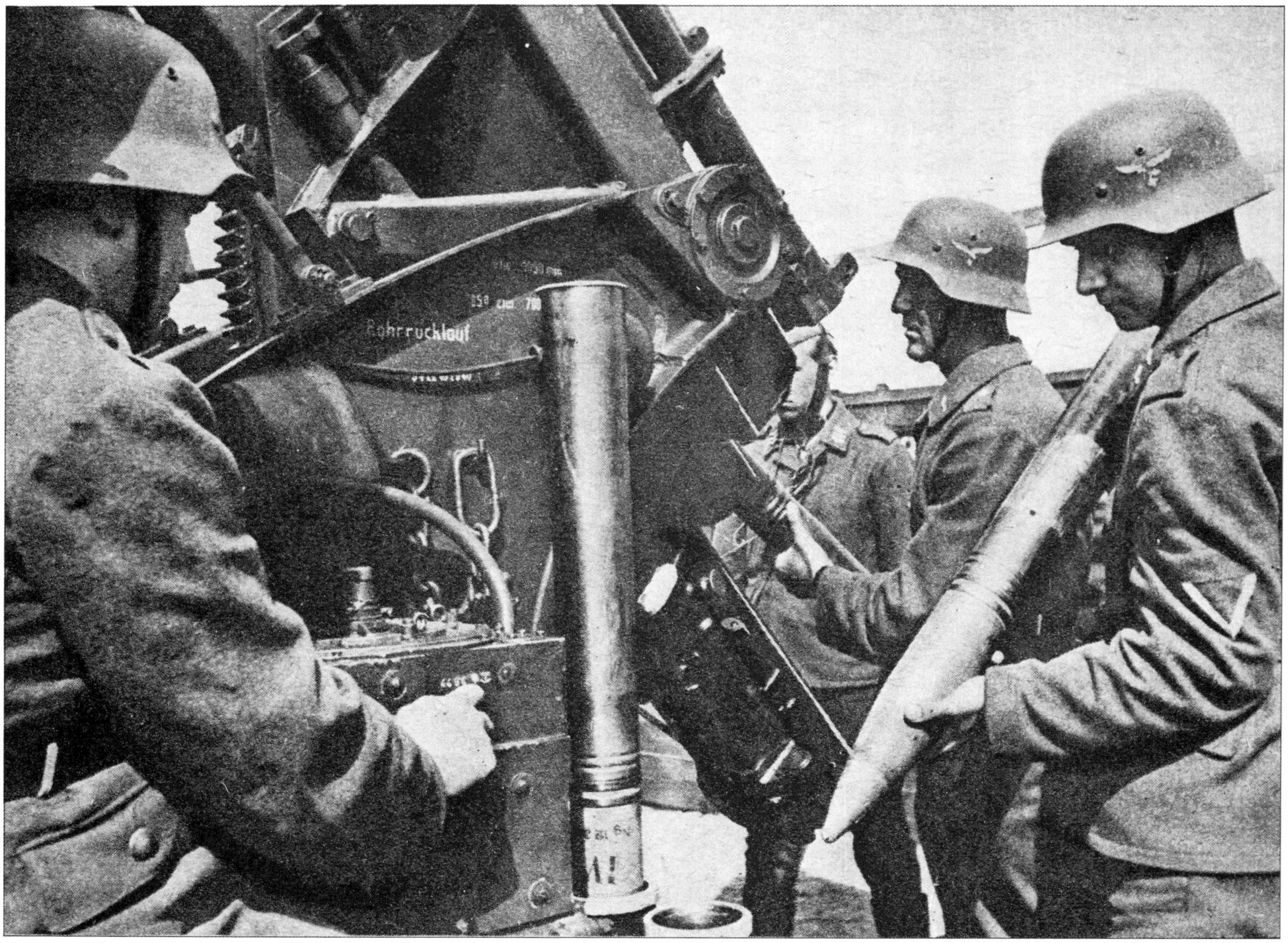

Above: German troops loading an 88mm anti-aircraft gun. But first the round is inserted, nose down, into a device which sets the time fuze.

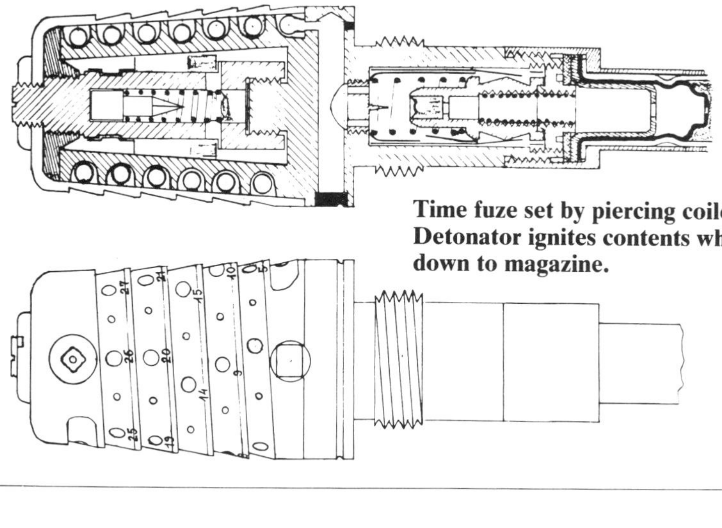

Time fuze set by piercing coiled tube. Detonator ignites contents which burn down to magazine.

be done by guesswork. The 'thinking fuze' uses a mechanism which arms the fuze on feeling the impact with the plate and then detonates it when it feels the slight acceleration of the shell as it pulls free of the plate on the inside of the target. Thus the timing of the delay is entirely self-regulating and the shell will always detonate at the correct point, irrespective of how thick or thin the target may be.

Attempts to make thinking fuzes for major-calibre naval and coast defence shells for the attack of warships were made for many years without much success. Present-day designs for use with small-calibre cannon against aircraft and helicopter armour appear to have been more successful, though there is little information published on their design.

TIME FUZE A fuze which is set into operation at the moment of firing from the gun, but which defers initiation of the shell until a pre-set time has elapsed.

Time fuzes are traditionally of two distinct types: combustion or mechanical. In recent years designs for a third type, based on electronic principles, have begun to appear though few have yet been put into service.

The operation of the combustion time fuze is described under that heading, above. It relies upon rings containing compressed black powder which burn at a pre-determined rate, and the time operation is adjusted by rotating one of the rings. A two-ring fuze has a maximum burning time of about 30 seconds, though this depends upon the type of powder used to fill the rings. In order to achieve longer times of burning, fuzes with three, four or even five rings were developed, though they were uncommon and their timing was not reliable.

An alternative method of using powder for timing was commonly used by the French with their 75mm gun M1897; this had a length of powder contained in a lead tube and wrapped spirally around a central pillar. To set the fuze, a hole was punched through the lead tube and into the central pillar; this allowed the initiating flash from a detonator to light the powder in the tube at the punctured point, and it then burned down until it ignited the magazine. Timing was achieved by simply punching the hole at a greater or lesser distance from the magazine.

Mechanical time fuzes can use pre-wound clockwork mechanisms or can be driven by centrifugal force derived from the spinning of the shell. The spring-driven mechanism is described under 'clockwork fuzes' and will not be repeated here. The centrifugally-driven mechanism is commonly known as the 'Junghans' mechanism (from the inventor) and uses two weights which revolve outwards under spin to drive a train of gears and a regulator. The gear train rotates a 'timing disc' which has a cut-out on its perimeter; when this cut-out aligns with a spring-loaded arm, the arm can move into the cut-out portion and so release a striker to fire a detonator. Fuze setting is achieved by turning the cap of the fuze which turns the timing disc in relationship to the gear drive; when the gun fires, the set-back force disconnects the setting cap from the disc and also engages a clutch which allows the gears to drive the disc.

Electrical time fuzes have been the subject of experiment since the early 1940s, and they are all based on the well-known resistor-capacitor delay circuit; in this a capacitor (condenser) is charged from a power source through a resistor. The voltage of the power source and the resistance of the resistor govern the speed at which the capacitor charges and thus the timing; when the capacitor is fully charged it automatically discharges through an electric detonator to fire the shell. In the early days the principal drawbacks to this system were the electrical power source, which tended to be cumbersome, and the method of connecting it to the fuze. A common idea was to use a loose wire dangling outside the gun muzzle which the fuze would brush as it was fired, making contact for sufficient time to charge part of the circuit. A better idea was simply to have a battery on the gun and make a direct plug connection to the fuze during the loading process. The principal difficulty was the physical size and fragility of early electronic components. In the 1970s, however, with the advent of micro-circuits and solid-state devices, these have been overcome and several designs of compact and reliable electronic time fuzes have been developed, though, so far as we know, few have actually entered military service.

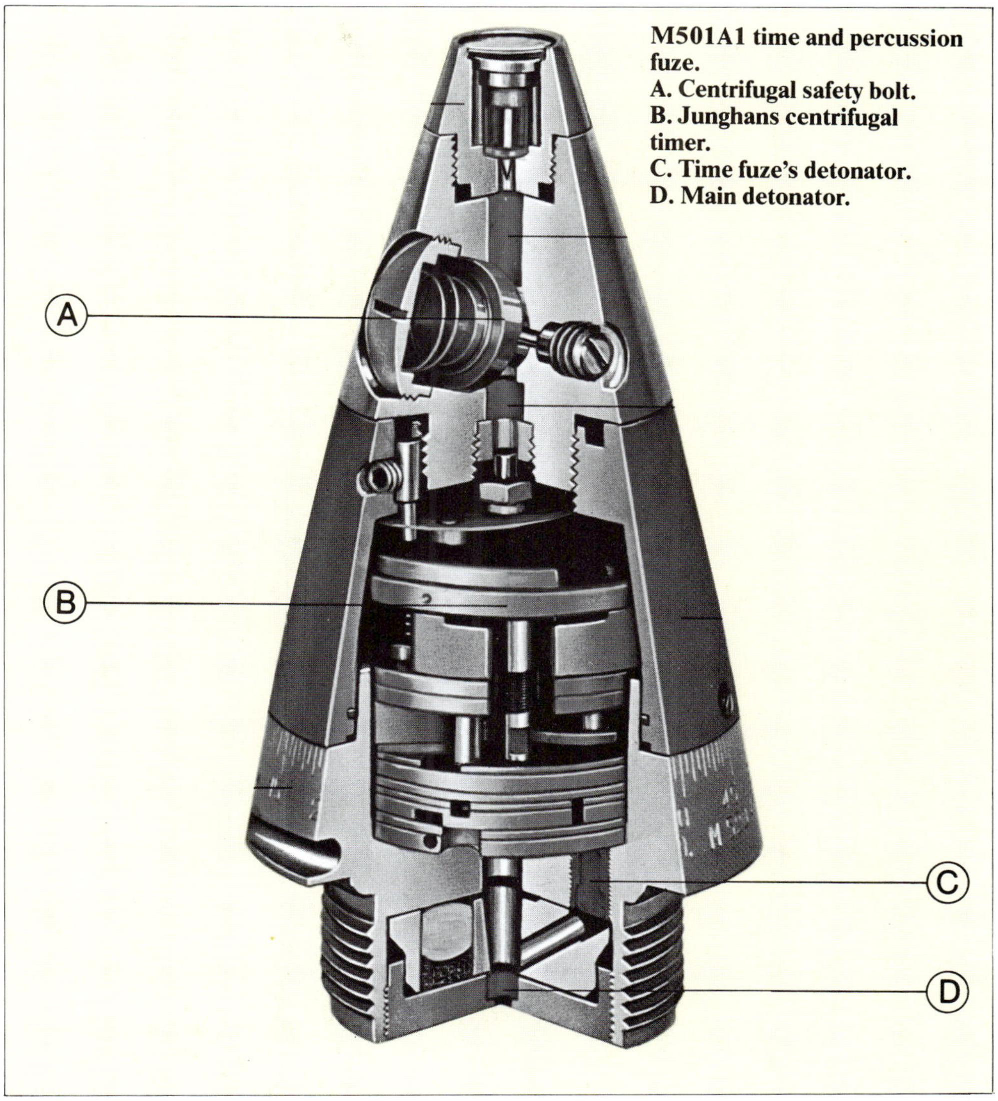

M501A1 time and percussion fuze.
A. Centrifugal safety bolt.
B. Junghans centrifugal timer.
C. Time fuze's detonator.
D. Main detonator.

TIME AND PERCUSSION FUZE A time fuze which also incorporates an impact element so that if the fuze is wrongly set, or the timing element fails and it passes the intended point of burst, it will at least detonate when it hits the ground and, it is hoped, do some damage to the enemy. Originally, these were almost entirely confined to use with ground artillery fire, since anti-aircraft fuzes were assumed to burst in the air whether or not they reached or passed their intended target. However experience soon showed that a small percentage of AA shell time fuzes failed to function and even if they failed to detonate when they fell to the ground the impact of 20–30kg of steel and explosive falling from several thousand feet could do considerable damage. Impact 'clean-up' elements were therefore added to time-only fuzes, and today a time-only fuze is a rarity.

TNT Tri-nitro-toluene (also Trotyl, Trinol, Tritol, Trilite and many similar names). One of the most widely used military explosives, as a shell filling and for mines and demolition charges. Favoured because it has a good resistance to shock, is prone to burn rather than detonate if accidentally ignited, is resistant to water, and has a low melting-point which makes filling of munitions easy. For further information see under *High Explosives.*

Left: A Swiss tank opens fire with tracer ammunition.

Above: Tracer in use in 1944 showing anti-aircraft fire against VI missiles.

TRACERS Pyrotechnic devices attached to the rear of projectiles to enable their flight path to be seen. Used only with direct-fire weapons (ie, weapons which are in sight of their targets) so that in the event of missing the target the observer can make a correction to bring the fire on to the target.

Tracers can be filled directly into a drilling in the base of the projectile (usually done only with solid AP shot) or can be filled into a tube which is then screwed to a threaded boss in the base of the projectile (usually done with composite shot of various types, eg APDS, APCR, etc.). 'Flat-base' tracers are flat discs with tubes running through the width of the disc and meeting in the centre; these are used on projectiles for separate-loading ammunition which must be rammed, and where a tubular tracer protruding from the shell base might be damaged by the ramming.

The composition used in the tracer is usually based on magnesium, to provide a bright flame, together with various salts to provide the desired colouring — usually strontium nitrate to give the flame a bright red colour, which experience has shown to be best for contrast against most sorts of background. The composition is pressed into the tube or cavity under considerable pressure to form a very hard-packed filling which then burns layer by layer; failing to press the mixture hard renders it liable to crack and burn too quickly, so that the trace goes out before the target is reached. This hard pressing makes the composition hard to ignite, and a layer of 'starter composition', usually a form of black powder, is pressed on top. This is ignited by the propellant

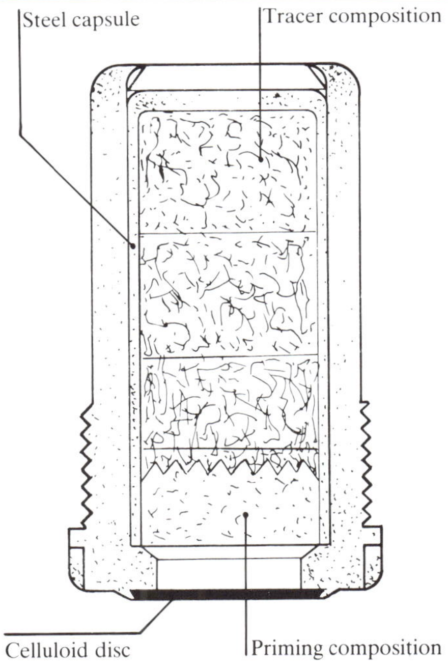

A typical tracer unit which would be screwed into the base of a projectile. The celluloid disc waterproofs the interior.

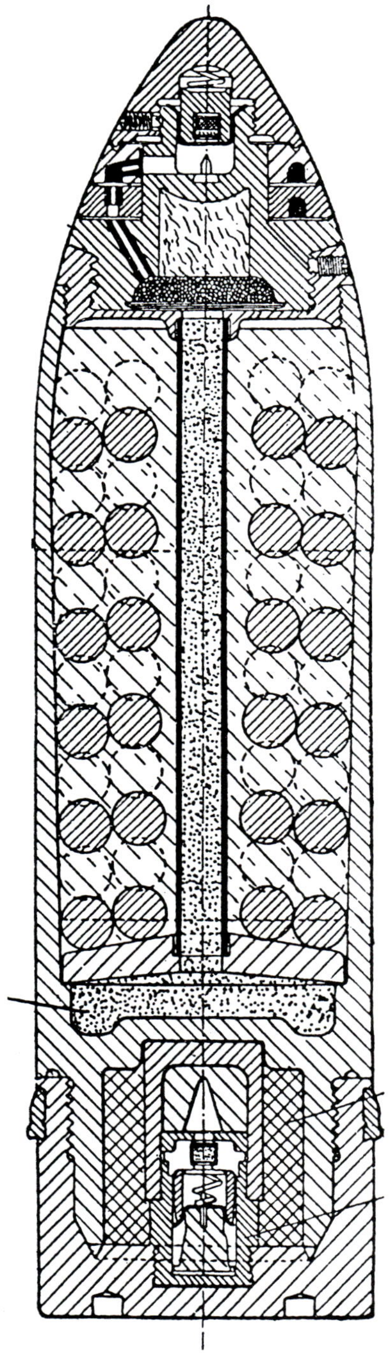

A Krupp universal shell of about 1910. This version uses a time fuze in the nose for normal shrapnel action and a graze fuze in the base to detonate the TNT packed around the bullets.

flash and transfers ignition to the tracer composition. It also defers the onset of the trace to some distance from the gun muzzle, which avoids blinding the gunlayer as the shell leaves the muzzle, and also prevents an enemy from following the trace back to its origin to discover the location of the gun.

U

UNIVERSAL SHELL Type of field-gun shell developed by Krupp of Germany in the early 1900s. It was essentially a *shrapnel shell* in which the balls were held in a matrix of TNT instead of resin. The fuze was a time and percussion fuze. When set for 'time' action, the fuze fired the expelling charge in the usual way and the balls and TNT were ejected, the TNT being ignited by the expelling charge flame so as to burn and give a puff of black smoke in the sky. When the fuze was set for 'percussion' the shell struck the ground and the impact action detonated the TNT, so that it then became a high-explosive shell with a very high proportion of fragments caused by the shell casing and the shrapnel balls. Though a successful design, it did not last long once the shrapnel shell was eclipsed by the more general use of high-explosive shell during World War I.

W

WALLBUSTER Early name for *squash-head* shell, since it was originally designed as a means of demolishing reinforced concrete defences and was intended for use against the Atlantic Wall defences during the 1944 invasion of Normandy. As events turned out, none of the weapons which used wallbuster shell were adopted for use by the invasion forces, and the wallbuster shell was redeveloped as an anti-armour projectile.

WATER SHOT Type of *paper shot* which instead of being filled with sand was filled with water prior to loading. On firing the paper casing burst open and the water was ejected from the muzzle of the gun, largely in the form of steam and spray, first having given sufficient resistance to movement to cause the gun to recoil.

An early wallbuster shell. This is the second version. The first had a wire-mesh interior container which carried the plastic explosive.

NEW DEVELOPMENTS

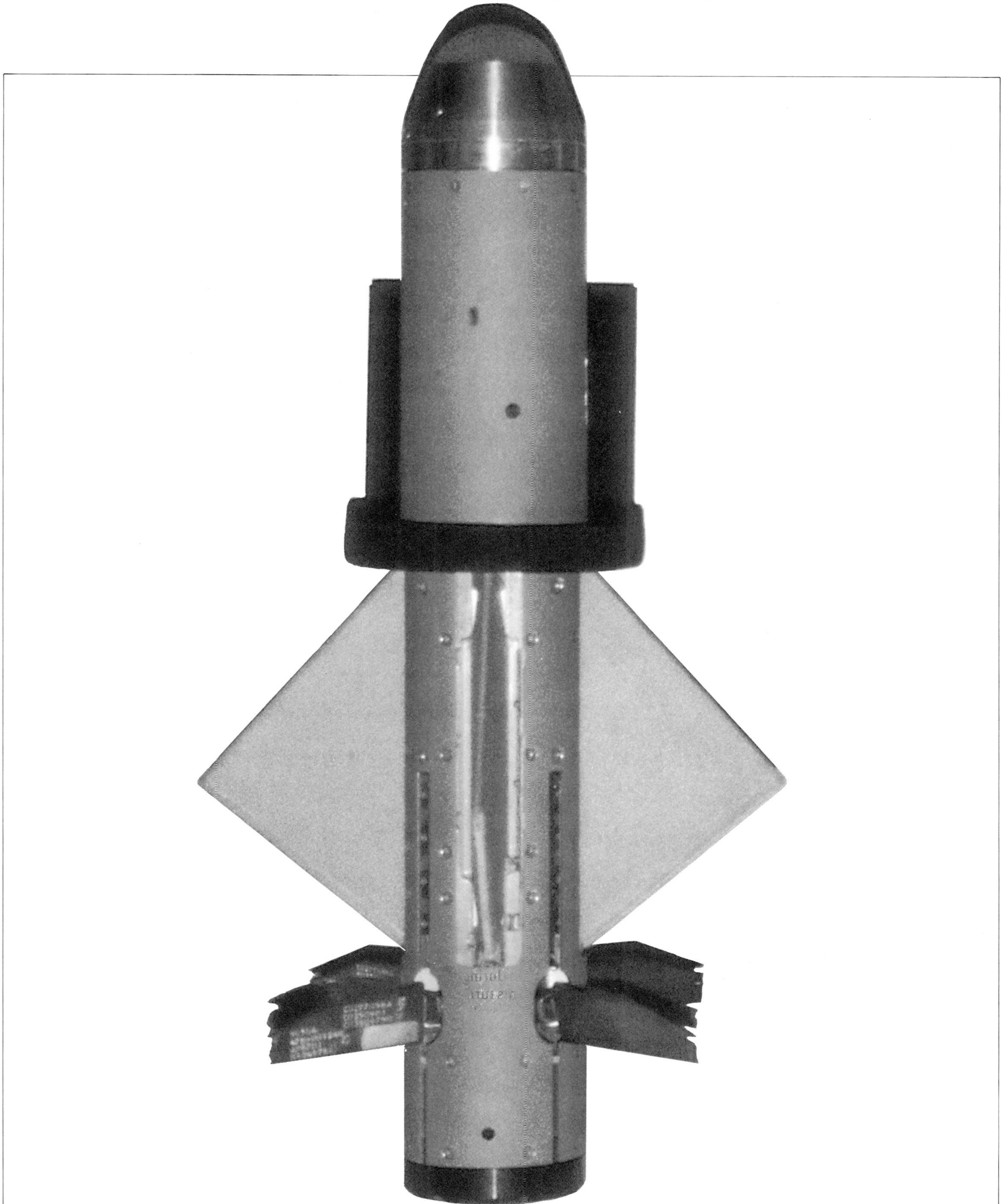

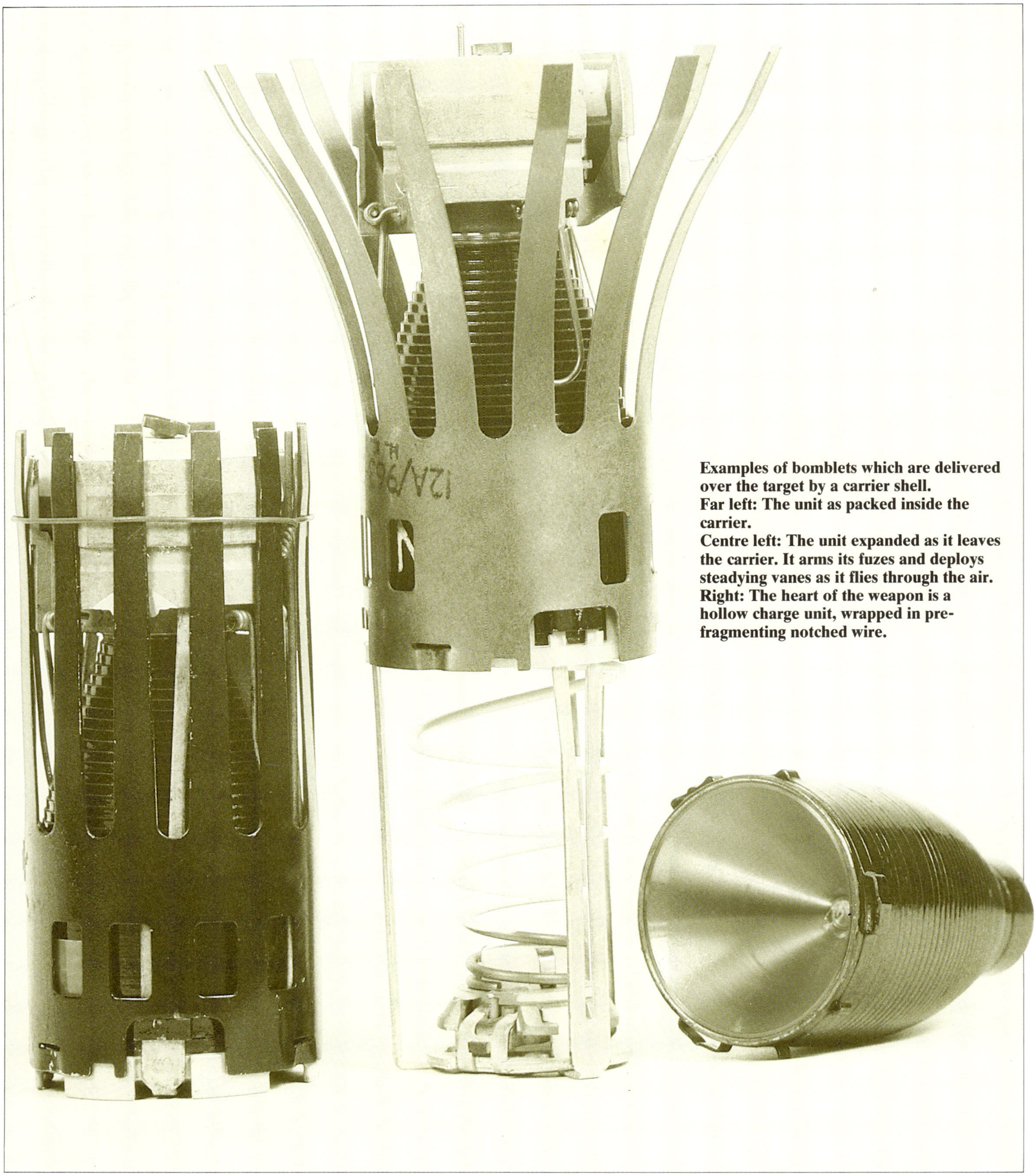

Examples of bomblets which are delivered over the target by a carrier shell.
Far left: The unit as packed inside the carrier.
Centre left: The unit expanded as it leaves the carrier. It arms its fuzes and deploys steadying vanes as it flies through the air.
Right: The heart of the weapon is a hollow charge unit, wrapped in pre-fragmenting notched wire.

B

BOMBLET A sub-projectile which is loaded into a type of carrier shell and released over the target. The bomblet is not a projectile and falls freely when released rather than attempting to follow the shell's trajectory. It is normally constructed with a warhead which contains a shaped charge surrounded by pre-formed fragments of some sort. Thus the bomblet is basically anti-armour, but also has an anti-personnel effect. It is intended to be used against armoured formations at long range, and the anti-personnel element is intended to deal with any accompanying infantry.

C

CONTINUOUS ROD Type of warhead used with air-defence missiles. It consists of a canister of high explosive, the wall of the canister being made of high-tensile steel rods, square in section, and welded together at the ends so that the tip of one rod is connected to the rod on one side and the bottom of the rod to the rod on the other side. When the explosive is detonated it blows the rods outwards; the welded ends remain connected and thus the linked rods expand in zig-zag fashion to form a connected circle of steel some 10–15 metres across, flying through the air at high speed. If this strikes an aircraft it will cut straight through it. Used in Western air defence warheads since the 1950s, and according to some authorities was used in the Soviet warhead which brought down the Powers U2 spy plane over Sverdlovsk in 1960.

COPPERHEAD Terminally-guided artillery projectile for use in 155mm howitzers, developed in the USA by the Martin Marietta Corporation and intended for NATO adoption.

Copperhead is officially known as the M712 Cannon-Launched Guided Projectile (CLGP). It is fired from conventional 155mm howitzers, using a normal propelling charge and takes up a conventional ballistic trajectory. The target, which has been selected by a forward observer, is then illuminated by a laser beam operated by the observer or carried in a remotely-piloted vehicle (RPV). In the forward section of the Copperhead is a laser seeker; this detects the reflected laser signals from the target, and by means of pop-out fins and control surfaces the laser seeker steers the Copperhead to impact with the target.

The warhead is of the HESH type and there is a 'fuzing' module' which has a direct-impact sensor and six graze sensors which ensure detonation even if the warhead does not make a head-on impact. The projectile weighs 63.5kg (140lb) and has a maximum range of 16 kilometres (10 miles).

Production of Copperhead began in 1980, but initial funding was later withheld since production models failed to reach the promised standards of accuracy. In 1983, after overhauling the production facilities, a success rate of over 90 per cent was achieved and by early 1984, 5,000 of the US Army order for more than 30,000 had been delivered. The British Army, however, have changed their minds and will not adopt Copperhead, since they consider that the requirement for the target to be laser-illuminated is unrealistic.

D

DISCRETE ROD Type of air defence missile warhead which is similar to *continuous rod*, but in which the individual rods making up the wall of the warhead cylinder are not joined, but are simply held together by a matrix of plastic material. As a result, when the explosive content is detonated the individual rods are blown outwards as separate (discrete) sub-missiles, capable of causing considerable damage to an aircraft target.

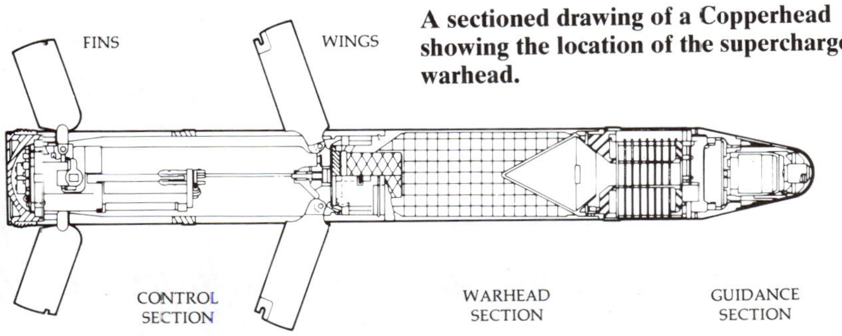

A sectioned drawing of a Copperhead showing the location of the supercharged warhead.

F

FLAT CONE CHARGE Type of *shaped charge* in which the hollowed face of the charge and its cone liner are opened to an angle of about 130°. This reduces the diffusion caused by spin, and makes a thicker and more coherent jet which blows a larger hole in the target, but which is not capable of such deep penetration as the conventional sharp cone pattern. Flat cone charges are being developed for small-calibre cannon shells which are required to cause severe damage to armoured aircraft targets, but which do not need to pierce great thicknesses of plate. They are also useful against lightly-armoured APCs and similar ground vehicles, and there is evidence that their use is contemplated for *top-attack* anti-tank weapons.

FUEL/AIR EXPLOSIVES Explosives which consist of a finely-divided combustible substance and air. It is the fineness of the combustible and its dispersion in the air which generates the explosive effect; the substance need not, of itself, be an explosive. A prime example of a fuel-air explosive which occasionally appears by accident is the mixture of flour dust and air which has been known to explode in empty grain silos, ignited by a casual flame or spark. Flour is not an explosive; but mixed with air in the right proportions it has considerable destructive ability. Mine explosions from a mixture of coal dust and air are also well-documented.

The military use of fuel/air explosives (FAE) began in World War II; there are persistent rumours, though unsubstantiated by any documentary evidence, that the German Luftwaffe used some type of FAE bombs against the Soviets in front of Leningrad. Both the British and American air forces also experimented with FAE bombs, since they promised immense destructive capacity against buildings. A properly-designed FAE bomb would, it was theorized, penetrate a building, disperse its payload of dust to mingle with the air in the building, and then detonate the FAE mixture. This would exert a powerful heaving effect against the walls, blowing them out and collaps-

A laser target designator. A laser beam from this instrument is directed at the target. The guided projectile detects the reflected laser light and homes in on the source of reflection.

ing the entire building. This would obviously have far greater effect against factories and similar establishments with large floor-spaces than the more concentrated force of a conventional high-explosive bomb.

Since then, FAE has been suggested for various roles and in various munitions; one idea which has considerable appeal is to fire an FAE projectile above a minefield, so that the detonation would exert a slow but considerable pressure downwards and trip the pressure-plates of all the mines. A few shots would thus clear a wide path through any form of minefield. However, no practical FAE munition exists yet.

L

LASER GUIDANCE Form of guidance proposed for various types of *remotely-delivered munitions*, and for the *Copperhead* cannon-launched guided projectile. The system relies entirely upon the target (usually a tank) being 'illuminated' by a laser projector operated by a ground observer or from the air — eg a helicopter, support aircraft or remotely-piloted vehicle (RPV). The 'laser designator' sends out a beam of laser light which strikes the target and is reflected off in several directions, just as ordinary white light would be. This reflection is detected by the incoming missile or munition and acts as a beacon to the guidance mechanism, causing the missile or munition to steer to hit the target. The laser emission is coded, and the responding munition is pre-set with the code, so that the munition will only respond to the correct laser reflection and will not be misled by another laser operating against some other target.

While sound in theory, the argument against this system is that it is not always convenient or practical to have someone in place to operate the laser designator. As a result there is a great deal of work going forward on devising weapons which will detect and steer towards targets without the necessity for a third party in the process.

LIQUID PROPELLANT Gun propellant which uses explosive liquids instead of conventional smokeless powder.

The attractions of liquid propellants are that they are likely to be more efficient, in the power-to-bulk ratio, than smokeless powders; they will leave no residue in the bore after firing and there will be no empty cartridge case; and that they will require less bulk in the logistic supply-line.

Against that are some formidable counter-arguments: that almost all the liquids suitable for this are so volatile or corrosive as to be dangerous to handle; that the sealing of the gun will be far more difficult than sealing against the explosion of smokeless powders; that precise measurement of the liquids will be extremely difficult; and that the ignition of mixed liquids proceeds irregularly, leading to unacceptable peaks of excessive pressure in the chamber.

Various ideas have been put forward, and some have been tried in experimental weapons. The most easy method is simply to inject some suitable liquid into the gun chamber, behind the projectile, and fire it with, say, an automobile sparking plug. Another method is to have two liquids, relatively harmless by themselves but explosive when mixed, and inject them simultaneously into the gun chamber. A third proposal is to use a conventional smokeless powder charge, but to increase its power by injecting some form of fuel into the chamber immediately before firing.

Among the more practical problems faced is the one of carrying such dangerous liquids safely on the battlefield, carrying them on guns or in tanks, feeding them to the gun at all angles of elevation, and refuelling the weapons when necessary.

Obviously, there are many things to be settled before liquid propellant becomes a practical proposition.

M

MILLIMETRIC WAVE RADAR Method of guiding some forms of remotely-delivered munition to their target. Millimetric wave radar is similar to conventional microwave radar, but operates at extremely high frequencies (10–400 gHz). The principal difference is that there is, as yet, no way of producing really powerful outputs. Current techniques produce about 100 watts, but there are hopes that this may be considerably improved in the near future.

However, millimetric waves can also be used in the passive role, a technique known as 'radiometry'. The device requiring guidance need not emit any radar signal because it takes advantage of the fact that high-frequency millimetric waves are approaching the far infra-red spectrum and tend to take on some of the characteristics of optical frequencies. At these frequencies, ground objects emit radiation which is largely due to their reflectance of the colour temperature of the sky; the earth emits a fairly regular figure, but objects of greater reflecting power — such as tanks — can be easily distinguished from their surroundings and can be detected by millimetric wave techniques.

MULTIPLE-SHAPED CHARGE Type of air-defence missile warhead which consists of a cylindrical core upon which are mounted a large number of small shaped charges facing outwards at various angles so as to provide all-round effect. At the altitudes at which these missiles are intended to function, the rarified air means that the jet from a shaped charge will travel for a considerable distance and retain its energy sufficiently to damage an aircraft or missile target. When the warhead is detonated (by a proximity fuze) all the shaped charges fire simultaneously, so generating a 'porcupine' of jets around the point of detonation.

N

NUCLEAR PROJECTILES Artillery shells containing a nuclear device as the explosive.

Nuclear shells are generally in the 2-kiloton yield area — ie the detonation is the equivalent of 2,000 tons of TNT. Details of their construction are not divulged, but it can be assumed that they rely upon fairly simple techniques of controlled explosion to bring together sub-critical masses of nuclear material. The principal design problem with this type of projectile is the ensuring of its absolute safety during transport, handling and firing, together with complete reliability in action at the target. The fuzing system of nuclear shells is known to be extremely complex, with three or more mechanically interlinked time fuzes controlling the various arming sequences within the shell and with a proximity fuze to detonate it at the correct height above the target. Multi-function proximity fuzes are used, capable of being set to varying burst heights or for impact burst, as dictated by the tactical requirement.

The complexity of nuclear shells, together

Loading the American 280mm nuclear cannon 'Atomic Annie' with a nuclear shell.

Above: An American 155mm carrier shell loaded with anti-tank mines. Below: Projectiles being developed for NATO howitzers. Loaded with shaped-charge bomblets, they can be fired against armour at long range. The lower projectile is fitted with a base-bleed motor for maximum range.

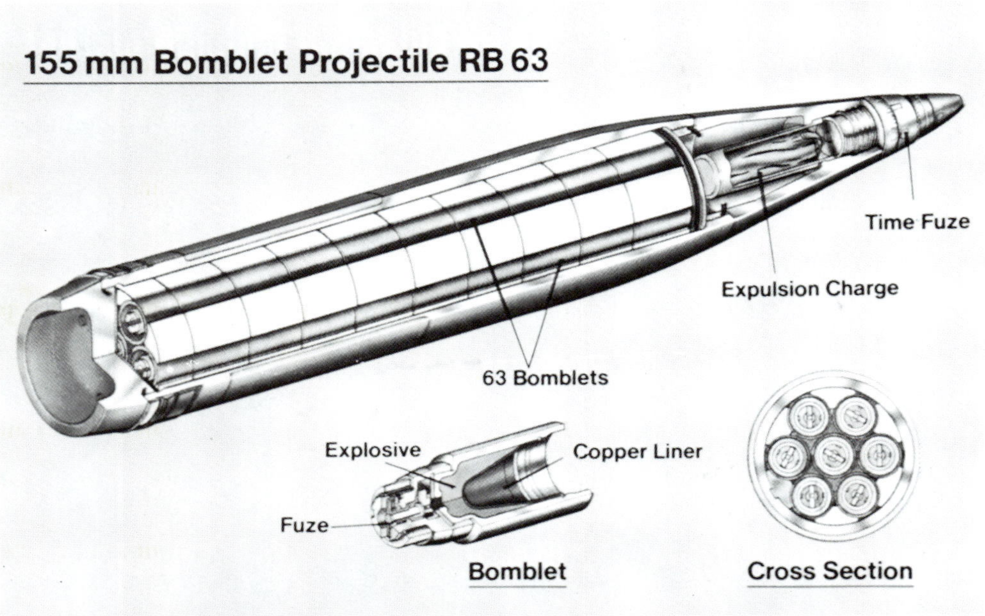

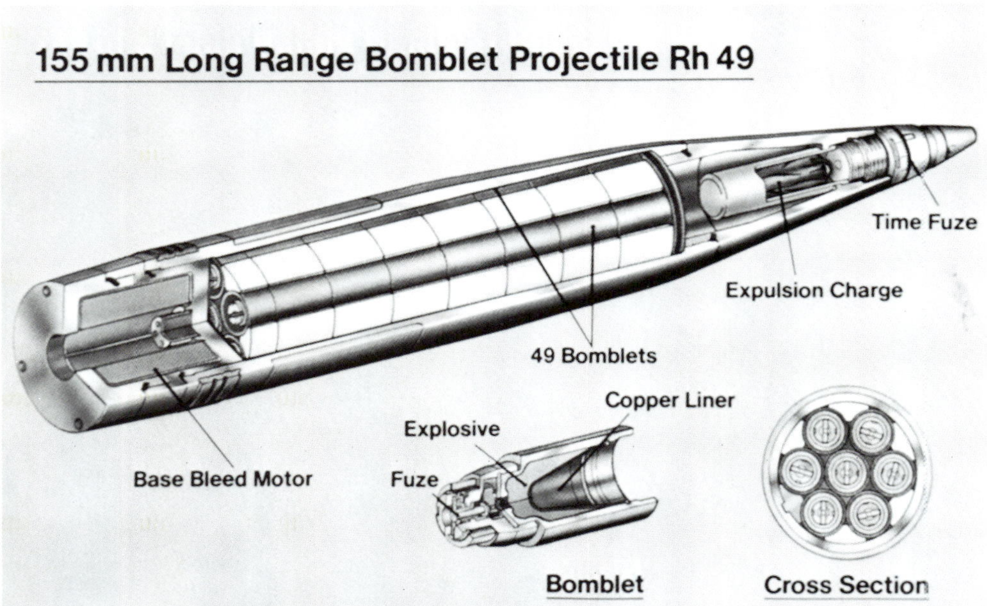

with their radius of destruction, means that there is a minimum calibre below which they become impracticable. At present this is 155mm, and a number of NATO and Warsaw Pact weapons of 155mm and larger calibres are understood to be provided with nuclear projectiles.

R

REMOTELY-DELIVERED MUNITIONS (RDMs) Munitions which are fired in a carrier shell to the target area, then ejected from the shell and allowed to seek and detect their own targets.

Most RDMs have been devised as a means of attacking tanks at long ranges. They consist either of bomblets carrying shaped charges, guided shaped-charge devices, both for the top attack of armour, or scatterable anti-tank or anti-personnel mines. Bomblets and mines are simply ejected from the shell and fall at random, the bomblets to strike any tanks which may be beneath, the mines to lie on the ground and, it is hoped, be run over by advancing tanks at some later time. Guided devices usually rely upon infra-red or millimetric wave seekers which detect the tanks either by their heat or by their different reflectance and then steer the sub-munition to impact.

S

SADARM Acronym for 'Seek and Destroy Armor', an American RDM designed to attack tanks at long range. Three SADARM munitions can be fitted into one 203mm or 155mm carrier shell and are ejected through the base in the vicinity of a target. Each munition deploys a parachute which is connected in such a way as to maintain the weapon canister at the constant angle of 30° to the vertical. While descending at about 9 metres per second (29fps), the canister spins at about 4 revolutions per second so that a *millimetric wave* sensor can scan the ground below in a reducing spiral pattern. Specific contrast characteristics of a tank target are incorporated in a memory circuit so that the sensor will ignore anything that does not look like a tank. Once such a target is detected the sensor fixes its position and then calculates the

An Aerojet Sadarm sub-munition detonates and fires a self-forging fragmentation device downwards at its target.

A model showing how a self-forging fragment develops and the damage it can do to armour plate.

precise point during the spin cycle at which to fire the explosive warhead so as to discharge a *self-forming fragment* at the target. Should the munition not detect a target, it falls to the ground and will then act as a shaped-charge mine if run over by a tank. Development of the SADARM system has gone as far as the firing of experimental weapons, but funding has been stopped pending the completion of other anti-armour system studies.

SELF-FORGING FRAGMENT (SFF) MUNITION The self-forging fragment is a development of the *Schardin effect.* It is a flat-cone shaped charge, but with a very heavy plate liner rather than the conventional thin shaped-charge liner. On detonation, the liner is not vaporized into a jet but is converted to a slug-shaped missile which is projected at about 1,200 metres per second (3,900fps) as a kinetic energy projectile. Over the short ranges (usually less than 50m) at which SFF munitions are proposed to be used, there is little velocity loss; penetration, though not as good as from a carefully designed discarding sabot sub-projectile, is good, and the disruptive effect inside the armour is enormous. The principal application of the SFF in artillery ammunition is as the warhead unit for various types of remotely-delivered munition. See, for example, *SADARM* above.

SUB-MUNITIONS General term for the explosive devices forming remotely-delivered munitions and taken to the target area inside carrier shells. Often, but not necessarily, guided in the final portion of their operation. See *SADARM*, *Remotely-Delivered Munitions* and *Bomblet* above.

T

TERMINAL GUIDANCE Guidance of a weapon during the final phase of its flight; the initial phase being a conventional ballistic trajectory. As, for example, in the *Copperhead* guided projectile which flies as a shell to the latter part of its trajectory, but then adopts a guidance mode.

TOP ATTACK Method of attack of armour which directs the attack against the upper surface of the tank hull and turret, since these are usually thinner than the front and sides. The principal threat to the tank has always been from ground weapons, and so the designers have concentrated on protecting the front and sides; as a result new designs of weapon are appearing which evade these thicker areas and go for the thinner top surfaces. Defence against this attack is, at present, somewhat difficult; simply adding extra armour will add excessive weight and lead to mechanical problems with transmissions and suspensions, and will also unbalance the tank. It can be expected that the next generation of tanks, for the 21st century, will adopt some form of top protection, by which time the weapons designers will have thought of something else.

W

WARHEAD General term for the explosive and fuze content of a missile or sub-munition.

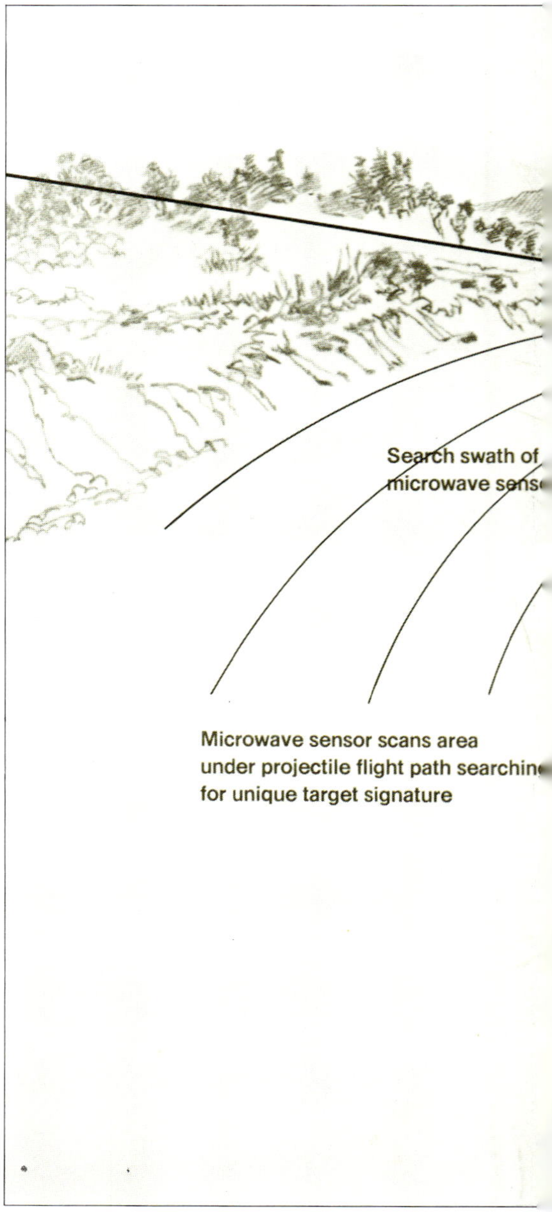

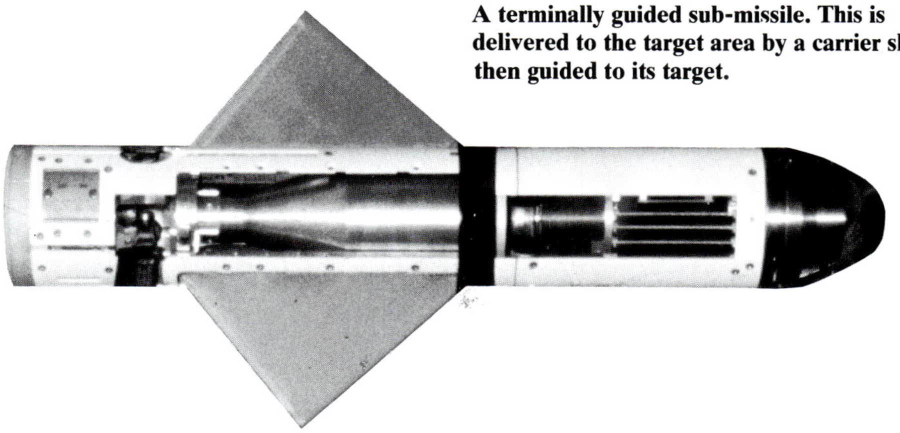

A terminally guided sub-missile. This is delivered to the target area by a carrier shell, then guided to its target.

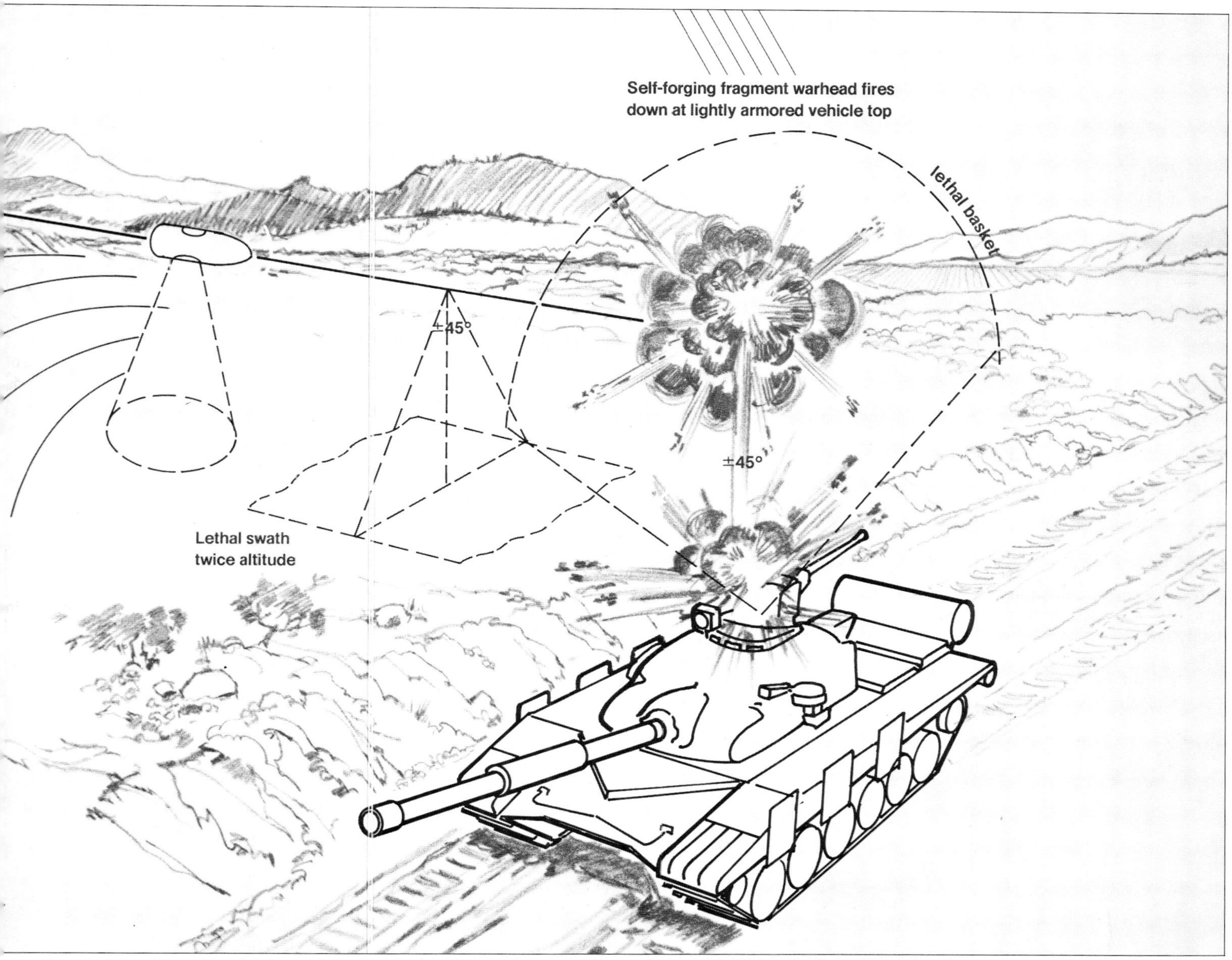

How a Smart Top Attack Fire and Forget projectile attacks a tank from above with high velocity self-forging fragments.

A US terminal guidance projectile under development. Guidance system is in the nose.

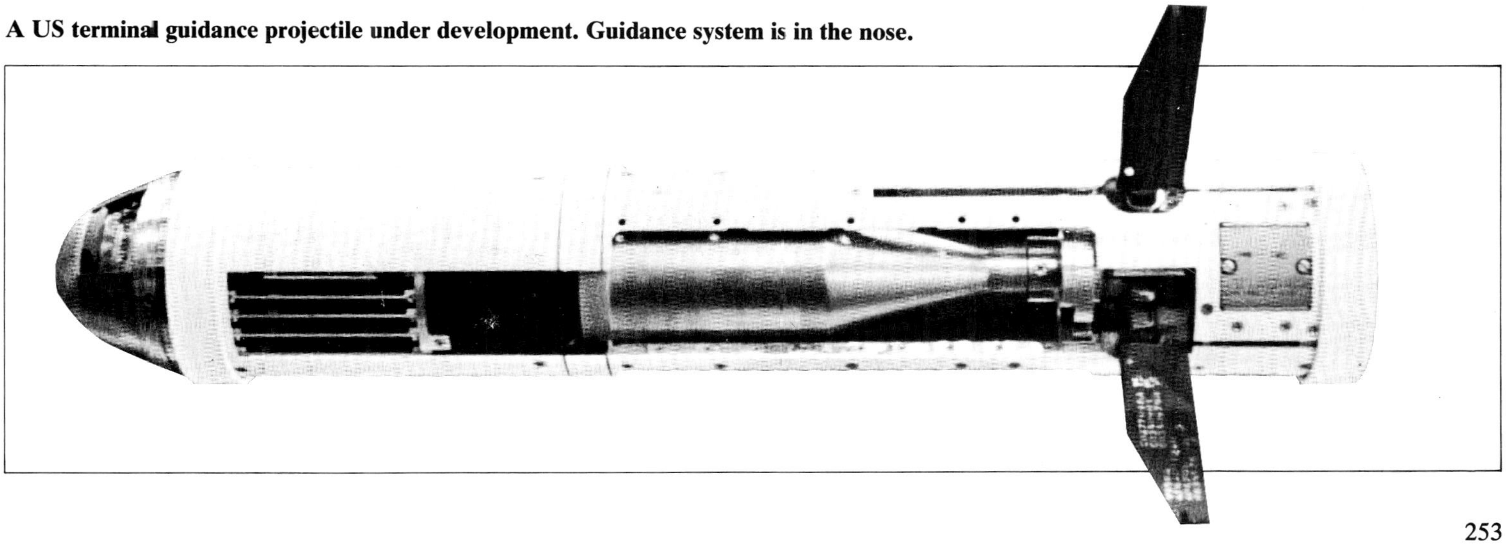

GENERAL INDEX

INDEX TO HEADWORDS

Acknowledgements

The author wishes to acknowledge the assistance given by various manufacturers in providing information and illustrative material, and particularly the Royal School of Artillery for their permission to photograph historical ammunition items in their Ammunition Display Room. Other illustrative material has been supplied by the following sources: Page 9 — Christ Church, Oxford; Page 10 — The Armouries, The Tower of London; Page 11 — Mary Evans Picture Library; Pages 12, 13, 14, 15 and 19, top — The Armouries, The Tower of London; Pages 19, centre and bottom, and 20, top — Mary Evans Picture Library; Page 21, top — The Armouries, The Tower of London; Page 30, top and bottom — National Maritime Museum, Greenwich; Page 37, top — The Armouries, The Tower of London; Page 51 — Mary Evans Picture Library; Pages 52 and 53 — Imperial War Museum, London; Page 69, bottom left — Dynamit-Nobel; Page 77, bottom — US Army; Page 84, centre — SMI; Page 87, bottom left — Accuracy Systems Inc; Page 88, centre right — Olin-Winchester; Page 92 — VEW AG; Page 94, bottom left — Brandt; Page 97, bottom — PRB SA; Page 98, top — Arges; Page 101, bottom right — Mecar SA; Page 104, top — Dynamit-Nobel Wien; Page 105, top — Esperanza y Cia; Pages 106/107 and 111 — PRB; Page 113, top — Diehl-Wehrtechnik; Page 119, bottom, left and right — Brandt Armements; Page 125 — Lacroix; Page 126, left — Losfeld; Page 127, bottom — MoD, UK; Page 130 — Steyr-Daimler-Puch; Page 133 — Mecar SA; Page 134, right — Soltam; Page 135, bottom, and page 136 — FFV Ordnance; Page 142, bottom right — Accuracy Systems Inc; Page 143 — Esperanza y Cia; Page 150, right — Rheinmetall; Page 152, left — GIAT; Page 155, centre — SNPE; Page 155, right — FFV Ordnance; Page 159, right, and page 160 — Rheinmetall; Page 165, top — Oerlikon Bürhle; Page 172 — Vincenzo Bernadelli SpA; Page 180/181 — T. Gander; Page 185, top — Diehl-Wehrtechnik; Page 188, left — Bofors Ordnance; Page 188, lower right — As Konsberg Vapenfabrik; Page 199 — Vincenzo Bernadelli SpA; Pages 210, 212, left, and 214, top and left — Diehl-Wehrtechnik; Pages 218 and 219 — Thomson-CSF; Page 225, right — General Defense Corp, Flinchbaugh Division; Page 241 — Vincente Bernadelli SpA; Page 248 — Texas Instruments; Page 249 — US Army; Page 250 — Rheinmetall; Page 251 — Aerojet Electro Systems.